Access to Destinations

Related Books

Eds. HENSHER & BUTTON	Handbooks in Transport 6-Volume Set

AXHAUSEN	Travel Behaviour Research
DOHERTY & LEE-GOSSELIN	Integrated Land-Use and Transport Models
GÄRLING et al	Theoretical Foundations of Travel Choice Modeling
HENSHER	Travel Behaviour Research: The Leading Edge
MAHMASSANI	Travel Behavior Research Opportunities and Applications Challenges
TIMMERMANS	Progress in Activity-Based Analysis

Related Journals

Transportation Research Part A: Policy and Practice
Editor: Phil Goodwin

Transport Policy
Editors: Moshe Ben-Akiva, Yoshitsugu Hayashi, John Preston & Peter Bonsall

For a full list of Elsevier transportation research publications visit:
www.elsevier.com/transportation

Access to Destinations

Edited by

David M. Levinson

University of Minnesota, USA

and

Kevin J. Krizek

University of Minnesota, USA

ELSEVIER

Amsterdam – Boston – Heidelberg – London – New York – Oxford
Paris – San Diego – San Francisco – Singapore – Sydney – Tokyo

ELSEVIER B.V.
Radarweg 29
P.O. Box 211, 1000 AE Amsterdam
The Netherlands

ELSEVIER Inc.
525 B Street, Suite 1900
San Diego, CA 92101-4495
USA

ELSEVIER Ltd.
The Boulevard, Langford Lane
Kidlington, Oxford OX5 1GB
UK

ELSEVIER Ltd.
84 Theobalds Road
London WC1X 8RR
UK

First edition 2005

Library of Congress Cataloging in Publication Data
A catalog record is available from the Library of Congress.

British Library Cataloguing in Publication Data
A catalogue record is available from the British Library.

ISBN-13: 978-0-08-044678-3
ISBN-10: 0-08-044678-7

∞ The paper used in this publication meets the requirements of ANSI/NISO Z39.48-1992 (Permanence of Paper).
Printed in The Netherlands.

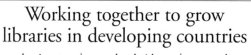

Dedication

This book is dedicated to our children.

ACKNOWLEDGEMENTS

This book would not have been possible without the initiative of the Center for Transportation Studies (CTS) at the University of Minnesota, who took the lead in planning and hosting an interdisciplinary conference in November 2004. The conference itself was funded by the University of Minnesota President's 21st Century Interdisciplinary Conference Series, with matching funds provided by CTS. Robert Johns, Gina Baas, and Stephanie Jackson at CTS and Teresa Washington of the College of Continuing Education were essential in organizing the conference. Representatives from several other organizations participated in planning the conference, including the Center for Urban and Regional Affairs (Tom Scott), Hennepin County (Gary Erickson), the Metropolitan Council (Connie Kozlak) and the Minnesota Department of Transportation (Abigail McKenzie). We would like to thank the army of almost 50 peer reviewers on whom we relied for detailed feedback on the scholarly contributions of the work presented herein. In addition, the chapter authors were responsive in addressing such comments and further improving their work. Donna Kern assisted with administrative matters related to the stringent review process. Most importantly, however, we would like to thank Peter Nelson at CTS for his diligent copy-editing and layout skills. Robert Johns, the director of CTS, was unwavering in his support of this book project throughout.

TABLE OF CONTENTS

PART III: APPLICATIONS

Access to Destinations
D.M. Levinson and K.J. Krizek (editors)
© 2005 Elsevier Ltd. All rights reserved.

CHAPTER 1

THE MACHINE FOR ACCESS

David Levinson, Kevin J. Krizek, University of Minnesota
David Gillen, University of British Columbia

In 1923, renowned architect Le Corbusier authored what is largely considered to be the best-selling architecture book of all time, *Vers Une Architecture* ("Towards a New Architecture"). Within the pages of his impassioned manifesto, he claimed that "A house is a machine for living." Architecture, he claimed, is able to improve people's lives. For Le Corbusier, Victorian cities were chaotic and dark prisons for many of their inhabitants; he was convinced that a rationally planned city could offer a healthy, humane alternative. Contemporary planning initiatives suggest the time is ripe to revisit this metaphor, albeit on a larger scale.

Much of land use-transportation planning today aims to reduce average vehicular delays, increase passenger throughput, and in general, keep traffic flowing smoothly and safety. The barometers used to measure such attributes include hours of delay, speed of traffic, number of cars in congestion. Such barometers have become accepted lore among populations from both the transportation industry and popular culture. Newspapers around the country wait eagerly for the well-known annual rankings from the Texas Transportation Institute to relay to their residents how well (or in a perverse sense of pride, how poorly) their city is performing.

Measures of congestion, however, have limited utility. They provide a snapshot of only a select dimension of a city's transportation system: the ability of residents to transport themselves under certain conditions (e.g., free flow travel times). Measures of mobility are merely concerned with the ability to move, but not with where one is going. In many respects, such measures fail to adequately capture other essential dimensions of a city's entire transportation environment—that is, how easy it is to get around. Transportation, concerned with the movement of matter (people and goods), is the machine for mobility.

Like a house is a machine for living, cities are the machine for accessibility. In a conference in 1993, Melvin Webber succinctly and eloquently outlined the motivation for and definition of 'accessibility' in a metropolitan context. He described the aim as a desire for greater connectivity and to maximize opportunities for social and economic interactions. The *ideal* city, he claimed, is "one that *maximizes* access among its interdependent residents and establishments." This central notion has been widely shared. For example, it is entirely consistent with Lewis Mumford's claim that the problem of urban transportation could be solved by "bringing a larger number of institutions and facilities within walking distance of the home" [i]

We agree with such statements even though these ideas are not necessarily novel. Many urbanologists have evoked the notion of accessibility. It has received considerable study over the years from researchers in land use, transportation, regional science, urban economics, and geography. However, the concept is receiving a resurgence of sorts from several professions. In response, at least three issues motivate this work:

- Quality of life and related policy dialogues are increasingly sensitive to the role of congestion throughout many metropolitan areas. The measures guiding such discussions—the Texas Transportation Institute's Congestion Indices—provide an incomplete picture,

- In depth transportation, land use, and city planning surrounding such issues deserve a balanced, and objective criterion on which to base future solutions. There is confusion over use of the various terms in such discussions, (see, for example, Susan Handy's paper in this volume on the misinterpretation of mobility versus accessibility),

- The time is ripe to revisit many of the concepts put forth during the novel work of the 1970's to suggest how matters have changed and progress that is being made over a quarter century later.

The idea of centering a research agenda—and a series of papers—around the concept of access is one that we have considered for some time in informal dialogue. There are several issues involved and questions that each of us have batted around for some time. For example, what are the implications for the different measures of accessibility. How do we know which one is right for which job? Is there a problem with providing excessive amounts of access? How do we know when a policy or project we have implemented achieves the *optimal amount* of access in a *real* city; that which is worth what it costs? There are ongoing debates (and more importantly, things taken for granted that should be debated) that the accessibility hypothesis can inform. First, what is the role of accessibility in the economy at large? Second, which, if any, mode of transportation (auto, transit, or other) should be favored? Third, what is the optimal city structure? And last, how do we best answer these questions and implement the answers given the fractures in the transportation field?

The optimal choice of accessibility today may be tomorrow's white elephant. Perhaps it is the failure to recognize that the optimum amount of access is not infinite, or permanent, and that access is not the result of a single technology or process that has fueled the debate of auto versus transit (low density versus high) as the appropriate vehicle for achieving access. The following section highlights additional matters.

ADDITIONAL CONSIDERATIONS

What creates accessibility, or why isn't economic and social activity ubiquitous? There are economies and externalities associated with the concentration of activity. These scale economies and spillovers allow greater choice ranging from consumer goods to entertainment to business specialization; how many small towns have a sushi bar, a symphony orchestra, or a stock market? One simply has to look at the differences in the range of opportunities between small towns and larger cities. Once you have economies of density (also called scale or agglomeration economies), it means there will be differences between places. If markets are working, in their attempts to be as well off as possible given the resources available (in the economist's jargon "maximizing utility under a budget constraint"), people will gravitate to those areas which offer them the greatest satisfaction. This is a very important point—small towns exist which have fewer services than larger towns, yet people move there, enjoy themselves, and die there. They have the amount of access they want.

Why should the private sector be concerned with accessibility? When individuals choose an activity, they must consider three elements: what they want to access, what everyone else wants to access, and what the people who run those places which are being accessed want. For instance: I want a bookstore nearby, my neighbors also want bookstores nearby, and somebody wants to sell books—great, we have ourselves a market. The owner, by locating his or her store and having extended operating hours, maximizes access to the most people. But if my neighbors didn't want books, I would be unable to support the bookstore myself. Or if nobody wanted to sell, we have the same problem. Or if the government set rules limiting store hours (for instance, "blue laws") or where bookstores could locate (zoning) we'd have less access.

Why must the public sector be concerned with accessibility? If all of the externalities from concentration were positive (e.g., increased market activity, access to specialized goods), then government should be in the business of encouraging higher density; if they were all negative (pollution, congestion, crime), government should discourage them. In the real world, there is a combination of the two, so there must be some attempt to optimize the level of density. That is the crucial issue; it is a central rationale for planning. But in the end it is a balancing act that reflects the varied preferences of people who want greater density and those who do not.

What constitutes the metric of accessibility, considering costs? In principle this is easy. Given relative costs it is the set of all goods and services (including social interaction) which the population could consume. Again in the parlance of the economist, it is the sum of aggregate consumer surplus. In practice, it is another story. Unlike most studies, accessibility analysis needs to consider variations in travel and cost over several time horizons (e.g., hour, day, week, month, and year), different modes and routes of travel; and activities in proportion to importance. It would be ideal to weigh each by their duration, frequency, or intensity. Moreover, the analysis should consider historical and forecasts of use patterns, not merely a single point in time, as well as all points in space. This is a much more difficult task to measure than simple congestion, but it will also explain far more. A proper metric for accessibility is essential to discovering the optimum mix of modes of access and optimal city structure.

SOME NORMATIVE PROPOSITIONS

We propose that a central role of cities is to help individuals and organizations maximize choice at the lowest cost. The government, and more specifically the public sector transportation profession, has been charged with providing certain components of the transportation system such as roadways, buses, and commuter rail. Further, the government regulates other components, including vehicles for safety and pollution. Also, communications can be considered a mode for many purposes. How often need we travel physically when all we really want to do is exchange information? But this provision is in response to people's desires. For example, the desire to have greater access to products led to the significant mail order business in America. With increasing demand, couriers supplemented and ultimately supplanted the Postal Service, for the most part, to reduce the transactions cost of shopping.

The parable of the mountains may be useful here. While some disagree, we view the collective goal of the city's residents as economic growth and progress. Analogous to mountain climbing they want get as high as possible, as quickly as possible. Assume there are two mountains, both very tall, named Mt. Auto and Mt. Transit. The final altitude of both mountains is obscured by the clouds. We (as society) had the choice of which mountain to climb. However, by climbing one, we travel further away from the valley that separates them and thus further from the peak of the other (not only do we have its height to overcome, we must retrace our steps). In the year 2005, we are committed to one mountain—Mt. Auto. There are those who claim that Mt. Transit is taller, and we should turn around, climb down our mountain and climb the other; but the farther up we go, the harder this alternative becomes. Moreover, we still have no assurances about the true height of either mountain , only some forecasts which are little better than astrology. Some prophets who preach "Pedestrian Friendly Design" or "Transit Oriented Development" warn that atop of Mt. Auto are dragons and monsters (e.g., Global Warming, environmental destruction, alienation from community). We climbed up Mt. Transit in the 1800s, in that bygone era of mass production and centralization, the era of a single downtown and high-density cities. We decided, from our vantage atop Mt. Transit, that Mt. Auto was higher. By 2005 we have long surpassed our peak on Mt. Transit. Maybe Mt. Transit is taller now than it used to be, the geology of complex systems changes suddenly by introducing new technology. Perhaps the fog will lift, we may yet find a bridge between the two, or more likely, we will see a third and taller mountain. All of this begs continued discussion—and research—about the form and function of our cities. Accessibility is central to such analysis.

If accessibility is paramount, what is the optimal city form to maximize access at the lowest cost? Clearly this depends on a number of factors: the cost of money, land, risk, construction, energy, information, transactions, and time come immediately to mind. When interest rates are high, short-term investments make more sense. If land is scarce (as on a peninsula such as Hong Kong or San Francisco), density makes more sense than dispersion. While construction is cheap, high rises have merit over low. A city which is particularly vulnerable to disaster (for instance, earthquakes) shouldn't put all its eggs in one basket, but rather spread the risk over more land. Where energy is expensive, one wants to conserve it. If the value of time is low, it is cheaper to make people commute longer rather than build new infrastructure. Further, these costs are not permanent over time. They tend to change from year to year, and in the case of interest rates, minute to minute, making impossible a definitive answer. There is a continuous struggle to find the optimal balance.

Asking the question another way: What should we be doing, maximizing opportunities for people (maximizing jobs, for example) or minimizing the transactions cost of consuming these opportunities? The latter, which transportation professionals have typically undertaken, wears the blinders of mobility; the former widens the view to include access. The optimal amount of access is not infinity; it is only that which is worth what it costs. So it comes back to the balancing of social costs and social gains. Both measures have to include private gains and positive and negative externalities. It is our proposition that the role of the city is to facilitate the market by providing maximum choice in destinations/activities at the lowest full cost (including individual travel time, as well as social cost of infrastructure, cost of externalities, etc. which should be charged to the individual).

While accessibility has as its goal something broader than reducing congestion, there is a place for congestion reduction. One means for maximizing choice includes reducing congestion where cost-effective through supply and demand management and facility expansion. Tools include pricing transportation—as communication and energy transmission are already, so that individuals can make trade-offs between time and money. Similarly, providing information so that people learn a shortest route, and if anyone else is going to the same place at the same time, will help manage the system—only in personal transportation do we rely on individuals to route their packages (themselves). Other utilities, be it the phone or power companies, or even the package delivery services, coordinate that function centrally.

The scientific program of the past few centuries has steadily divided the fields of knowledge into more and more specialized disciplines. In this reductionist approach, each discipline established its own language, journals, goals, and research agendas. On occasion, there were syntheses: Smith's Invisible Hand, Darwin's Theory of Evolution and Natural Selection, Einstein's Space-Time Continuum. Transportation has not escaped the reductionist trend, we have traffic signal controllers, highway engineers, transit planners, traffic modelers, travel demand managers, transport economists, and the like. Moreover these groups are distinct from urban designers, community planners, growth managers, land use forecasters, and demographic modelers. Each discipline has its own objectives, which are coordinated through the visible hand of government that employs them, and the invisible hand of the market that directs government action.

A CALL FOR ADDITIONAL RESEARCH AND METRICS

Aside from political expediency (the maximization of voter satisfaction), there is no overarching goal of government action in the city. There are the occasional platitudes such as the transportation mobility goal of "safe and efficient movement of people and goods," but nothing on which to evaluate such aspirations. Urban designers are adept at formulating "theories"—manifestoes really, which purport to instruct us how to shape our urban environment. Environmentalists establish their own core beliefs. Engineers adopt various rules, the AASHTO Green Book, the Highway Capacity Manual, which inform and instruct the design of transportation facilities. Planners have various service standards to measure the quality of systems, level of service at intersections, the travel-time response of fire departments, etc. All of these rules lack a core. They are assertions little better than religion in some cases, or averages of (questionable) practice in others.

In the theory of evolution, it is suggested that those individuals (or even their genes) who adopt the best rules are more likely to survive and propagate than those who have inferior behaviors. The same may be true for cultures and societies. We have the advantage over other species that we can rewrite our rules quickly and consciously, with an aim to perfect them. We need hypotheses, which can be refuted and revised, not merely asserted, followed and believed. We need evaluation of policies, not mere promulgation of rules. There is a saying in science "Theories Destroy Facts," an overarching theory explains the accumulation of data in a field, such that once a theory is corroborated, it is essential to learn the theory rather than each specific case. Once a theory is established, the specific observations and rules of behavior can be tested against whether they are consistent with the greater theory. Is "The City is the Machine for Access" such a theory?

The issue of the interaction of location and travel is central to community design, urban economics, regional planning, and transportation engineering. Maintaining, or attaining, job/housing balance is becoming a goal of many regional and local plans. However, this goal conflicts with the restrictive zoning policies mandated by many communities and with the gains from agglomeration that results in central *business* districts. Similarly, growth management schemes are often based on the desire to restrict traffic congestion. Traffic engineers attempt to maximize flow at the highest speed, and have established level of service standards to identify congested areas, and restrict trip-making (land development) in their vicinity. But by looking only in the immediate area, these traffic impact studies ignore more regional effects—restricting development in one area, thereby pushing it into another, may only increase the total amount of travel and traffic. Traveling 30 minutes at Level of Service C is viewed as preferred to traveling 15 minutes at LOS F. We believe these goals or rules are misguided. If the city is the machine for access, then policies need to be evaluated against that metric—policies that are based on empirically grounded research.

The chapters in this volume provide a focus on which to base such metrics and policies around the central concept of accessibility. They are wide-ranging, including issues of transit, network growth, definitions, and modeling. The papers presented herein are the result of an interdisciplinary conference held at the University of Minnesota in November, 2004. The conference was funded by the University of Minnesota President's 21st Century Interdisciplinary Conference Series as well as the Center for Transportation Studies. Each of the papers presented in this volume was subject to a critical peer review of between two and four experts in their respective field. Because of the broad spectrum of chapters contained within this volume, there is something for everyone. The practicing planner will find several chapters directly applicable to their work; there is theory; there is applied research; and even advanced econometric modeling.

Part I—Overview—begins with Bertini's account of alternative definitions of traffic congestion in U.S. metropolitan areas (Chapter 2). The chapter discusses current definitions of metropolitan traffic congestion and ways it is currently measured and describes the accuracy and reliability of these measures, and reviews how congestion has been changing over the past several decades. The results of a survey among transportation professionals are summarized to assist in framing the issue. Additional analysis of recent congestion measures for entire metropolitan areas is provided, using Portland, Oregon and Minneapolis, Minnesota as case examples. Some discussion of the stability of daily travel budgets and alternative viewpoints about congestion are provided.

In Chapter 3, Gifford complements the above by investigating society's discontent with congestion. He examines the history of congestion and the effectiveness of travel demand management programs and operations programs. He considers the prospects of congestion pricing and High Occupancy/Toll lanes.

Miller argues that traditional, place-based measures of accessibility need to be enhanced and complemented with people-based measures that are more sensitive to individual activity patterns and accessibility in space and time (Chapter 4). He reviews place-based measures, suggests what new tenets need to be incorporated, and offers several strategies for measuring people-based accessibility.

Knaap and Song show the effects of policy on land use and transportation, and ultimately on behavior. After reviewing the literature on the effectiveness of infrastructure on shaping land use they then turn to policy, showing that market-oriented policies will produce conventional density gradients, with the highest densities in the city center, while non-market regimes (e.g. Brasilia, Johannesburg, and Moscow) have much less traditional urban forms. As part of their description in Chapter 5, they investigate current U.S. land use regulations and incentives, including urban growth boundaries, subdivision regulation, land use plans, transferable development rights, and priority funding areas.

Krizek reviews various perspectives that exist in the literature on both neighborhood and regional accessibility (Chapter 6). He urges the need to consider not only what modes are available, but which ones are attractive in assessing the inter-relationship between infrastructure, urban form, and travel behavior.

Chapter 7 consists of Handy's description of how mobility and accessibility are distinct concepts with vastly different implications for planning. This chapter describes the concepts in theory and in practice by examining the language in a sample of regional transportation plans in Northern California.

The second part of the volume—Behavior and Measures—begins with Kim and Morrow-Jones challenging the assumption that access to the workplace matters in residential location decisions (Chapter 8). Working with a longitudinal dataset based in the fast growing city of Columbus, Ohio, they show empirically how other, more pressing matters of accessibility (such as school quality) dominate most residential location decisions.

Yang and Ferreira (Chapter 9) compare three measures relating the spatial separation of jobs and housing: ratio of jobs to employed residents, accessibility, and minimum required commute (MRC) for both Boston and Atlanta to commuting costs. While concluding that MRC has the most explanatory power, the authors go on to suggest some alternatives such as proportionally matched commuting that should be considered in future analyses.

Horner and Mefford, using a case study of Austin, Texas, examine how an urban bus transit system provides accessibility to residents (Chapter 10). They apply accessibility indices at two different scales: disaggregate measures for the neighborhood level and aggregate measures at the system-wide level.

Relying on detailed parcel level data from Seattle, in Chapter 11, Lee develops some very precise and accurate measures of transit travel time, including both in-vehicle and access times, which can be used to assess accessibility. He applies his method to grocery stores in Seattle and identifies significant gaps in service.

The volume then turns to the third part—Development—where Levinson and Chen use a Markov chain model to analyze the spatial co-evolution of transportation and land use (Chapter 12). They find that existing agricultural and recreational zones that contain highways are much more likely to convert to employment and residential zones than those without highways, suggesting infrastructure drives development. Similarly, highways were more likely to be added to developed areas.

In Chapter 13, Woudsma and Jensen apply transportation and land use data for Calgary, Alberta, Canada to test whether transportation system performance has a quantifiable influence on the timing and spatial character of Distribution-Logistics-Warehousing (DLW) land use development. They find there is a positive relationship between DLW land development and transportation accessibility—those locations with better accessibility (smaller congestion influence) were developed more often over those which did not offer better accessibility. However, there are inconsistencies whereby some destinations are more influential than others suggesting that accessibility is spatially uneven. But they also show that there is a 5-10 year temporal lag in the relationship.

The final contribution under the domain of development comes from Ottensman who describes how matters of accessibility are incorporated into land use-transportation forecasting models (Chapter 14). The LUCI model, he demonstrates, easily allows a user to create and compare of future development scenarios reflecting a wide range of dimensions related to accessibility such as policy choices and assumptions affecting future development.

The final part of the book—Applications—describes how the concept of accessibility has been applied in various contexts. This part begins with Chapter 15 where Primerano and Taylor argue that the accessibility of a location does not necessarily reflect the accessibility of an individual to that destination. They suggest the focus should be on accessibility to activities, which can be satisfied by a number of destination alternatives. They reviews existing measures of accessibility from three different perspectives: the traveler (individual or group); the transport system (mode, roads and traffic characteristics); and land-use (characteristics of land-uses at origins and destinations).

Abdel-Rahim and Ismail present a graph-based approach to model accessibility in urban transportation networks under different scenarios (Chapter 16). Rather than based on the assumption of optimization, their approach incorporates issues of connectivity and functionality.

Using longitudinal data covering entire Switzerland from 1950 to 2000, Tschopp, Fröhlich, and Axhausen analyze the development of spatial accessibility and models its impacts on demographic and economic change by using a multi-level regression approach (Chapter 17). They maintain that Switzerland has been developing a more dispersed form over this period.

Finally, Ferguson considers the relationship between parking and accessibility in Chapter 18. Parking, by consuming space, reduces the space available for other activities, and thus negatively affects most measures of accessibility. However, parking that is difficult to access from destinations also has consequences. More attention needs to be paid to parking pricing, supply, and access costs from parking.

The end result is a collection of papers that serves to broaden our understanding of the meaning of accessibility. As we learn more, we discover more of what we don't know. Many of the papers raise questions that provide a research agenda for years to come on understanding the relationship between infrastructure, location, behavior, and accessibility.

NOTES

[i] Lewis Mumford (1968), p.70 in *The Urban Prospect*

Access to Destinations
D.M. Levinson and K.J. Krizek (editors)
© 2005 Elsevier Ltd. All rights reserved.

CHAPTER 2

CONGESTION AND ITS EXTENT

Robert L. Bertini, Portland State University

INTRODUCTION

"You're not stuck in a traffic jam, you are the jam." – *German public transport campaign*
(Kay, 1997)

Congestion—both in perception and in reality—impacts the movement of people and freight and is deeply tied to our history of high levels of accessibility and mobility. Along spatial and temporal dimensions, traffic congestion has been around since ancient Rome (Downs, 2004); it wastes time and energy, causes pollution and stress, decreases productivity and imposes costs on society equal to 2–3% of our gross domestic product (GDP) (Cervero, 1998). In terms of technology, it was noted that an automobile is "a conveyance which is capable of moving 1.6 km (a mile) a minute, yet the average speed of traffic in large cities is of the order of 17.7 km/h (11 mph)." (Buchanan, 1963). For 2002, it was estimated that congestion "wasted" $63.2 billion in 75 metropolitan areas during 2002 because of extra time lost and fuel consumed, or $829 per person. (Schrank and Lomax, 2004) Some refer to these kinds of estimates as misleading since the prospect of eliminating all congestion during peak periods is "only a myth; congestion could never be eliminated completely." (Downs, 2004).While some research emphasizes that "rush hour is longer than an hour in the morning and an hour in the evening and few people are 'rushing' anywhere," others say that "gridlock is not going to happen because people change what they do long before it happens." (Garrison and Ward, 2000) Some view congestion as a "problem" that individual drivers are subject to, while others emphasize that the users of transportation networks "not only experience congestion, they create it." It has been shown that most people make travel decisions based on an expectation of experiencing a certain amount of congestion; while "few consider the costs their trips impose on others by adding to congestion." (Mohring, 1999) The objective of this chapter is to discuss current definitions of metropolitan traffic congestion and ways it is currently measured. In addition, the accuracy and reliability of these measures will be described along with a review of how congestion has been changing over the past few decades.

FRAMING THE ISSUE

Congestion measurement can be thought of as focusing on system performance and measures of people's experiences. In order to assist in framing the issues, with a focus on how people think about congestion, an unscientific survey about metropolitan area congestion was distributed by email to more than 3,500 transportation professionals and academics, and a total of 480 responses were received. The survey, conducted specifically for this chapter, asked four qualitative questions:

- How do you define congestion in metropolitan areas?
- How is congestion in metropolitan areas measured?
- How accurate or reliable are traffic congestion measurements?
- How has metropolitan traffic congestion been changing over the past two decades?

Respondents were provided with an opportunity to comment on congestion in general. The survey results are described below and are used to motivate later elements of the chapter.

Definition of Congestion

In attempting to define congestion, a total of 557 responses were provided since many responses included separate definitions for freeways and signalized intersections. As shown in Figure 1, survey respondents mentioned time, speed, volume, level of service (LOS) and traffic signal cycle failure (meaning that one has to wait through more than one cycle to clear the queue) as the primary definitions of congestion. Respondents who used the term "LOS" were not more specific; typical LOS measures include volume/capacity, density, delay, number of stops, among others. The majority of the responses included a "time" component—travel time, speed, cycle failure and LOS are all related to the fact that users experience additional travel time due to congestion. It is clear from these responses, that some definitions of congestion rely on point measures (e.g., volume and time mean speed) and some rely on spatial measures (travel time, density and space mean speed).

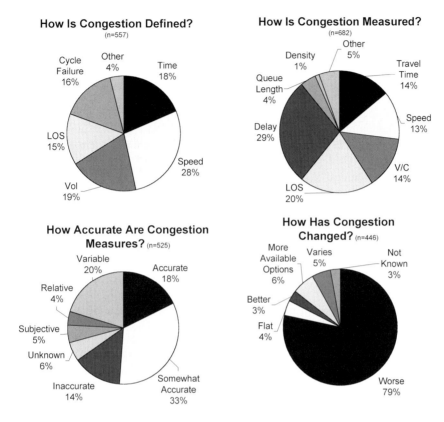

Figure 1. Congestion Survey Results

The definitions of congestion that are related to point measures include vehicle count (expressed as flows) and time mean speed extracted from point detectors (often extrapolated over a segment to estimate link travel time). A point-based travel time estimate can be compared to a free flow travel time for a link, where the difference between actual and free flow travel times is defined as the delay. The product of the number of vehicles to pass a point during a given time interval and the length of an associated segment will provide the vehicle-kilometers traveled (VKT). The product of the number of vehicles passing the point during a time interval and the travel time for the associated link results in the vehicle-hours traveled (VHT). Definitions of congestion related to spatial measures include density, queue length and actual segment travel time (recorded by a probe vehicle). Delay can be calculated as actual travel time over a segment minus the free flow travel time for that segment.

Several survey responses included important comments. It was pointed out that "if we want to reduce congestion we need to be able to define it and quantify it." Some were willing to define congestion as "anything below the posted speed limit," or below some "speed threshold (e.g., <35 mph)." Others noted that "congestion is relative," "a perception" and, "I know it when I see it." One response pointed out that congestion was not so much of a concern anymore since "fortunately we have had an economic collapse." Other surveys have found similar results. In the U.K. respondents defined congestion as stop-start conditions (38%); a traffic jam with complete stops of 5+ min at a time (24%); having to travel at less than the speed limit (19%); and moving very slowly at less than 10 mph (16 km/h) (17%). (U.K. Department for Transport, 2001) A conference of European Transport Ministers concluded that there is no widely accepted definition of road congestion, but that an appropriate definition might be: "the impedance vehicles impose on each other, due to the speed-flow relationship, in conditions where the use of a transport system approaches its capacity." (European Conference of Ministers of Transport, 1998) Finally, a survey conducted by the National Associations Working Group for ITS included a response describing the difficulty in defining congestion: "you know it when you see it—and the severity of the problem should be judged by the commonly accepted community standards." (Orski, 2002)

Measurement of Congestion

There were 682 responses to the question asking how congestion is measured (multiple response). As shown in Figure 1, most responses were related to time: delay, speed, travel time and LOS, all of which include the notion that actual travel time can be a primary measure of congestion. Other measures included volume/capacity (point measure) and queue length and density (spatial measures). The small number of responses in the "Other" category included such measures as number of stops and travel time reliability. One responded summed up the issue by stating "it is never truly measured." The literature includes a wide array of possible congestion measures including: volume/capacity (disregards duration),VKT, VKT/lane km, speed, occupancy, travel time, delay, LOS and reliability. In the U.K. survey several helpful measures of congestion were identified: delay (51%); risk of delay (20%); average speed (18%); and amount of time stationary or less than 10 mph (16 km/h) (11%).

Accuracy and Reliability of Congestion Measurements

Next, respondents were asked, "how accurate or reliable are traffic congestion measurements?" There were 525 responses to this question, indicating that respondents had mixed feelings about the accuracy/reliability of congestion measurements. As shown in Figure 1, about half of the responses indicated that the measurements are accurate or somewhat accurate, while the other half indicated that they are inaccurate or variable. Many comments indicated that congestion measurements are often based on very small sample sizes, that they are relative and variable and that they should be presented with confidence intervals rather than purely deterministic values. One comment stated that congestion measurements are "reasonably accurate despite the fact that they measure the wrong things," while another stated that "congestion perception is a personal thing based on personal experiences and anecdotes." Finally, one comment indicated that "the result is really just a snapshot in time."

Changes in Congestion

The final survey question aimed to assess the extent to which urban traffic congestion has changed over the past two decades; most respondents indicated that congestion has worsened. Some respondents indicated that more transportation options (such as transit and intelligent transportation systems) are now available and some indicated that congestion has gotten better. The response is not surprising since the U.S. population increased by 24% to 282 million from 1980-2000, the number of highway vehicles has increased 39% to 225 million and the passenger car VKT has increased 44% during the same period. Respondents commented that the impact of congestion has increased in both spatial and temporal dimensions, including the spread of peak periods. Others pointed out that change has been relative, depending on the area and on the user. For example, "western cities are experiencing increasing congestion, as both population and per capita kilometres of travel increase. Some rust belt cities, on the other hand, are experiencing decreasing population and thus are experiencing decreases in congestion." Some respondents indicated that drivers have been conditioned to tolerate more congestion, using an analogy about how to boil a frog: "... start out by putting him cold water, gradually raise the temperature, and he won't figure it out (and thus escape) so he stays 'til he's cooked! We're all gradually getting cooked!"

Several responses pointed out that the congestion will always be a by-product of a healthy, vibrant urban area. In this context, it was pointed out that traditional traffic/transportation engineering antidotes to congestion have been reactive in nature, and that roadways are not improved until there is a problem. Another comment stated that "current conceptions of congestion have more to do with preserving the world as it was, rather than preparing us for the world as it will be," and another stated that there is "too much focus on congestion, there should be more attention to accessibility. What can people get to in a reasonable period of time (20-30 minutes)?" Several respondents indicated that there has been a transformation from the mentality that you can "build your way out of congestion" to the point where other options such as HOV lanes and reversible lanes" are available.

Other Survey Comments

Survey respondents offered comments and suggestions. These included sometime conflicting suggestions to consider alternative transportation strategies such as: congestion pricing, more transit, demand management, regional/statewide approaches, more funding, higher gas taxes, more frequent systematic assessment of traffic operations, driver and citizen information and education, probe vehicles, multi-jurisdictional traffic signal operations, incident reduction, better land use policies, consideration of user vs. system goals, consideration of impacts on health and air quality, focus on moving people during the peak hour, more operations improvements, ITS improvements, more understandable measures, consideration of how trucks affect capacity and simply building more lanes. On the other hand, other respondents suggested: consider equity impacts of congestion pricing, don't build more lanes, don't waste money on ITS, don't waste money on transit and "don't spend too much time on it." Now that the input from a survey of experts has been considered, it is important to examine the literature.

LITERATURE REVIEW

The survey results motivated a comprehensive literature review in order to explore current federal, state and local efforts to define and quantify congestion.

Federal Definition and Monitoring

The Federal Highway Administration (FHWA) defines traffic congestion as: "the level at which transportation system performance is no longer acceptable due to traffic interference." Because there is a relative sense to the word "congestion," the FHWA continues their definition by stating that "the level of system performance may vary by type of transportation facility, geographic location (metropolitan area or sub-area, rural area), and/or time of day," in addition to other variations by event or season. (Lomax, Turner and Shunk, 1997) The definition of congestion is imprecise and is made more difficult since people have different perceptions and expectations of how the system should perform based on whether they are in rural or urban areas, in peak/off peak, and as a result of the history of an area.

Congestion can vary since demand (day of week, time of day, season, recreational, special events, evacuations, special events) and capacity (incidents, work zones, weather) are changing. Most researchers agree that recurrent congestion (due to demand exceeding capacity (40%) and poor signal timing (5%)) makes up about half of the total delay experienced by motorists, while nonrecurrent congestion (due to work zones (10%), incidents (30%) and weather (15%)) makes up the other half. It has been shown that four components interact in a congested system (Lomax, Turner and Shunk, 1997):

- Duration: amount of time congestion affects the travel system.
- Extent: number of people or vehicles affected by congestion, and geographic distribution of congestion.
- Intensity: severity of congestion.
- Reliability: variation of the other three elements.

Because of user expectation, one proposal is to define "unacceptable congestion" as the travel time in excess of an agreed-upon norm, which might vary by type of transportation facility, travel mode, geographic location, and time of day. (Lomax, Turner and Shunk, 1997) "A key aspect of a congestion management strategy is identifying the level of 'acceptable' congestion and developing plans and programs to achieve that target." (Lomax et al., 2001) Based on U.S. Census data, an extensive analysis of commuting patterns has been conducted (Pisarski 1996). In this analysis of journey to work data, there seem to be several thresholds for unacceptable congestion occurring: if less than half of the population can commute to work in less than 20 minutes or if more than 10% of the population can commute to work in more than 60 minutes. It is apparent that several agencies use the term "acceptable congestion," but clearly this can mean different things to different people and at different times and locations. In this context, it has been argued that individuals and firms may choose to locate in a congested area due to easier access to other individuals and firms. (Taylor, 2002) This highlights the need to consider the interaction between transportation and land use when attempting to define congestion.

The FHWA has initiated a Mobility Monitoring Program based on measured travel time in which they are trying to answer a mobility question: "how easy is it to move around?" and a reliability question: "how much does the ease of movement vary?" The measures used include (Turner, et al., 2002; Lomax, Turner and Margiotta, 2001; Jung, et al., 2004; FHWA, 2004):

- *Travel time index*: ratio of travel conditions in the peak period to free-flow conditions, indicating how much longer a trip will take during a peak time (a travel time index of 1.3 indicates that the trip will take 30 percent longer). The calculation of this index assumes a capacity of 14,000 vehicles per freeway lane per day and 5,500 vehicles per principal arterial lane per day and compares measured VKT to these assumed capacity values.
- *Average duration of congested travel per day* (hours): "How long does the peak period last?" Trips are considered across the roadway network at five-minute intervals throughout the day. At any time interval, a trip is considered congested if its duration exceeds 130% of the free-flow duration. When more than 20% of all trips in a network are congested in any five-minute time interval, the entire network is considered congested for that interval. The total number of hours in which the network is designated as congested is reported in this measure.
- *Buffer index*: this measure expresses the amount of extra time needed to be on-time 95 percent of the time (late one day per month). Travelers could multiply their average trip time by the buffer index, and then add that buffer time to their trip to ensure they will be on-time for 95 percent of all trips. An advantage of expressing the reliability (or lack thereof) in this way is that a percent value is distance and time neutral.

A recent synthesis examined more than 70 possible performance measures for monitoring highway segments and systems (NCHRP, 2003). From users' perspectives, key measures for reporting the quantity of travel included: person-kilometers traveled, truck-kilometers traveled, VKT, persons moved, trucks moved and vehicles moved. In terms of the quality of travel, key measures included: average speed weighted by person-kilometers traveled, average door-to-door travel time, travel time predictability, travel time reliability (percent of trips that arrive in acceptable time), average delay (total, recurring and incident-based) and LOS.

Other Congestion Definitions and Monitoring

A review of the literature reveals that transportation agencies have adopted particular definitions of congestion for their purposes. INCOG, the regional council of governments in Tulsa, Oklahoma defines congestion as "travel time or delay in excess of that normally incurred under light or free-flow travel conditions." (INCOG, 2001) INCOG applies their policy of identifying recurring congestion and documenting its magnitude. To do so, traffic counts are compared to capacity and then the ratio of volume/capacity is expressed as a level of service. Tulsa uses traffic counts (and traffic volume forecasts) as an initial screen to locate congested routes and future problems. In Rhode Island, the state DOT recognizes that "congestion can mean a lot of different things to different people." As a result, the state attempts to use objective congestion performance measures such as percent travel under posted speed and volume/capacity ratios. In Cape Cod, Massachusetts, a traffic congestion indicator is used to track average annual daily bridge crossings over the Sagamore and Bourne bridges. (Cape Cod Center for Sustainability, 2003) This very simple measure was chosen for this island community since it is appropriate, easy to measure, and since historic data are available to monitor long-term trends. In the State of Oregon, the 1991 Transportation Planning Rule (TPR) uses VKT as a primary metric, with a goal of reducing VKT by 20% per capita in metropolitan areas by 2025.

In Minnesota, freeway congestion is defined as traffic is flowing below 45 mph (72 km/h) for any length of time in any direction, between 6:00 a.m. and 9:00 a.m. or 2:00 p.m. and 7:00 p.m. on weekdays. Michigan defines freeway congestion in terms of LOS F, when the volume/capacity ratio is greater than or equal to one. Since the function of the transportation system is to provide transport of people and goods, and its benefits are a function of the number of trips served, in California "congestion" is defined as the state when traffic flow and the number of trips are reduced. The California Department of Transportation (Caltrans) defines congestion as occurring on a freeway when the average speed drops below 35 mph (56 km/h) for 15 minutes or more on a typical weekday. There is currently a proposal to change the definition of congestion to be measured as the time spent driving below 60 mph (97 km/h), based on analysis of 3363 loop detectors at 1324 locations as part of the California Performance Measurement System (PeMS) database (Varaiya, 2002). The State of Washington DOT aims to provide congestion information (in plain English) that uses real time measurements, reports on recurrent congestion (due to inadequate capacity) separately from nonrecurrent congestion (due to incidents). This includes the measurement of volumes, speeds, congestion frequency, and geographical extent of congestion, travel time and reliability. The Washington DOT also focuses on travel time reliability and predictability by presenting a "worst case" travel time for a set of corridors such that commuters can expect to be on time for work 19 out of 20 working days a month (95 percent of trips), if they allow for the calculated travel time. (Washington State Department of Transportation, 2004)

The Urban Congestion Report

The Urban Mobility Report (UMR) (Schrank and Lomax 2004), sponsored by a consortium of state departments of transportation and several interest groups, has been conducted by the Texas Transportation Institute since 1982 (FHWA 2004). The very popular UMR (see http://mobility.tamu.edu) tracks congestion patterns in the 75 largest U.S. metropolitan areas. The main mission of the UMR is to convert traffic counts to speeds, so that delay can be computed. Since 2002, the UMR has also reported on the contributions of operational strategies (such as incident management and ramp metering) and public transportation have on reducing delay (FHWA 2004).

The UMR uses several measured variables reported as part of the Highway Performance Monitoring System (HPMS). The HPMS was developed by the FHWA and the states in 1978 to promote a systematic, national approach for identifying highway conditions, estimating capital investment needs, and measuring changes in highway conditions over time (Hill et al. 2000). In support of the HPMS, states are required to report 70 data elements on pavement condition, traffic counts, and physical design characteristics for a statistical sample of about 100,000 highway sections. For some segments, traffic count data are available from continuous (usually hourly) automatic traffic recorder systems, while on other segments these data are measured over 48 hour periods on a triennial basis. The UMR uses the following measured and reported variables for its analysis (for facilities defined in the HPMS as freeways and principal arterials):

- Population: U.S. Census data are obtained for metropolitan areas. The census definition of a metropolitan area may or may not coincide with city, county or metropolitan planning organization (MPO) limits or urban growth boundary.
- Urban Area Size: this variable is based on census definitions of metropolitan areas.
- Segment Length: length of each freeway or principal arterial segment.
- Number of Lanes: number of lanes for each freeway or principal arterial segment.
- Average Daily Traffic (ADT): total daily traffic volume of freeway or principal arterial segment.
- Directional Factor: estimate of directional split for average daily traffic volume.

The UMR takes these measured parameters and follows some well-documented procedures toward the production of the performance measures listed above. In order to complete the process, a number of assumptions and constants are used, including:

- Vehicle Occupancy: 1.25
- Working Days Per Year : 250.
- Consumer Price Index (CPI): taken from the U.S. Department of Labor.
- Value of Time: $13.45 (2002 value; adjusted using CPI).
- Commercial Vehicle Operating Cost: $71.05 (2002 value; adjusted using CPI).
- Vehicle Mix: 5% commercial vehicles.

- ·Fuel Cost: taken from the American Automobile Association.
- Peak Periods: assumed to be 6:00-9:30 AM and 3:30-7:00 PM.
- Percent of Daily Travel in Peak Period: assumed to be 50%.
- Uncongested "Supply" (vehicles per lane per day): assumed to be 14,000 for freeways and 5,500 for principal arterials.
- Relation between Road Congestion Index (RCI) and Percent of Daily Travel in Congested Conditions: the RCI is a ratio of daily traffic volume to the supply of roadway, and is applied using one piecewise linear relation (see Exhibit B-4 in Schrank and Lomax 2004).
- Relation between ADT and Speed for Freeway (Peak and Off Peak Direction) and Arterial (Peak and Off Peak Direction): applied using equations (see Exhibits B-7 and B-8 in Schrank and Lomax 2004).
- Free Flow Speed: assumed to be 96 km/h (60 mph) for freeways and 56 km/h (35 mph) for principal arterials.

Given the measured or estimated traffic counts, data describing the length and numbers of lanes for each freeway and principal arterial segment and the constants described above, the UMR then computes nine derived variables for each metropolitan area:

- Daily VKT by Facility Type: based on Segment Length and ADT data.
- Lane Miles by Facility Type: based on Segment Length data.
- Road Congestion Index (RCI): a ratio of daily traffic volume to the supply of roadway, based on Daily VKT and Lane Kilometer data.
- Percent of Congested Travel During Peak Period: based on RCI Relation.
- VKT by Congestion Level and Direction: based on Percent of Congested Travel and Directional Factor data.
- Segment Speed by Congestion Level and Direction: based on Relation between ADT and Speed.
- Delay: based on Speed and Free Flow Speed.
- Travel Rate (minutes/km) by Facility Type (actual and free flow): based on calculated actual Speed and Free Flow Speed.
- Travel Rate Index (TRI): based on VKT and Travel Rate.

Given the count based estimates of speed, and assuming free flow speeds by facility type, the UMR reports four primary performance measures:

- Annual Delay per Traveler: Extra travel time for peak period travel during the year for freeways and principal arterials.
- Travel Time Index (TTI): The ratio of travel time in the peak period to the travel time at free flow conditions.
- Travel Delay: Extra travel time for peak period travel above that required for travel at free flow conditions.

- Excess Fuel Consumed: Increased fuel consumption due to travel in congested conditions rather than free-flow conditions.
- Congestion Cost: Value of delay and excess fuel consumption converted to dollars for person and commercial vehicle travel.

It is clear that the UMR results are based on traffic count data that were originally collected for system monitoring. No actual traffic speeds or measures extracted from real transportation system users are included, and it should be apparent that any results from these very limited inputs should be used with extreme caution. To its credit, the UMR does leverage existing data sources (using 6 measured variables and 13 constant values or relations) and produces a document that provides a basis for drawing some limited conclusions. The annual release of the UMR results in widespread media coverage and often major headlines describing worsening congestion and comparing one metropolitan area to another. However, the UMR authors caution against comparisons between cities, emphasizing that the UMR is more appropriate for comparisons of trends for individual cities, rather than focusing on any value for a particular year. Further, the performance measures should be viewed with skepticism since in any one year, there are traffic counts that are three years old, very limited data (48 hour counts) are extrapolated over the entire freeway and principal arterial system, and the U.S. Census definitions of metropolitan area populations and areas may be at odds with a region's actual boundaries or urban growth boundary. The Census definition of "urban" may be different than that of a particular region, and traffic volumes actually accommodated in a metropolitan area may be excluded from the UMR due to boundary definition inconsistencies. It is also worth noting that there may be a lag in reporting changes to metropolitan area population and size data coupled with lags in reporting changes in the extent of the freeway and arterial system. This may result in spikes in the reported UMR performance measures.

A number of other seemingly arbitrary assumptions are also worth noting. The UMR applies uniform definitions of vehicle occupancy, vehicle mix, lengths of peak periods, amount of travel occurring during peak periods, daily capacities of freeway and principal arterial lanes, and the relations between volume and speed. It is probable that the combination of measurement error noted above plus the errors introduced by these constants end up producing results with some quantifiable error distribution—but this is not reported. The UMR also defines congestion as travel occurring at anything less than a pre-defined national value of free flow speed. As noted earlier in this chapter, there may be diverging opinions about what an appropriate speed threshold for congestion should be—again this is not included in the UMR. With this review of the literature in mind, there are some basic theoretical issues that should be explored, so that various metrics and their derivations can be better understood.

SOME BASIC THEORY

Having reviewed the literature, it is clear that many common traffic measurements are derived from the basic traffic flow parameters—flow, density and speed. This section describes how these fundamental measures can be applied at the level of the roadway segment, a corridor and over an entire door-to-door trip.

Segment Level

Figure 2 illustrates some basic points about traffic flow. In Figure 2(a), a set of vehicle trajectories on a time-space plane is shown in the context of a roadside observer (or detector) at location x. (Daganzo, 1997) During time interval t, an observer would count 7 vehicles passing point x. Flow, a point measure, is defined as the number of vehicles that pass a point during a particular time interval; in this case $7/t$, usually expressed in vehicles/hour. Under certain circumstances the "capacity" of the highway at point x might be estimated, and the actual measured volume could be compared to that theoretical capacity value in the form of a volume/capacity ratio. Speed could also be measured at point x, for example by a radar gun. If the arithmetic average of the speeds measured at a point is taken over a measurement interval t, this is called the time mean speed.

Figure 2(b), which also shows a set of vehicle trajectories on a time-space plane, illustrates that some key traffic flow parameters are measured over distance. For example at time j, the number of vehicles on the segment d at that instant would be counted as six vehicles. The density at time j is the number of vehicles on the section at that time divided by the section length, in this case $6/d$, usually expressed in vehicles/km. The actual travel times of vehicles can also be recorded over space; in this case for vehicle i, its travel time is shown as v_i. The free flow travel time for segment d might be assumed to be v_f. Therefore, for vehicle i on this roadway segment the delay is defined as v_i-v_f.

Depending on what data collection system is available, sometimes a point measure, such as speed, can be applied over a roadway segment. As shown in Figure 2(c), if a roadway is equipped with measurement sensors, a sensor's area of influence can be assumed to be the distance between the upstream and downstream midpoints between each detector pair. In Figure 2(c) this would be equal to $0.5s_1+0.5s_2$. For federal reporting purposes, as part of the HPMS, limited point-level count and speed measurements are taken on a sampling of urban roadway locations for one 48-hour period every three years and extrapolated over the entire roadway network. Figure 2(a) also shows how a point speed measured at location x can be extrapolated to determine segment travel time, v_e for vehicle i. This can be used to estimate the delay for vehicle i, v_e-v_f.

Figure 2(d) illustrates the basic relation between fundamental traffic flow variables on a density-flow plane for the hypothetical road segment shown in Figure 2(c). (Coifman and Mallika, 2004) The relation is approximated as a triangle, where zero flow occurs when there are no vehicles on the facility—density and flow are both zero. Zero flow also occurs when density increases to a level such that all vehicles must stop—the speed and flow are zero. (FHWA, 2003) Figure 2(d) illustrates four distinct traffic states: 1, 2, 3 and C. Traffic states 1, 3 and C fall on the unqueued branch of the flow-density relation, while state 2 falls on the queued branch. Figure 2(e) illustrates several issues related to the measurement of traffic parameters along a highway segment where an incident of some kind occurred at a bottleneck location *b* at time t_1. (Coifman and Mallika, 2004) Prior to time t_1, traffic along this

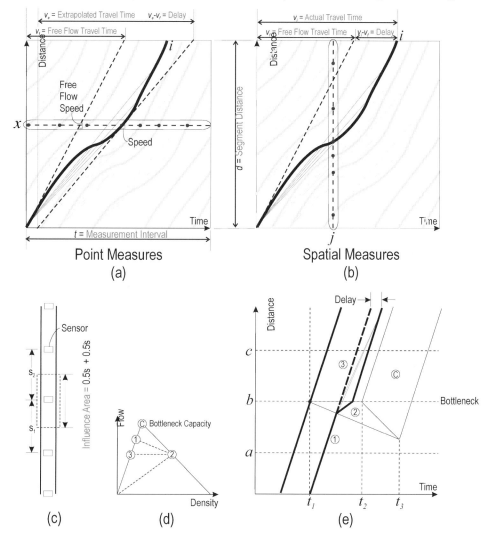

Figure 2. Segment Level Measures of Congestion

hypothetical road segment was in state 1. At t_1 traffic transitioned abruptly to state 2, with lower flow, lower speed and higher density. A shock passed backwards. The traffic state just downstream would be characterized by state 3. At some time t_2, the bottleneck was deactivated, so a backward moving recovery wave passed upstream until it intersected with the initial shock at time t_3. During the recovery, traffic flowed at state C until it returned to state 1. It is very important to understand how, when and where bottlenecks occur on a highway. For example, in Figure 2(e) if a detector were located at location a, it never would have "seen" queued traffic and furthermore if a vehicle entering the road section at time t_1 had received traffic speed information recorded at locations a, b or c prior to t_1 that motorist would have had no way of predicting the actual delay that was later experienced. Also, if a detector were located at point c, traffic speeds would remain high throughout the entire time period shown, falsely characterizing the segment conditions as unqueued. Thus, depending on where traffic conditions are monitored, it is possible to misreport actual conditions; this would adversely affect congestion measures for a segment.

Corridor Level

It is possible to compute congestion-related measures over a larger freeway corridor where more detection locations are available. Travel time can be calculated from real-time or archived freeway sensor data by extrapolating a measured speed value over an influence area (segment). For example, Figure 3(a) shows travel time versus time for one day on northbound Interstate 5 in Portland, Oregon. This was performed over this 35.2 km (22 mi) corridor using data from inductive loop detectors at 25 locations that recorded count, occupancy and speed at 20-sec intervals. The figure also illustrates the cumulative travel time and free-flow travel time (dashed line) throughout the day. As the cumulative line deviates from the cumulative free-flow travel time the travel time increases can be clearly observed. At 7:05 the travel time increased from 23 min. to 28 min. Similarly, at 19:42 the travel time decreased from 49 min. to 24 min. The free-flow travel time on this day was approximately 24 min.

One of the costs of congestion is delay, defined as the excess time required to traverse a section of roadway compared to the free flow travel time. As shown in Figure 3(b), the average delay was calculated for northbound Interstate 5 over five weekdays. Delay was estimated based on the difference between actual travel time and the free-flow travel time on the freeway segments. Total delay for each detector station, defined as the sum of all delay at that station throughout the day, is shown on a three-dimensional plot in Figure 3(c) for the southbound direction. For locations that indicate higher delays, as an example, a DOT can focus its incident response efforts to reduce further delays. From this plot one can see several spikes of delay that occurred at key bottlenecks along the corridor.

Figure 3(d) shows a speed plot for northbound Interstate 5 on one day, where the greyscale variation represents the average speeds measured at 20-second intervals at six detector stations. In addition, 20 express bus trajectories recorded from an automatic vehicle location (AVL) system have been superimposed over the speed plot, indicating that the loop detectors can provide a good indication of mean travel time for a corridor. The slopes of the trajectories changed at nearly the same locations where the freeway speed declined (darker grey). This method was used to show how accurately the speed is reported by the loop detectors.

Statistical analysis was used to validate that there was no evidence of difference between the means at the 95% level of confidence.

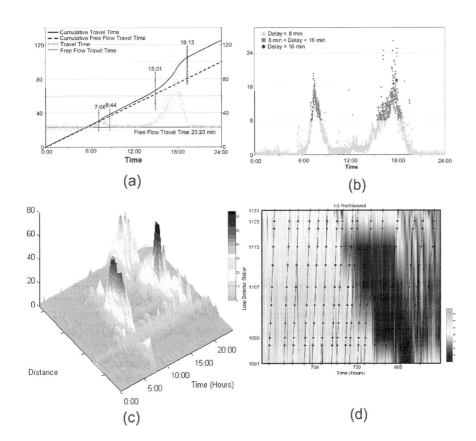

Figure 3. Corridor Level Congestion Indicators

Table 1. Percentage of U.S. Lane Kilometers and Vehicle Kilometers Traveled by Facility Type, 2000

	Total Lane Km	% Lane Km	Total VKT (Million)	% VKT
Freeway	401,400	3.0	1,353,100	30.6
Arterial	1,523,500	11.6	1,840,000	41.5
Other	11,258,400	85.4	1,234,600	27.9
Total	13,183,400	100.0	4,427,700	100.0

Consideration of Total Trip

As shown in Table 1, freeways comprise about 3% of the lane km in the U.S., but they carry more than 30% of the traffic. (FHWA, 2002) Most congestion measures are relevant for a particular link, with a focus on freeways since that is where the most traffic is located and where sensors are in place. Some researchers point out that what is relevant for the traveler is the entire door-to-door trip (Taylor, 2002). For example, Figure 4 is based on Taylor (2002) and illustrates a hypothetical vehicle trajectory (solid line) on a time-space plane. For this trip, the traveler walks to her car, travels on a local street, collector and arterial, followed by a freeway segment, an arterial, a parking lot, and finally walks from her car to her workplace. This trip took 36.1 min and traversed 17 km (10.6 mi). As shown on the *x*- and *y*-axes of Figure 4, the congested freeway component of the trip (at 40 km/h) accounted for 57% of the distance and 40% of the total travel time. On this trip 60% of the travel time occurred off of the freeway. If we focus on the freeway segment and imagine a solution that would return freeway speeds to free-flow conditions (96 km/h), the trip time would be reduced to 27.7 min, as shown by the dashed line. As shown on the *x*-axis, this would reduce the freeway segment's share to 22% of the travel time; now 78% of the trip time would occur off the freeway. It is worth noting that improving freeway conditions may impact many more trips than improvements to other links in the network.

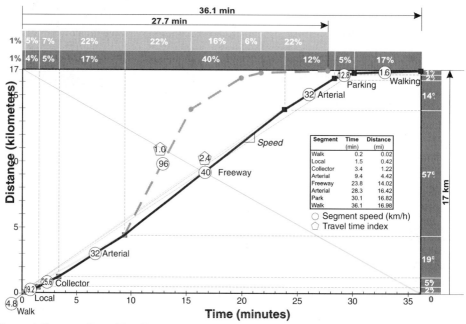

Figure 4. Door-to-Door Trip Times

METROPOLITAN LEVEL MOBILITY MEASURES

When thinking about ways to measure congestion at the metropolitan scale, it is important to remember that our current perceptions are strongly influenced by what happened during the 1960s and 1970s in the U.S. This period (within the memory of many of today's drivers) was one of relatively low congestion since the Interstate system construction era provided much greater expansion in travel capacity than the growth in travel during the same period. (Lomax, Turner and Shunk, 1997)   The result was that in many large urban areas traffic congestion actually decreased. This recent experience frames the debate in that some would like to try to return mobility levels to those earlier conditions

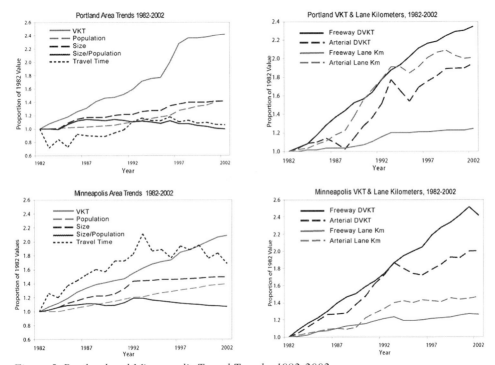

Figure 5. Portland and Minneapolis Travel Trends, 1982-2002

Using data from the 2004 UCR (Schrank and Lomax 2004), a recent study was conducted to begin tracking transportation performance in Portland, Oregon (Portland State University, 2004; Gregor, 2004) for the past 20 years. Keeping the caveats mentioned above regarding the UCR in mind, as an example, Figure 5(a) shows trends in the proportional change in VKT, population, metropolitan area size (sq. mi.), the ratio of size to population and average travel time in peak periods for the Portland, Oregon-Vancouver, Washington urbanized area since 1982. For example, the plot indicates that the VKT in 2002 was 2.4 times that recorded in 1982 while the population and size were nearly 1.4 times their 1982 values. The travel time, on the other hand is nearly the same as it was in 1982. Focusing on freeways and principal arterials, Figure 5(b) shows that daily VKT on Portland-Vancouver area freeways more than doubled between 1982 and 2002, and has also doubled on arterials. Lane km on arterials have been added at a rate greater than the increase in VKT. However, lane km on freeways have increased by only 25 percent over the past 20 years. The gap between VKT and lane km on freeways may explain the declining speeds on Portland-Vancouver freeways. Figures 5(c) and 5(d) show similar data for Minneapolis-St. Paul, Minnesota. As shown in Figure 5(c), population and metropolitan area size have increased to 1.4 times their values in 1982, while the ratio of size/population has remained constant. VKT has approximately doubled and peak period travel time has reportedly increased to more than 1.6 times its 1982 value. As shown in Figure 5(d), Minneapolis-St. Paul's freeway VKT and lane km grew at similar rates to those in Portland-Vancouver, but the notable difference is that the extent of the arterial system has not kept pace in Minneapolis-St. Paul.

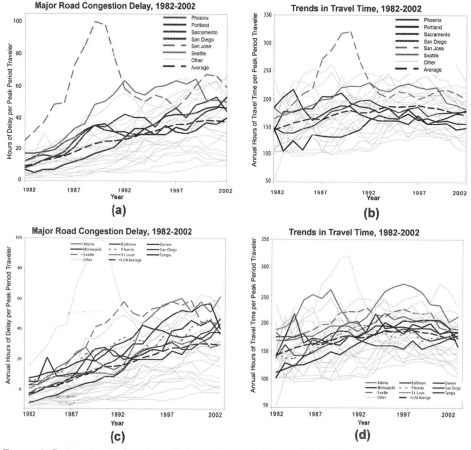

Figure 6. Comparing Urban Area Delay and Travel Times, 1982-2002

As part of the Portland performance measures analysis, the Portland-Vancouver urbanized area was compared to 26 other urban areas with populations between 1-3 million. (Portland State University, 2004) Despite the caveats and limitations in the UMR data mentioned above, as shown in Figure 6(a), when comparing the 20-year trends for Portland-Vancouver with other urban areas, the highlighted lines are for six western peer cities: Phoenix, Sacramento, San Diego, San Jose and Seattle, plus Portland-Vancouver. The lighter grey lines are for the remaining cities the 1-3 million population category, and the dashed black line represents the average value measured across all 27 Large cities. Figure 6(a) is a comparison of the hours of delay per peak period traveller as reported by the UMR. Delay in all cities has been increasing. This figure shows that annual congestion delay estimated for peak period travelers in Portland-Vancouver reportedly increased from 7 hours per year in 1982 to 46 hours per year in 2002, and has remained close to the mean value for similarly sized cities. It had been below the average before 1992, and exceeded the average after that. Portland-Vancouver's peer cities have larger estimates of delay in 2002. In Portland-Vancouver, it appears that shorter-than-average travel distance coupled with lower-than-average travel speed has leveled off the delay actually experienced by travelers.

Figure 6(b) compares the average peak period freeway and major arterial travel time reported by the UMR. This figure shows that Portland-Vancouver's annual travel time per peak period traveler has remained below the average of all large metropolitan areas and has actually declined slightly in recent years. The reported mean travel time for Portland-Vancouver's peer cities also tended to be higher than average. Again, for Portland-Vancouver, shorter-than-average travel distance, coupled with operational improvements, some capacity improvements, and expansions of the public transportation system has eased the impact of congestion on travel time. It is difficult to draw conclusions from these comparisons—and some of the trends are easier to explain than others. For example, the large drop in delay and travel time for San Jose might be explained by the opening of several new freeways in the 1990s. Other fluctuations (such as for Seattle) may be explained by economic conditions.

Similarly, Figure 6(c) shows 20-year trends on delay with a focus on Minneapolis and 8 other cities with populations between 2-3 million: Atlanta, Denver, Phoenix, Seattle, Tampa, Baltimore, San Diego and St. Louis. The average trend in reported delay is shown for all 27 cities with populations between 1-3 million. As shown, Minneapolis began somewhat below average in the delay measure but now falls above average. As shown in Figure 6(d), the trend in mean travel time is similar, yet the more recent data illustrate that the travel time for Minneapolis is about average among all cities in the 1-3 million population range.

Using comparisons as in Figure 6, Figures 7(a) and 7(b) illustrate the Travel Time Index (TTI) trends between 1982-2002 for Portland and Minneapolis, respectively. As noted, based on limited traffic count data, the TTI is the ratio of travel time in the peak period to the travel time at free-flow conditions for freeways and principal arterials. A value of 1.35 would indicate that a 20-minute free-flow trip took 27 minutes in the peak. These figures show that the TTI for both cities is comparable with other peer cities in this category. Figure 7(c) shows a scatter plot of population vs. peak period travel for the 27 cities with populations between 1-3 million. Portland's population is 13th out of the 27 large cities (25th out of all 85 cities), and the amount of travel per peak period traveler is 19th out of the 27 large cities. The population of Minneapolis is 5th out of the 27 large cities, yet the annual hours of delay per peak period traveler is only the 16th highest. Figure 7(d) shows that the annual amount of travel per peak period traveler in Portland is among the 9 lowest when compared to other large cities, while the Travel Time Index for Portland is among the top 6 out of the 27 large cities. In the case of Minneapolis, the Travel Time Index is 11th among the 27 large cities.

The presentation of comparative plots using the UCR data leads to the question of whether the measures shown are the correct measures, or whether the proper variables are actually being measured. For example, would it be better to measure actual speeds, consider reactions from actual travelers, or compute confidence intervals for the reported performance measures? This concern for considering other issues beyond traffic counts at discrete points motivates the discussion presented in the next section.

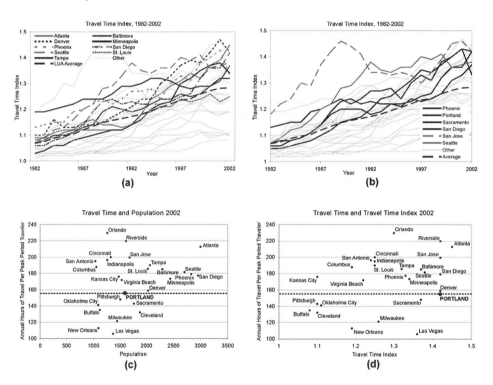

Figure 7. Comparing Urban Area Travel Time Index, Population and Travel Times, 1982-2002

GOING BEYOND CONGESTION MEASURES

The earlier discussion of the congestion survey results, the presentation of the literature review with a focus on understanding the UCR and other definitions of congestion, and the comparative results just described lead to more questions. Do we need to go beyond the congestion measures currently used? The comparisons just drawn are aggregate in the sense that they are derived from very general UCR measurements that were designed for one purpose and used for another. It is thus important to consider going beyond these measurements when attempting to grasp the issues related to congestion. It is also important to use caution when using these results for informing policy decisions—it is possible that performance is being measured incorrectly and that the wrong things are being measured. First, the concept of a travel time budget will be examined briefly.

Travel Time Budget

It has been mentioned above that the journey to work time has remained stable in recent years despite large increases in VKT. Some authors have stated that "average journey to work time changed little as suburban road and highway expansion has accommodated growing number of trips." (Garrison and Ward, 2000) It is said that typical commutes are actually becoming shorter. (Garreau, 1991) The concept of a "travel time budget" was conceived and has been investigated over many years. (Zahavi and Ryan, 1980; Zahavi and Talvitie, 1980; Ryan and Zahavi, 1980) When looking back at development patterns of even ancient cities, the average travel time to work locations has been relatively stable for a few thousand years, and that people have maintained a total daily travel time budget of roughly one hour over the past 5,000 years. (Lomax, et al., 2001; Ausubel and Marchetti, 2001; and Shafer and Victor, 1997) This point has been linked to an assessment of the size of cities as transportation technology has evolved. (Crawford, 2000) For example, if we consider that the maximum accepted (average) commute time is about 45 minutes, it has been reported that the city of Istanbul was approximately 9.6 km in diameter in the sixteenth century. In that case and if people walked at about 6.4 km/h it would have taken 45 minutes to walk from the city's periphery to the city center. (Garreau, 1991)

Several studies have attempted to document these phenomena more specifically. In one study it was shown that the average weekly commuting time for males and females in the U.S.S.R. between 1910-1990 ranged between 4.2 and 5.8 hours week, which would have translated to 50-70 minutes per day with a five day work week. (Grübler, 1990) Another study analyzed aggregate survey data from 1958 and 1970 in Washington, D.C. and Minneapolis-St. Paul, and found that in both cities and at both times, travelers averaged approximately 1.1 hours of travel per day. (Ryan and Zahavi, 1980) A more recent paper has provided more up to date values of travel budget for Washington, D.C. and the Minneapolis-St. Paul regions. (Levinson and Wu, 2005). The authors rejected the notion of a travel time budget, found increasing commute times in Minneapolis and concluded that commute time clearly depends on the spatial structure of an urban area.

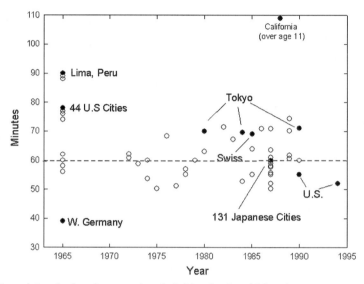

Figure 8. Travel time budget (source: Ausubel, Marchetti and Meyer)

Finally, a comprehensive analysis of daily travel between 1965-1995 in various U.S. and international cities, it has been noted that it appears that people travel about one hour per day. As technology has evolved, the distance traveled per day in the U.S. has increased approximately 2.7% per year from 4 km (walking) in 1880 to approximately 80 km per day in 1990. (Ausubel, Marchetti and Meyer, 1998) As an extract from this work, Figure 8 shows representative data for studies of the U.S. and a dozen other countries since 1965. Despite one obvious outlier from a California study, the figure does indicate that much of the available data centers on the 60 minute range. With this in mind, in order to provide some balance against the measures described up until this point, a short discussion of some other viewpoints on congestion measures will now be presented.

Other Viewpoints

A number of authors have been presenting slightly different views about traffic congestion. Some have noted that successful cities are places where economic transactions are promoted and social interactions occur, and that traffic congestion occurs where "lots of people pursue these ends simultaneously in limited space." (Taylor, 2002; Downs, 2004) It has also been stated that congestion is not necessarily all bad, since it can be a sign that "a community has a healthy growing economy and has refrained from over-investing in roads." (Cervero, 1998) Similarly, it has been noted that unpopular places rarely experience congestion (Garrison and Ward, 2000) and that declining cities have actually experienced reductions in congestion. (Taylor, 2002)

Given the limitations of metro level congestion indices, some alternative techniques have been proposed. For example, a congestion burden index (CBI) was proposed to account for the presence of commute options. (Surface Transportation Policy Project, 2001) The CBI is the travel rate index multiplied by the proportion of commuters who are subject to congestion by driving to work. For example, the 1999 Portland travel rate index was 1.36 (rank 8), and the transit share was 0.14. So the CBI was 1.36×(1-.143) = 1.16 (rank 14). As another indicator that the provision of transportation choices in an urban area is helpful, the transportation choice ratio was also proposed (Surface Transportation Policy Project, 2001), which is calculated by dividing the hourly km of transit service per capita by the lane km of interstates, freeways, expressways and principal arterials for each metro area. It has also been recognized that there is an interaction between personal lifestyles and traffic congestion. Some have noted that during peak periods, only one-third to one-half of all trips are work trips (Lomax, et al., 2001).

Knowing that congestion is often poorly measured, there are few standard indices and it is difficult to compare congestion across metro areas and years, a capacity adequacy (CA) has been proposed. (Boarnet, Kim and Parkany, 1999) The CA system establishes six capacity levels for highway classifications between principal arterials in rural areas to major urban expressways, based on peak hour traffic flow rather than daily VKT. The CA is calculated as $100 \times$ (capacity/volume during present design hour). In this equation, the capacity is estimated and design hour volume is based on the 30th highest hour (rural) or 200th highest hour (urban). This analysis was performed at a county level for California counties; the CA for each highway was weighted by ADT and summed for the entire county. This measure would be in contrast to the UMR calculations, where the capacities of freeway and arterial lanes are fixed at 14,000 and 5,500 vehicles per hour per lane respectively. Given these other research efforts aimed at improving the way congestion is measured toward providing better inputs to policy makers, it is perhaps not surprising that current research is under way as well.

Current Research

With many decades of work to define and monitor congestion as a backdrop, this section discusses some current research efforts that are revealing some fundamental changes in how vehicles use the highway system. For example, very high sustained freeway flows have been measured, more than 20% greater than was once considered to be a theoretical maximum (Lomax, et al., 2001; Cassidy and Bertini, 1999). This means that drivers are accepting very short headways, such that one vehicle's hesitation can cause other vehicles to brake suddenly (Lomax, et al., 2001). This also has immense safety implications. In addition, it has been shown that under some circumstances, freeway flows drop when congestion forms. On Canadian, German and U.S. freeways, this drop is in the range of 2-11% (Cassidy and Bertini, 1999; Bertini, et al., 2004, Zhang and Levinson, 2004), while in Los Angeles, there are reports that the uncongested flows of 2,000 to 2,500 vehicles per hour per lane (VPHPL) drop to about 1,400 to 1,600 VPHPL. (Garrison and Ward, 2000). Finally, earlier studies have mentioned the need to protect transit vehicles from congestion (Buchanan, 1963). Recent developments in bus rapid transit and transit signal priority are taking advantage of opportunities to exploit some gains in improving person travel through congested corridors. (Byrne, et al., 2004) Finally, there is increasing evidence shows that travelers and shippers, when making travel-related choices, consider not only the absolute extent of congestion, but also its temporal variation. As one means to examine this, a regional transportation data archive has been established in Portland, Oregon, and includes systems for automatically generating system performance measures (Bertini et al. 2005). Figure 9 illustrates a sample monthly report for one freeway segment (northbound Interstate 5) for April 2005. This automatically generated plot shows the mean travel time (by 5-minute time slice), the 95th percentile travel time and the congestion frequency for each time interval (right hand axis). There is a need to continue these and other research programs in order to improve our understanding of how the transportation system operates at both microscopic and macroscopic levels.

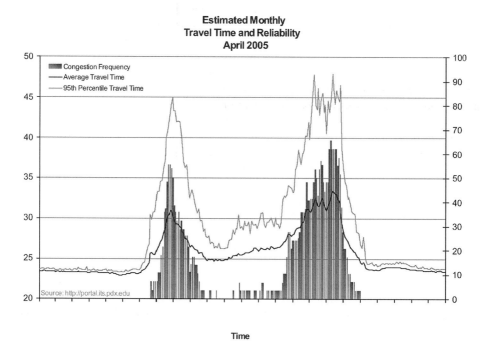

Figure 9. Travel time reliability.

FINAL REMARKS

"Congestion is people with the economic means to act on their social and economic interests getting in the way of other people with the means to act on theirs!" (Pisarski, 2005)

This chapter has covered a lot of territory, and it would be difficult to provide a definitive conclusion, since work in this area is continuing on many fronts. Instead, this section mentions a few important points that should be considered when thinking about congestion. It is generally felt that congestion on the nation's highway system continues to increase, both in reality and in people's perceptions. The notion of simply expanding capacity is limited due to constraints in transportation finance and in public acceptance because of environmental impacts. Efforts to reduce congestion and improve safety through operational means and with improved public transportation should and will continue. One implication of this situation is that since congestion cannot (and some would argue should not) be eliminated the standard methods for measuring and reporting system performance in those terms are no longer very useful. We can no longer simply evaluate the effects of road widening projects on vehicles using limited, aggregate measures such as traffic counts, VKT, the volume/capacity ratio and LOS, nor is it helpful to apply arbitrary speed or volume thresholds across all facility types. These limited measures are usually derived from simple, limited data (e.g., average volumes, number of lanes) extrapolated over large segments of the network and do not consider the impacts on different types of users. The current poor measurements may also be clouding our thinking and leading to irrational policy actions. These factors limit the specificity of performance reporting to large areas and generalized effects. Given new developments that allow for more robust data collection and demands for reporting actual system performance, we can no longer rely on the old way of system performance measurement.

Improvements or changes to the transportation system will impact different users differently—and the magnitude of that impact depends on the type of travel (e.g., freight, commute, recreation) and when their travel needs occur. Therefore, we now need to develop the ability to assess how different system users and society in general are affected by congestion and how that would change with different congestion mitigation actions. For example, reducing congestion on a highway serving a retail center might not be as beneficial as reducing congestion on a freight route because shoppers may be less sensitive to congestion delay than manufacturers and shippers, especially where just-in-time delivery is an important business practice.

In order to reliably estimate how congestion affects different travelers we need three things. First we have to know who is on the congested highway links and how and why they're traveling. Second we need to understand the trip characteristics that are important to travelers (e.g., travel time, reliability). Third, we need data that can be used to estimate these important trip characteristics. For example, if truck movements to and from a high tech manufacturing area are occurring on a congested highway segment, and if travel time reliability is an important travel characteristic, then we must be able to collect performance data that can be used to estimate travel reliability. Future efforts to define and measure traffic congestion should include these important principles. These efforts may include the development and expansion of transportation data archiving systems, the extraction of detailed travel behavior data using floating car data (probe vehicles), and the improvement of tools provided to transportation and land use decision-makers.

ACKNOWLEDGEMENTS

The author appreciates the dedicated support and input of Brian Gregor, Oregon Department of Transportation. Tim Lomax of the Texas Transportation Institute generously supplied the advance 2004 Urban Mobility Report (2002 data). Sonoko Endo conducted the comparative analysis of Urban Mobility Data. Chris Monsere, Jennifer Dill and Jacob Baglien also assisted with data analysis. Matt Lasky, Steve Hansen, Alex White, Aaron Breakstone, Erin Qureshi and Abram VanElswyk assisted with the literature review. Thanks to Prof. Joe Sussman for the quote. The author also appreciates the valuable suggestions provided by four anonymous reviewers, Kenneth Dueker and Alan Pisarski. This research has been supported by the Oregon Department of Transportation and the Portland State University Center for Transportation Studies. Any views presented here, or any errors or omissions are solely the responsibility of the author.

REFERENCES

Ausubel, J.H., and C. Marchetti. (2001). The Evolution of Transport. *The Industrial Physicist*, American Institute of Physics.

Ausubel, J.H., C. Marchetti and P. Meyer. (1998). Toward Green Mobility: The Evolution of Transport. *European Review*, Cambridge University Press, U.K.

Bertini, R.L., S. Hansen, A. Byrd, and T. Yin. "PORTAL: Experience Implementing the ITS Archived Data User Service in Portland, Oregon." *Transportation Research Record: Journal of the Transportation Research Board*, Washington, D.C., 2004. (In Press).

Bertini, R.L., R. Lindgren, D. Helbing and M. Schonhof. (2004). Empirical Observations of Dynamic Traffic Flow Phenomena on a German Autobahn. *Research Report, Portland State University, Center for Transportation Studies*.

Boarnet, M., E. Kim and E. Parkany. (1999). Measuring Traffic Congestion. *Transportation Research Record 1634*, TRB, Washington, D.C., 93-99.

Buchanan, Colin. (1963). *Traffic In Towns*. Her Majesty's Stationery Office, London.

Byrne, N., P. Koonce, R.L. Bertini, C. Pangilinan and M. Lasky. (2004). Using Hardware-in-the-Loop Simulation to Evaluate Signal Control Strategies for Transit Signal Priority. *Transportation Research Record: Journal of the Transportation Research Board*, Washington, D.C., (In Press).

Cape Cod Center for Sustainability. (2003). *Cape Cod Sustainability Indicators Report*, http://www.sustaincapecod.org/SIR03/EnvTrafficTransit.htm

Cassidy, M.J., and R.L. Bertini. (1999). Some Traffic Features at Freeway Bottlenecks. *Transportation Research*, **33B**, 25-42.

Cervero, R. (1998). *The Transit Metropolis*. Island Press, Washington, D.C.

Coifman, B. and R. Mallika. (2004). "Distributed Surveillance on Freeways with an Emphasis on Incident Detection," Proceedings of the 2004 IEEE Intelligent Transportation Systems Conference, Washington, D.C., USA, October 3-6, 2004, 773-778.

Crawford, J.H. (2000). *Carfree Cities*. International Books, Utrecht, the Netherlands.

Daganzo, C.F. (1997). *Fundamentals of Transportation Engineering and Traffic Operations*. Elsevier Science, Oxford, U.K.

Downs, A. (2004). *Still Stuck in Traffic*. The Brookings Institution, Washington, D.C.

European Conference of Ministers of Transport. (1998). *The Spread of Congestion in Europe*, Conclusions of Round Table **110**, Paris.

Federal Highway Administration. (2002). *Status of the Nation's Highway, Bridges, and Transit: Conditions & Performance*. U.S. Department of Transportation. http://www.fhwa.dot.gov/policy/2002cpr/index.htm Accessed Oct. 29, 2003.

Federal Highway Administration. (2003). *Freeway Management and Operations Handbook*, FHWA Report No.: FHWA-OP-04-003.

Federal Highway Administration. (2004). *Traffic Congestion and Reliability: Linking Solutions to Problems*. U.S. Department of Transportation. http://www.ops.fhwa.dot.gov/congestion_report/congestion_report.pdf

Garreau, J. (1991). *Edge City: Life on the New Frontier*. Anchor Books, New York.

Garrison, W.L., and J.D. Ward. (2000). *Tomorrow's Transportation: Changing Cities, Economies, and Lives*. Artech House, Norwood, MA.

Gregor, B. (2004). *Statewide Congestion Overview*. Oregon Department of Transportation. http://www.odot.state.or.us/tddtpau/papers/cms/CongestionOverview021704.PDF

Grübler, A. (1990). *The Rise and Fall of Infrastructures*. Physica-Verlag, Heidelberg, Germany.

Hill, M.C., B.D. Taylor, A. Weinstein and M. Wachs. (2000). Assessing the Need for Highways. Transportation Quarterly, **54**(2), 93-103.

INCOG. (2001). *Congestion Management System*, Tulsa, Oklahoma.

Jung, S., K. Wunderlich, J. Larkin, and A. Toppen. (2004). Rapid Nationwide Congestion Monitoring: The Urban Congestion Reporting Program. *Proceedings of the 2004 IEEE Intelligent Transportation Systems Conference*, Washington, D.C., 286-291.

Kay, J.H. (1997). *Asphalt Nation: How the Automobile Took Over America, and How We Can Take It Back*. Crown Publishers.

Levinson, D. and Wu, Y. (2005). The Rational Locator Reexamined: Are Travel Times Still Stable? *Transportation*, 32(2), pp. 187–202.

Lomax, T., S. Turner and G. Shunk. (1997). *Quantifying Congestion*, NCHRP Report **398**, National Academy Press, Washington D.C.

Lomax, T., S. Turner and R. Margiotta. (2001). *Monitoring Urban Roadways in 2000: Using Archived Operations Data for Reliability and Mobility Measurement*, Federal Highway Administration.

Lomax, T., S. Turner, M. Hallenbeck, C. Boon and R. Margiotta. (2001). *Traffic Congestion and Travel Reliability*.

Meyer, M. (2000). Measuring That Which Cannot Be Measured—At Least According to Conventional Wisdom, *TRB Conference Proceedings 26*, Conference on Performance Measures to Improve Transportation Systems and Agency Operations, Irvine, California.

Mohring, H. (1999). Congestion. In: *Essays in Transportation Economics and Policy*. Brookings Institution Press, Washington, D.C.

NCHRP. (2003). *Performance Measures of Operational Effectiveness for Highway Segments and Systems*, Synthesis **311**, Transportation Research Board, Washington D.C.

Orski, K. (2002). *ITS Forum*.
    http://www.nawgits.com/itsforum/apco/index.cgi?noframes;read=1348

Pisarski, A. (1996). *Commuting in America II*. Eno Foundation, Washington, D.C.

Pisarski, A. (2005). Personal communication.

Portland State University Center for Transportation Studies. (2004). *Portland Metropolitan Region Transportation System Performance Report*, Portland.

Ryan, J.M., and Y. Zahavi. (1980). Stability of Travel Components Over Time, *Transportation Research Record 750*, TRB, Washington, D.C.

Schafer, A., and D. Victor. (1997). The Past and Future of Global Mobility. *Scientific American*.

Schrank, D. and T. Lomax. (2004). *Urban Mobility Report*. Texas Transportation Institute.

Surface Transportation Policy Project. (2001). *Easing the Burden: a Companion Analysis of the Texas Transportation Institute's Congestion Study*.

Taylor, B. (2002). Rethinking Traffic Congestion. *Access*, University of California Transportation Center, **21**, 8-16.

Turner, S., T. Lomax, R. Margiotta and V. Pearce. (2002). *FHWA's Mobility Monitoring Program: Transforming Gigabytes of Archived Operations Data into Mobility and Reliability Performance Measures*.

U.K. Department for Transport. (2001). *Perceptions of and Attitudes to Congestion*.

Varaiya, P. (2002). California's Performance Measurement System: Improving Freeway Efficiency Through Transportation Intelligence. *TR News*, **218**, Washington, D.C.

Washington State Department of Transportation. (2004). *WSDOT's Congestion Measurement Approach: Learning from Operational Data*, Seattle.
    http://www.wsdot.wa.gov/publications/folio/MeasuringCongestion.pdf

Zahavi, Y. and J. Ryan. (1980) Stability of Travel Components Over Time, *Transportation Research Record 750*, TRB, Washington, D.C.

Zahavi, Y. and A. Talvitie. (1980) Regularities in Travel Time and Money Expenditures, *Transportation Research Record 750*, TRB, Washington, D.C.

Zhang, L. and D. Levinson. (2004). Some Properties of Flows at Freeway Bottlenecks, *Transportation Research Record 1883*, TRB, Washington, D.C., pp. 122-131.

Access to Destinations
D.M. Levinson and K.J. Krizek (editors)

CHAPTER 3

CONGESTION AND ITS DISCONTENTS

Jonathan L. Gifford, George Mason University

INTRODUCTION

The title of this chapter is taken from Sigmund Freud's *Civilization and its Discontents*, (1930) which is one of his most famous works . The book's central theme is that "the conflict between sexual needs and societal mores is the source of mankind's propensity for dissatisfaction, aggression, hostility and ultimately, violence" (Davidson, 2002). After the horrors of World War I, Freud rejected the "common Enlightenment view of human beings as naturally sociable and of social life as a reflection of the spontaneous harmony of a natural world governed by laws established by God and discoverable by reason…. [H]uman beings, he believed, are driven by extremely powerful instincts, the full satisfaction of which is incompatible with social life" (Garrard, 1996).

What can Freud's views on the conflict between sexual needs and societal mores contribute to our understanding of congestion? Congestion, I believe, has a tendency to bring out in us powerful instincts the full satisfaction of which is incompatible with social life.

Highway congestion is an endemic problem of American cities. Yet the remedies to congestion are not obvious. Should we do nothing, build more roads and highways, institute pricing, reorganize the layout of our communities so that we are less dependent on our automobiles, induce a shift from single-occupant automobile to carpooling and public transit, encourage telecommuting, revise the green-field development and the redevelopment processes to encourage and/or enable carpooling and public transit?

These questions have been at the center of American urban transportation policy for a half century or more. More fundamental questions are who should decide and on what basis? What outcomes will maximize general welfare? How much weight should be given to the conservation of open space, farmland, fossil fuels, historic preservation, and other resources? How much to individual preferences?

This paper examines how capacity, operations and demand strategies have been used to address congestion. It begins with a discussion of various perspectives on congestion, followed by some historical background on urban congestion. The next section discusses alternatives to highway capacity expansion. [1] The paper concludes with a discussion of the prospects for the application of these strategies for addressing congestion in the future.

PERSPECTIVES ON CONGESTION – PSYCHOLOGY, ACCESSIBILITY, SYMBOLISM AND SOCIAL CAPITAL

While the author is no expert in psychology, Freud's dark view of humanity's primal instincts seems to resonate with the behavior of many of today's drivers. Anger was cited as the cause of 28,000 road deaths in 1996. Aggressive driving, or "road rage," has become a matter of increasing concern and research (Ross and Antonowicz, 2004; Parker *et al.*, 2002; Lawton and Nutter, 2002; Shinar, 2004; Miller *et al.*, 2002; Wells-Parker *et al.*, 2002; Smart, 2004; Smart, 2002; Sharkin, 2004; Rathbone and Huckabee, 1999; Sarkar, 2000). In a 2000 incident that has become almost legendary, an angry driver, whose bumper a woman had tapped after he pulled in front of her, reached into her car, grabbed her 18-pound bichon frise dog and threw it into oncoming traffic, killing it (Hewitt *et al.*, 2001). Other sensational cases include an angry driver who threw a hot cup of coffee through the driver's window of another car, scalding an infant, and an angry driver who fatally shot a 14-year-old passenger in a car driven by his father (UPI NewsTrack, 2004; Linkous, 2005). A discussion of road rage in the author's undergraduate general education class elicited a torrent of personal accounts of road rage incidents, most as victim and some as perpetrator.

Many states have responded to concerns about aggressive driving by launching public safety campaigns. A 1999 survey identified programs in 9 states in the U.S. and 2 Canadian provinces (Rathbone and Huckabee, 1999).

Does congestion cause aggressive driving? Certainly some claim that the private vehicle occupied by a single person stuck in an endless line of congestion fosters anger and social alienation.

A contrasting view to Freud's dark vision comes from Heinz Kohut's "psychology of the self." Kohut rejected Freud's sexually-charged ego in the development of "self" for a more empathetic, personal self-development, where individuals attempt to nourish themselves with shared experiences with others, testing and thus finding one's talents and character, and building a foundation of self-esteem (Kohut, 1971).[2] How does congestion play into self psychology? Does congestion impair incipient desires for social harmony? The author can offer no definitive answer to these questions. But engineers and planners may wish to bear them in mind as they consider how best to respond to concerns about congestion.

A related body of transportation research focuses on access as the proper objective for public policy, rather than the more traditional focus on mobility. This work challenges the conventional notion that mobility is a basic human desire, indeed one that is highly income elastic (that is, as incomes rise, demand for mobility rises disproportionately). The alternative perspective is that the true human need is for access to goods and services. Mobility is simply one way to achieve access. The central theme in this work is that access may also be achieved through improved urban design (Hanson, 2004). Bicycle- and pedestrian-oriented communities with mixed land uses and traditional grid street patterns afford access without requiring automobile-based mobility, which has also been called "automobility" (Tierney, 2004).

To argue that access is about more than mobility seems sound. Yet it is also the case that mobility is about more than access. One need only observe the new 7-ton International CXT (commercial extreme truck) from International Truck and Engine to recognize that vehicles are about much more than utilitarian mobility. The CXT, initially viewed as a niche vehicle with limited market appeal and expected sales of 150 for 2005 had already sold more than 200 in January (Hakim, 2005).

Observations about the automobile as a symbol are not new, of course. Critics of the automobile have long viewed it as being as much symbolic as instrumental. Lewis Mumford, an early and fierce critic of the Interstate system, wrote of the automobile as a "compensatory device for enlarging an ego" (Mumford, 1963). James Flink, another critic, wrote that "[m]otoring [in the 1920s] had a hedonistic appeal rooted in basic human drives" (Flink, 1970). In fact, there has long been an interest in the automobile by cultural studies scholars in the automobile's symbolic value (see, e.g., Brownell, 1972).

A final perspective on congestion is that it is part and parcel of our "social capital." Social capital is the collection of rules, habits, customs and norms that govern social behavior. A fundamental element of social capital is trust, which influences everything from the nature of business transactions to tradeoffs between family and professional management of companies (Fukuyama, 1995).

Driving behavior is also a form of social capital. Casual observation suggests that the "culture" of driving differs markedly from place to place. The same undergraduate class mentioned above readily cited sharp differences in the culture of driving in the Washington, D.C., metropolitan area and their home towns, including reluctance to give way to merging vehicles and lane changes, and high speeds.

One of Fukuyama's key observations about social capital is that it is difficult to create, and once an element of social capital is eroded, it is very difficult to rebuild. How does one regenerate healthy families that have been destroyed by misguided social policies that discourage marriage and responsible paternity, he asks?

Has congestion eroded the quality of driving behavior. And if so, how can transportation officials rebuild a culture of courteous and cautious driving behavior?

HISTORICAL BACKGROUND ON URBAN CONGESTION

Congestion is as old as the city itself. Ancient Rome is said to have suffered from chariot gridlock. Dickens' London was an intensely congested place. And many American cities of the nineteenth century were chronically congested as well. But cities had more chronic problems than congestion, such as poor housing and unsanitary water and waste treatment. Prevailing beliefs about crowding as a source of disease led to continuing efforts to reduce the intense concentrations found in central cities. Other initiatives included the construction of urban parks (Wachs, 2002; Foster, 1981).

Improving transportation speeds was a critical part of this effort, and the latter half of the nineteenth century witnessed a series of experiments with various new forms, including horse cars (steel-wheeled wagons on tracks drawn by horse), elevated steam railways, cable cars, and electric traction street cars. The street car era, beginning in the 1880s and continuing into the 1920s, was a powerful force for expanding the functional radius of cities, allowing decentralization and suburbanization. With the emergence of the private automobile and truck in the early twentieth century, congestion took on a whole new dimension, eventually leading to the invention of the traffic signal, and its widespread implementation in the 1920s (Foster, 1981; McShane, 1994).

Pressure for suburbanization arose not only out of concerns for public health and transportation innovations. It also arose out of a desire by the upper and middle classes to remove their families from the filth and corrupting influence of the central city (Fishman, 1987).

Pleasure driving also exploded in the mid-nineteenth century. Urban cemeteries complete with carriageways for pleasure driving were so popular that non-lot-holders were banned on Sundays to control crowding. Public urban parks with carriageways soon followed, with New York City's Central Park becoming the prototype. It provided an extensive network of carriageways for the elite. Omnibuses, hacks and street railways were excluded from the park and restricted to underpasses (McShane, 1994).

A parallel development was the landscaped urban parkway, which also excluded wagons and common carriers. An 1868 proposal by Frederick Law Olmsted provided a traveled way abutted on both sides by a landscaped strip, which not only afforded pleasant views but also prevented abutting landowners from legal claim to access. Common law provided for access to landowners of all abutting streets, but not to parks (McShane, 1994).

Large scale metropolitan highway system development began to emerge in the 1920s and 1930s as a remedy for urban congestion. Street cars and street railways, which had begun to decline in market share as early as the 1890s, had become intensely unpopular. Cities often refused to allow fare increases, and owners adopted a policy of continued disinvestment, which reduced service quality, leading to a downward spiral. Los Angeles opted against expanding its public transit system in the 1920s, in favor or an extensive network of urban highways (Foster, 1981; Jones, 1985; Wachs, 1996).

During the Depression, municipal funds for road building plummeted. But federal programs, under the auspices of the Works Progress Administration, provided extensive funds, and enterprising cities were able to exploit these funds to expand their urban highway systems (Gifford, 1983). Robert Moses in New York is perhaps the most well known example (Caro, 1974).

Highway spending paused during World War II as materials were required for the war effort. The federal Bureau of Public Roads was not idle during this time, and was engaged in a bold proposal to address what it saw as a major national problem – burgeoning urban highway congestion. The bureau's primary responsibility to date had been the development of a large intercity highway system, which it had begun in the 1920s and largely completed by the early 1930s. The bureau had only limited experience with urban highways, mostly gained during the Depression.

The bureau's proposal for an urban highway program ran into difficulty in the White House and Congress, however. President Roosevelt was concerned about a large, expensive new program at a time when he was seeking to preserve signature programs like the Works Progress Administration and the Reconstruction Finance Corporation. Congress, then still dominated by rural interests, was also less interested in an urban program. Further, the bureau's proposal for urban highways was squarely at odds with a competing proposal for a national system of national superhighways, inspired in part by a mix of enthusiasm and alarm over Germany's successful autobahns (Seely, 1987; Gifford, 1983, 2003; Weingroff, 2000).

After the war, auto ownership, suburbanization and congestion exploded, and public pressure for highway expansion grew accordingly. The creation of the federal Highway Trust Fund in 1956 finally provided the funding needed to pay for the Interstate highway system – which had actually been created in law in 1944 (Rose, 1990). The ensuing two decades constituted an intense period of urban interstate highway design and construction, along with a comparably ambitious rural interstate program (Schwartz, 1976).

The aggressive urban interstate highway program had many impacts, not least of which was that it built a lot of urban highways. Cities across the country developed highway plans and undertook construction. For better or worse, many of those highways were built through low income areas. Prevailing theories of poverty of the day held that urban "blight" could be halted by removing slums, which were considered "decadent" areas prone to high crime, juvenile delinquency, tuberculosis and other health problems, and disproportionately low contributions to city tax revenues (Weingroff, 2000).

In the 1950s, the prevailing view of how to deal with congestion was characterized as "traffic service," which held that transportation facilities should be designed to accommodate the demands of traffic, within reason and to the extent possible (Rose and Seely, 1990; Brown, 2005). Indeed, the prevailing design guidance specified that facilities should be sized to accommodate the thirtieth busiest hour of the design year, usually twenty years after construction (Gifford, 1983). There was little appreciation at the time for the "induced demand" problem, that is, that traffic in urban areas often responded very quickly to new highway capacity and filled it up much sooner than expected. More recently, critics have dubbed this philosophy "predict and provide," suggesting that it inappropriately places concerns about traffic over other considerations, such as urban design and community preservation (see, e.g., Great Britain Department of the Environment Transport and the Regions, 1998).

The twenty-year horizon was an outgrowth of concerns in the 1930s that highways designed in the 1920s had become functionally obsolete due to geometric and capacity limitations. Those highways could not accommodate the increased speeds that became prevalent, and increased vehicle usage led to high levels of congestion, especially in metropolitan areas. Twenty-year forecasts, it was hoped, would avoid such functional obsolescence in the future (U. S. Bureau of Public Roads, 1939).

The traffic-service philosophy was behind the development of the standard four-step transportation planning process, beginning in the late 1940s. It was recognized that simple extrapolative forecasting was not adequate for a metropolitan area because traffic typically had multiple paths through the road network to reach its destinations. Thus, adding a new facility or expanding a new one could cause traffic to redistribute itself on the network. The idea of the four-step model was to predict land use twenty years into the future, then the trips such land use would generate (step 1, called trip generation), where they would go (step 2, called trip distribution), the modes those trips would use (step 3, mode choice), and finally the routes each trip would take (step 4, traffic assignment).

The development of these models was seen as a highly challenging technical exercise that would require the best that then-emerging computers could provide. Few recognized at the time that it was at least as much a policy challenge as a technical challenge. Predicting land use twenty years into the future was a hugely uncertain enterprise, and of course, land use depended as much on transportation decisions as transportation performance depended on land use.

An additional issue was the design of the interstate highway facilities. The debate over whether the system should be metropolitan or interregional had continued through the 1940s, and the resolution was to make it both. Thirteen percent of the 40,000-mile (64,000-km) system initially allocated in 1947 was for urban routes, about half for radial routes and half for beltways. All of these urban interstates, however, were designed to accommodate high-speed traffic safely, to have limited access, and to have no at-grade intersections. The high-speed geometry was very important for interregional trips. They were less beneficial for local trips, however, which tended to be shorter. In addition, the high speed geometry also required long acceleration and deceleration lanes, which limited the spacing between interchanges to several miles (Schwartz, 1976).

As it turned out, many interstate routes were quickly swamped with local trips, which were not well suited to the travel. The long interchange spacing meant that drivers experienced long waits to reach an exit when encountering unexpected delays. The high speed geometry provided few benefits when congestion slowed traffic to a crawl.

This was not terribly surprising in retrospect. Existing traffic on nearby arterials moved quickly to take advantage of the new travel capacity on the new routes. In addition, the resulting access to open land convenient to outlying interstate exits was quickly exploited by real estate developers eager to serve the exploding demand for suburban housing.

The initial reaction to this unexpected congestion was to increase the number of lanes on new facilities to accommodate the trips that they attracted. But it soon became apparent that the congestion problem was more complicated than a simple shortage of highway lanes. The express highways had to tie into the local street system, which in turn had to provide access to adequate parking spaces, either on the street, which competed for road space, or in parking lots or parking structures. Expanding the expressway without dealing with the complementary local streets and parking simply would not solve the problem.

Furthermore, initial efforts to address congestion by increasing the size of expressways required significantly more right of way, which in turn required the condemnation of more property, which increased the number of households and businesses displaced. Such displacements increased costs, and in many places prompted a political backlash, which was exacerbated by the disproportionate number of poor and minority communities being affected.

A final exacerbating influence was the fact that federal interstate highway funds could only be spent on interstate highways, and the centerline length of interstate highway routes was fixed in law at 44,000 miles (70,800 km). States could add lanes to a mile of interstate highway and be reimbursed for ninety percent of the cost as long as the federal government agreed that the capacity was required under its twenty-year forecast requirement. But if a state or city wanted to add lanes to another facility to address congestion, it had to pay fifty or one hundred percent of the cost itself (Taylor, 2000).

The ensuing backlash against many urban interstates fed a growing national concern with cities and the environment, and the passage of the National Environmental Policy Act of 1969 (NEPA). NEPA required any project receiving federal funds to undergo an environmental impact assessment and receive a finding that there was no reasonable alternative. Other environmental laws enacted near this time included the National Historic Preservation Act of 1966 and the Clean Air Act Amendments of 1970.

Collectively, these acts raised the hurdle for capacity expansion as a remedy for congestion, and led the nation to begin to examine where and whether expanding highway capacity still made sense. Alternatives to predict and provide that began to receive consideration included transportation demand management and improved transportation operations, which are the subjects of the next section.

ALTERNATIVES TO HIGHWAY CAPACITY EXPANSION

As concerns about the impacts of new highway capacity grew in the 1970s, attention turned to two alternative approaches. The first was transportation demand management (TDM), which is a set of measures intended to reduce demand for peak hour road space, such as car- and van-pooling and flexible work hours. The second approach was to improve transportation operations. These are discussed in turn below.

Transportation Demand Management

Transportation demand management (TDM) is an umbrella term that encompasses a broad range of measures aimed at reducing or modifying the demand for transportation (Ferguson, 1998, 2000). The term itself arises out of a desire to consider alternatives to "supply-side" measures such as those arising from the traffic-service philosophy. In the 1970s, it was also viewed as an energy conservation measure that would help the U.S. respond to the OPEC oil embargos of 1973 and 1979.

Interest in TDM arose out of a recognition that supply-side solutions to congestion—that is, increases in capacity—were becoming increasingly difficult to effect. Exclusive attention to supply, without a thoughtful treatment of induced demand, could lead to unexpected and un-desirable community impacts. Researchers today have begun to understand that induced de-mand has multiple components. When a new facility opens, some existing trips are redistrib-uted to utilize it, either from other routes or from other times of day. Some increase in demand arises from overall economic growth in a region. Some increase in demand is attributable to land development attracted to the newly available capacity. And some increase in demand is due to new trips by existing travelers who wish to take advantage of the new capacity (Cervero, 2003; Noland and Lem, 2002).

In practice, TDM has come to mean almost any alternative to a general purpose highway lane (see *Table 1*). It is also viewed as a useful complement to new capacity in order to manage in-duced demand. Measures include the following (Gifford and Stalebrink, 2001):[3]

1. Value pricing (also known as congestion pricing) utilizes tolls and other fees to raise the apparent price of travel in order to induce travelers to shift the time of day they travel, the number of people with whom they travel, or their mode of travel.

2. Ridesharing (i.e., carpooling and vanpooling) programs match travelers with proximate origins and destinations in order to reduce travel in single occupant vehicles. Ridesharing programs sometimes organize shared rides to originate in park-and-ride lots or other locations, and may encourage ridesharing by providing preferentially located or priced parking.

Table 1. TDM Measures

Program/Objectives	Strategy	Measures/Technologies
Improve mobility between two points, during peak-periods	• Reduce travel • Stimulate alternative routes and modes • Promote ride-sharing	• Peak-period road-pricing (congestion or "value" pricing) • Traveler information technologies • Remote parking lots
Reduce environmental impacts from travel	• Stimulate ride-sharing • Improve non-motorized travel conditions • Increased vehicle use fees	• High occupancy vehicle (HOV) lanes • Bike lanes • Transit • Public financial telecommuting incentives
Reduce number of deaths and injuries	• Reduce speed/traffic calming • Improve road conditions	• Increased police visibility • Speed cameras • Traffic calming
Corporation driven TDM programs	• Financial subsidies • Regulation	• On-site employee transportation coordination • Alternative work schedules. • Telecommuting
Source: Adapted from Gifford, J.L., and Stalebrink, O.J. 2001. Transportation Demand Management. In: Hensher, D., Button, K. (Eds.), Transport Systems and Traffic Control. Elsevier, Oxford: Elsevier, pp. 199-208.		

3. Park-and-ride lots provide parking for travelers to leave their private vehicles in order to switch modes to either transit or shared rides.

4. High occupancy vehicle (HOV) lanes are highway lanes that are reserved for the use of vehicles containing two or more travelers (the minimum varies among facilities). Lanes are sometimes divided from general purpose travel lanes by physical barriers, or sometimes denoted with "diamonds" or other markings.

5. High occupancy toll (HOT) lanes are HOV lanes that also allow access for a price to single occupant vehicles.

6. Employer commute option programs utilize on-site or dedicated transportation coordinators to work with employees to encourage ridesharing and other TDM measures.

7. Telecommuting measures allow employees to work at home or at remote locations some or all of the time instead of commuting to a distant office.

8. Alternative work schedules allow employees to work schedules that differ from standard 40 hour per week, fixed schedules (e.g., 9 to 5, or 8:30 to 5 five days per week). Such alternatives include working four 10-hour days with three-day weekends, nine 9-hour days with alternating three-day weekends, etc.

9. "Cashing out" free parking provides an option to employees who have access to free parking to "cash out" their access to free parking in exchange for an employer-subsidized transit pass. Recent changes in U.S. law allow employers to treat such programs as a tax-deductible employee benefit up to $100/month.

10. "Family friendly" transportation services are those that attempt to accommodate the needs of commuters with children by, for example, providing child care, grocery, pharmacy and other services at places of employment, near transit stations, or park-and-ride lots.

11. Traffic calming measures attempt to reduce traffic speeds through a variety of physical measures, such as narrowing lanes and speed humps.

TDM measures have been applied widely since the 1970s. Most major metropolitan areas have adopted at least some TDM measures. Yet comprehensive data on the nature and extent of the adoption of TDM is spotty at best, as is comprehensive evaluation data. At the metropolitan level, however, much more data is available, both ex post evaluations and ex ante planning studies.

Today, there are approximately 1210 miles (1947 km) of HOV lanes nationwide (Transportation Research Board, 2001). HOV travel for the journey to work in 2000 comprised 12.2%, or one out of eight, down from 19.7% in 1980, or one out of five (Pucher and Renne, 2003). Thus, while the share of HOV is significant, it is declining nationally. Part of the shift is towards 2-person carpools that are often family based, that is, a husband and wife who commute together. However, recently there appears to be a resurgence in 3- to 4-person carpools that is not yet well understood.[4]

In Minneapolis, two HOV lanes on an 11-mile stretch of Interstate 394 carry about the same number of people as two parallel general purpose lanes, 5175 persons/lane in 1674 vehicles in the HOV lanes vs. 5324 persons/lane in 5267 vehicles in the general purpose lanes during the AM peak (Pratt (R. H.) Consultant *et al.*, 2000). These lanes were recently converted to high occupancy toll (HOT) lanes (see below).

Comprehensive evaluations of the impact of TDM measures other than HOV are quite limited. Given the diversity of the measures, and the range of places where they are located, it is difficult to derive a single measure of impact. The Transit Cooperative Research Program has since 1997 been sponsoring the updating of a handbook, *Traveler Response to Transportation System Changes,* originally published in 1977 and subsequently updated in 1981, which is synthesizing information from a range of applications of TDM nationally (Pratt (R. H.) Associates, 1977; Pratt, 1981; Pratt (R. H.) Consultant *et al.*, 2000; Pratt, 2003). The revised work has begun to appear. One indicator of the complexity of the overall enterprise is that it will be published in nineteen volumes.

The Central Puget Sound Region evaluated a suite of TDM measures to address congestion on State Route 520, which includes a bridge crossing of Lake Washington. Their assessment found that a combination of 14 TDM measures used in various combinations, including parking charges and land-use based measures applied in 10 communities, would yield trip reductions ranging from 0.9% to 11.3%. The most effective measures in their assessment were (DKS Associates, 2003):

1. Parking pricing at employment sites. An additional parking fee of $2 to $3 reduced commute trips up to 9.8%.

2. Increased infill and densification would result in reductions of up to 4.4%.

3. Increased mixed-use development would result in a reduction of up to 2.8%.

4. Multi-employer transportation management associations would lead to a reduction of up to 1.8%.

5. Alternative mode subsidies would reduce trips by up to 1.6%.

While specific to the particular region, these indicators suggest a range of impacts that might be achievable with a well-designed and implemented TDM program. However, the most effective strategy – parking pricing – is often very difficult to accomplish politically, and hence its benefits in trip reduction could be difficult to achieve.

Congestion Pricing

One category of TDM that deserves special mention is the use of pricing on highway facilities. The use of a pricing mechanism to meter demand has been a matter of considerable interest to economists and other transportation specialists for decades (see, e.g., Small *et al.*, 1989; Gómez-Ibáñez, 1997, 1999; Mohring, 1999; Friel, 2004; Downs, 2004). Congestion pricing has been successfully implemented in a number of world cities. One of the highest profile applications occurred in central London in 2003. Other significant implementations include Singapore (beginning in 1975) and Rome (2001).

In the London case, the city imposed a charge of £5 for any private vehicle entering an area of 8 square miles (13 square km) in central London. The restricted zone employs 1 million workers and, prior to the charge, had average traffic speeds of 9 mph, with vehicles spending half their time in queues. As of late 2003, the impact within the zone was a reduction in traffic delays of 30%, a reduction in travel times across the zone of 15%, and an increase in travel time reliability of 30%. Outside of the zone, travel on the orbital surrounding the zone changed within a threshold of 7% plus or minus. Most ex-car users transferred to public transport. Transport officials claim that only 4,000 people per day are no longer entering central London as a result of the charge. However, the reduction of 60,000 car trips per day into the zone has coincided with a reduction of 60,000 to 80,000 fewer people entering the zone daily. Transport officials have attributed the reduction to factors other than the congestion charge (Iraq War, SARS, economic downturn, etc.). However, it is still too early to determine the long term impact on economic activity in central zone. Currently the city is planning to expand the congestion charge to an adjacent zone (Cheung, 2004).

Other significant congestion charging initiatives have been launched in a few other major metropolitan areas, including Singapore and Rome. The Singapore system was instituted in 1975, and remains one of the most successful. Car traffic into the central business district during the morning peak dropped about 80% after it was introduced. The scope was expanded to include the evening peak in 1988, and to the entire day in 1994 (Cheung, 2004). The Rome system was institute in 2001 and restricted access to the historic center, leading to a reduction from 90,000 to about 72,000 vehicles per day (Di Carlo, 2004).

High Occupancy Toll (HOT) Lanes

An implementation of congestion pricing that has begun to be implemented in the U.S. is the high-occupancy toll, or HOT, lane. HOT lanes allow access to a facility – usually an underutilized HOV lane – for a price. As usually conceived, the price varies in order to ensure that the tolled facility is always free flowing. A driver then has the choice whether to pay the toll and travel unimpeded or travel for no toll on (usually congested) general purpose lanes (see, e.g., Poole and Orski, 2000; Perez, 2003; Gómez-Ibáñez, 1997; Surowiecki, 2004; Swope, 2005).

California has implemented HOT lanes in the State Route (SR) 91 corridor in Los Angeles (Sullivan, 2002) and Interstate 15 in San Diego. The San Diego lanes are currently being extended (San Diego Association of Governments, 2005). Houston, Texas, and Minneapolis, Minnesota, also both have HOT lane implementations. Virginia is considering the addition of HOT lanes to its portion of the Capital Beltway (Safirova *et al.*, 2003).

A recent study of traveler behavior on SR 91 estimated that its users value their time at the rate of $21.46/hour and value travel time reliability at a rate of $19.56/hour. The sum of the two rates – travel time and reliability – closely matches the peak rate charged on the express lanes during the time of the study. At peak hour the average commuter would pay $1.89 to realize travel savings of 5.6 minutes and an additional $0.98 to avoid the possibility a 3-minute delay, totaling $2.85, which is somewhat less than the peak toll of $3.30.[5] This suggests that somewhat less than half the peak traffic would use the express lanes, which is consistent with observed behavior. The median motorist in the study sample would pay $0.42/trip to reduce the frequency of 10-minute delays from 0.2 to 0.1 (Poole, 2004; Small *et al.*, 2002).

While only deployed in a few places in the U.S., HOT lanes have received considerable attention as mechanisms for expanding highway capacity, or making better use of underutilized HOV lane capacity.

Transportation Operations

Another option for reducing congestion is improving the operations of highway facilities. A substantial amount of urban congestion—perhaps as much as half—arises from so-called "non-recurring" incidents, such as vehicle crashes and breakdowns. The ability to detect such incidents quickly and respond quickly with the necessary resources has received considerable attention in the last decade. In addition, optimizing traffic signal timing can significantly reduce congestion.

An early effort in this regard was TOPICS (Traffic Operations Program to Improve Capacity and Safety), which received explicit federal funding from 1970 to 1973. Later, the term "transportation systems management" (TSM) came into favor, and became a required element of metropolitan transportation improvement programs (Weiner, 1997).

Today, improved highway operations have in part been made possible by the improved detection and communication capabilities afforded through intelligent transportation systems (ITS), which have received substantial funding since the early 1990s. The range of ITS technologies is quite broad. An illustrative list of technologies with particular relevance for congestion reduction are:

1. Incident detection and management systems – Such systems use a variety of detectors, including video monitors, inductive loops buried in the pavement, and acoustic sensors, to detect when traffic flow is interrupted. The prevalence of cellular telephones among highway users has also proved to be a valuable source of detection information. Once detected, such systems can allow reconnaissance to determine what resources are needed on the incident scene and the prompt deployment of those resources. In addition, highway users can be warned of incidents through variable message signs, AM

and FM radio and, in more advanced systems, automated in-vehicle traffic monitoring and advisory systems.

2. Advanced traffic management systems – Such systems allow the coordination of traffic signals along a corridor, in a local jurisdiction, or at the regional level. More advanced systems use traffic detectors to adjust traffic signal timings in real time. More basic systems allow a range of operating time plans (morning peak, afternoon peak, midday, night, etc.).

3. Electronic toll collection systems – In areas with tolls, the collection of tolls can be a significant source of congestion. Electronic toll tags are now widely deployed in the U.S. and abroad that allow toll collection at highway speeds.

While the technology of transportation operations is a significant source of potential congestion reduction, the implementation of such systems is usually subject to a number of institutional considerations, including (Horan and Gifford, 1993):

1. Jurisdictional issues – Most metropolitan areas consist of a number of separate jurisdictions and the facilities that serve them are owned and operated by a number of separate operating agencies. These include state departments of transportation, cities, counties, townships, toll road authorities, public transit agencies, and so forth. In addition, police, fire and emergency medical service organizations are often responsible for responding to incidents occurring on or around transportation facilities. Collaborating across jurisdictional boundaries is often extremely challenging. As technology has improved and made such coordination and collaboration technically feasible, such jurisdictional issues have often become barriers to effective operations. While regional coordination in the planning and programming of transportation funds has been coordinated at the metropolitan level through metropolitan planning organizations (MPOs), no comparable organization dedicated to transportation operations has generally existed.

2. Organizational issues – The organizations that own and operate transportation systems often face serious limitations on their capacity to plan, acquire and operate some of the more advanced systems that are available. These limitations can arise from funding shortages, the technical capability of staff and procurement requirements.

3. Behavioral issues – Transportation users themselves may undermine the capability of ITS technologies to address congestion through behavioral changes, such as ignoring HOV restrictions and requirements, ignoring parking restrictions, purchasing anti-

detection devices (e.g., radar warnings). In addition, concerns about privacy may inhibit adoption of certain systems.

All of these issues are, in some sense, boundary issues: spatial and territorial boundaries, organizational boundaries, and behavioral boundaries.[6]

Concern over jurisdictional and organizational issues has given rise in the last decade to new approaches to transportation operations. The current nomenclature for this effort is "regional operations collaboration and coordination" (ROCC).[7]

Significant research into regional transportation collaboration and cooperation has been conducted in the last 5 years that aims to understand models and best practices at a regional scale (Briggs and Jasper, 2001a, b, c, d, e, f, g; Briggs, 1999; Gifford and Stalebrink, 2002; U.S. Department of Transportation, 2002). Successful models include:

1. A virtual organization: a voluntary group of agencies that agree, often through a memorandum of understanding (MOU). Typically virtual organizations own no assets and employ no staff. All resources are provided by the partners.

2. A private corporation: such as a non-profit 501(c)(3), which is legally established to perform some aspect of operations.

3. A regional government or authority, which is usually a preexisting organization such as an MPO that takes on the responsible for some aspect of operations.

The scope of such operating responsibilities may include any or all of the following:

1. sharing of resources, such as staff, funds, equipment;

2. sharing of information, such as construction schedules or operating status; and/or

3. sharing of operational control, such as the ability to control another agency's VMS during night operations when the owner agency's staff is not on duty.

While improving operations is not a panacea, it can significantly reduce congestion in a region. However, successful transportation operations require a combination of technological and institutional development.

DISCUSSION

What can we learn from the past century of experience with urban highway congestion and efforts to combat it? First, of course, it is important to remember that congestion is not an entirely bad thing, and absence of congestion should not necessarily be the objective of public policy. Chernobyl, after all, has no congestion (Taylor, 2002; Wachs, 2002; Stopher, 2004; Mokhtarian, 2004; Taylor, 2004). While highway congestion can be frustrating to those who are caught up in it, much if not most congestion is routine, that is, it occurs in the same places day after day. Yet people routinely make decisions about where to live and where to work and how to travel between them recognizing that their trips are going to occur in congestion. There are alternatives to such lifestyles. This tells us that the benefits they derive from living and working where they do more than offset the frustrations associated with congestion. Otherwise they would reorganize their lives to reduce their exposure to it.

The fact that our systems are not operating at an optimal engineering level of service may be frustrating to engineers. That they do not appear to maximize consumer surplus may frustrate economists. But travelers may be making choices on the basis of other considerations that fall outside of these frameworks (Levinson, 2003). Indeed, some travelers report enjoying their commutes as periods of privacy, reflection and relaxation between home and office (Mokhtarian, 2005; Ory and Mokhtarian, 2005; Handy *et al.*, 2005).

Yet experiences such as those with HOT lanes in Los Angeles and San Diego suggest that travelers will exercise an option to pay for a premium service when it is available, a clear indicator that they view such options as improving their wellbeing.

A second observation is that the policies that led us to where we are unfolded across more than a century, beginning in the streetcar suburb era and continuing to the present (Adams *et al.*, 1999). Our urban congestion "problem" is not going to be solved overnight. Much of the stock of facilities and buildings is already in place. Our options are to add new stock, or to modify or retire existing stock. It is a problem that is faced by almost all of America's large cities. Major change at the national level is unlikely, given the nature of Washington politics and the momentum of current programs. The strongest trend seems to be toward earmarking of projects, which is a threat and opportunity that most major cities have surely already recognized.

Whatever policy remedies we decide to adopt must be sustainable for many decades. Structural market incentives are in a way much more sustainable than tax-funded initiatives. Coercion and public taxation have sharply limited potential, and much of the public resource is already spoken for, for the foreseeable future (Johnson, 2003). Sustaining a publicly funded program over the span of decades is extremely difficult, especially without its becoming diluted with political earmarks.

Expectations play a key role. "The relative attractiveness for investment (of time or money) in a location is influenced as much by the expectation of future change in the value of the capital stock as it is by the current value of the capital stock" (Ward, 2003). If owners or prospective property owners in a particular location believe it is likely to become less appealing, their behavior and willingness to invest will be reduced accordingly.

A third point is that in large part our dilemma is a product of enormous wealth. A study of the Minneapolis-St. Paul, Minnesota, region observed, "The decentralization of jobs and population are, in large measure, the consequences of increasing affluence. Decentralization will continue as long as income levels increase and current incentives are in place" (Ward, 2003). As it is in the Twin Cities, so it is in most other prospering U.S. cities.

Our error, then, has been a failure to adequately anticipate increasing affluence and – not a separate issue – entry of women into the workforce. So we might think about the problem as affluence management, rather than congestion management. The problems posed by affluence are similar but slightly different. Reducing affluence is not a winning strategy. Rather, the challenge is managing the consequences of affluence, including distributional impacts on those unable to access a car. A similar point can be made about female labor force participation, that is, that managing the consequences of female labor force participation is well within the scope of policy objectives, but reducing it is not.

Furthermore, while cross-subsidies are clearly present in the urban transportation domain, many estimates of underpayment by highway users are flawed and probably overstated (Gómez-Ibáñez, 1997). A long-term study in the Minneapolis-St. Paul, Minnesota, region concluded that "[e]ighty-four percent of known transportation costs are internal costs, borne by the people who travel. Nevertheless, the absolute size of governmental and external costs, plus a likelihood that external costs are considerably underestimated, calls for public and policy makers' attention" (Ward, 2003).

A fourth point relates to the so-called "non-correspondence" problem (Gifford and Pelletiere, 2002). While it is sometimes appealing to wish for strong regional governments that are well-matched to the scale and scope of transportation problems, the logic of technical systems will rarely correspond to the structure of our governmental institutions. And why should they? Technical systems are transient, and governmental institutions, at least those embodied in constitutions, are meant to abide indefinitely. Moreover, while many would not agree (Peirce *et al.*, 1993; Rusk, 1993; Orfield, 1997, 2002), there is a credible argument that the coexistence of distinct city and suburban entities provides citizens and businesses a choice of jurisdictions within a region, leads jurisdictions to compete, and improves welfare (Tiebout, 1956).

It would be naïve to design our technical systems with the assumption that our governmental institutions will yield to their internal logics. Rather, we must conceive and implement our transportation systems and policies in recognition of our constitution, and the powers and responsibilities it allocates.

These observations suggest that our discontent with congestion should be tempered by a recognition that its causes are at least in part healthy ones: robust communities competing for residents and businesses, and individuals and businesses using the transportation system to afford themselves the locational amenities they desire. Indeed, the diversity of our modern metropolitan communities affords and extraordinary range of choice of where to shop, worship, study, work and live (Fishman, 1990). We can do a better job of developing our communities and transportation systems so as to reduce the level and duration of congestion. HOT lanes and other forms of congestion charging, in conjunction with efficient system operations and demand management, hold a great deal of promise.

REFERENCES

Adams, J. S., J. L. Cidell, L. J. Hansen, and B. VanDrasek (1999). Synthesizing Highway Transportation, Land Development, Municipal and School Finance in the Greater Twin Cities Area, 1970–1997. In: *Transportation and Regional Growth Study,* University of Minnesota, Center for Transportation Studies, Minneapolis, MN.

Briggs, V. (1999). New regional transportation organizations. *ITS Quarterly,* **7,** 35-46.

Briggs, V. and K. Jasper (2001a). *Organizing for regional transportation operations: an executive guide.* U.S. Federal Highway Administration, Washington, D.C.

Briggs, V. and K. Jasper (2001b). *Organizing for regional transportation operations: Arizona AZTech.* U.S. Federal Highway Administration, Washington, D.C.

Briggs, V. and K. Jasper (2001c). *Organizing for regional transportation operations: Houston TranStar.* U.S. Federal Highway Administration, Washington, D.C.

Briggs, V. and K. Jasper (2001d). *Organizing for regional transportation operations: New York/New Jersey/Connecticut TRANSCOM.* U.S. Federal Highway Administration, Washington, D.C.

Briggs, V. and K. Jasper (2001e). *Organizing for regional transportation operations: San Francisco Bay Area.* U.S. Federal Highway Administration, Washington, D.C.

Briggs, V. and K. Jasper (2001f). *Organizing for regional transportation operations: Southern California ITS Priority Corridor.* U.S. Federal Highway Administration, Washington, D.C.

Briggs, V. and K. Jasper (2001g). *Organizing for regional transportation operations: Vancouver TransLink.* U.S. Federal Highway Administration, Washington, D.C.

Brown, J. (2005). A tale of two visions: Harland Bartholomew, Robert Moses, and the development of the American freeway. *Journal of Planning History,* **4,** 3-32.

Brownell, B. (1972). A Symbol of Modernity: Attitudes Toward the Automobile in Southern Cities in the 1920s. *American Quarterly,* **24,** 20-44.

Caro, R. A. (1974). *The power broker,* Knopf, New York.

Cervero, R. (2003). Road expansion, urban growth, and induced travel: A path analysis. *Journal of the American Planning Association,* **69,** 145-163.

Cheung, F. (2004). Road Pricing in Singapore. In: *Managing Transport Demand through User Charges: Experiences to Date.* The Chamber, London City Hall.

Davidson, A. L. (2002). Sigmund Freud: Civilization and Its Discontents. Vol. 2004 PageWise.

Di Carlo, M. (2004). Road Charging in Rome. In: *Managing Transport Demand through User Charges: Experiences to Date.* The Chamber, London City Hall.

DKS Associates (2003). *Modeling TDM Effectiveness: Developing a TDM Effectiveness Estimation Methodology (TEEM) and Case Studies for the SR 520 Corridor.*

Downs, A. (2004). *Still stuck in traffic: coping with peak-hour traffic congestion,* Brookings Institution Press, Washington, D.C.

Ferguson, E. (1998). *Transportation demand management,* American Planning Association, Chicago.

Ferguson, E. (2000). *Travel demand management and public policy,* Ashgate, Aldershot.

Fishman, R. (1987). *Bourgeois utopias: the rise and fall of suburbia,* Basic Books, New York.

Fishman, R. (1990). America's New City: Megalopolis Unbound. *Wilson Quarterly,* **14,** 24-48.

Flink, J. J. (1970). *America adopts the automobile, 1895-1910,* MIT Press, Cambridge, Mass.

Foster, M. S. (1981). *From streetcar to superhighway: American city planners and urban transportation, 1900-1940,* Temple University Press, Philadelphia.

Freud, S. and J. Riviere (1930). *Civilization and its discontents,* L. & Virginia Woolf at the Hogarth Press, London.

Friel, B. (2004). Are tolls the answer? *National Journal,* **36**, 3754.

Fukuyama, F. (1995). *Trust: social virtues and the creation of prosperity.* Free Press, New York.

Garrard, G. (1996). Josep de Maistre's Civilization and its Discontents. *Journal of the History of Ideas,* **57,** 429-446.

Gifford, J. L. (1983). *An analysis of the federal role in the planning, design and deployment of rural roads, toll roads and urban freeways.* University of California, Berkeley CA.

Gifford, J. L. (2003) *Flexible Urban Transportation,* Elsevier Sciences, Oxford.

Gifford, J. L. and D. Pelletiere (2002). New Regional Transportation Organizations: Old Problem, New Wrinkle? *Transportation Research Record***, 1812**, 106-111.

Gifford, J. L. and O. J. Stalebrink (2001). In: *Transport Systems and Traffic Control*, Vol. 3 (D. Hensher, and K. Button, eds.), pp. 199-208, Elsevier, Oxford.

Gifford, J. L. and O. J. Stalebrink (2002). Remaking Transportation Organizations for the 21st Century: Consortia and the Value of Learning Organizations. *Transportation Research, part A: Policy & Practice,* **36**, 645-657.

Gómez-Ibáñez, J. A. (1997). In *The full costs and benefits of transportation: contributions to theory, method and measurement* (D. L. Greene, D. W. Jones, and M. A. Delucchi, eds.). Springer, Heidelberg, pp. 149-172.

Gómez-Ibáñez, J. A. (1999). Pricing. In: *Essays in transportation economics and policy: a handbook in honor of John R. Meyer* (J. R. Meyer, J. A. Gómez-Ibáñez, W. B. Tye, and C. Winston, eds.), pp. 99-136, Brookings Institution Press, Washington, D.C.

Great Britain Department of the Environment Transport and the Regions (1998). *A new deal for transport: better for everyone.* Stationery Office, London.

Hakim, D. (2005). New Way for Stars to Keep Truckin'. *New York Times* (January 29).

Handy, S., L. Weston and P. L. Mokhtarian (2005). Driving by choice or necessity? *Transportation Research Part A: Policy and Practice,* **39,** 183-203.

Hanson, S. (2004). In *The geography of urban transportation* (S. Hanson and G. Giuliano, eds.). The Guilford Press, New York, pp. 3-29.

Hewitt, B., E. Bazar and M. Schorr (2001). Collared: Fourteen Months after the Road-Rage Killing of a Dog Named Leo, a Tenacious Detective Brings in a Suspect. *People Weekly.*

Horan, T. A. and J. L. Gifford (1993). New Dimensions in Infrastructure Evaluation: The Case of Non-Technical Issues in Intelligent Vehicle Highway Systems. *Policy Studies Journal,* **21,** 347-356.

Johnson, C. (2003). Market Choices and Fair Prices: Research Suggests Surprising Answers to Regional Growth Dilemmas. In *Transportation and Regional Growth Study,* University of Minnesota, Center for Transportation Studies, Minneapolis, MN.

Jones, D. W. (1985). *Urban transit policy: an economic and political history,* Prentice-Hall, Englewood Cliffs, N.J.

Kohut, H. (1971). *The analysis of the self.* International Universities Press, New York.

Lawton, R. and A. Nutter (2002). A comparison of reported levels and expression of anger in everyday and driving situations. *British Journal of Psychology,* **93,** 407.

Levinson, D. (2003). Perspectives on Efficiency in Transportation. *International Journal of Transportation Management,* **1,** 145-155.

Linkous, J. (2005). Boy, 14, Killed in Road-Rage Shooting. The America's Intelligence Wire (Jan 31, 2005).

McShane, C. (1994). *Down the Asphalt Path: The Automobile and the American City,* Columbia University Press, New York.

Miller, M., D. Azrael, D. Hemenway, and F. I. Solop (2002). 'Road rage' in Arizona: armed and dangerous. *Accident Analysis & Prevention,* **34,** 807-814.

Mohring, H. (1999). In *Essays in transportation economics and policy: a handbook in honor of John R. Meyer* (Meyer, J. R., J. A. Gómez-Ibáñez, W. B. Tye, and C.Winston, eds.) Brookings Institution Press, Washington, D.C., pp. 181-221.

Mokhtarian, P. L. (2004). Reducing road congestion: a reality check--a comment. *Transport Policy,* **11,** 183-184.

Mokhtarian, P. L. (2005). Travel as a desired end, not just a means. *Transportation Research Part A: Policy and Practice,* **39,** 93-96.

Mumford, L. (1963). *The highway and the city,* Harcourt Brace & World, New York.

Noland, R. B. and L. L. Lem (2002). A review of the evidence for induced travel and changes in transportation and environmental policy in the U.S. and U.K. *Transportation Research Part D: Transport and Environment,* **7,** 1-26.

Orfield, M. (1997). *Metropolitics: a regional agenda for community and stability,* Brookings Institution Press, Washington, D.C.

Orfield, M. (2002). *American metropolitics: the new suburban reality,* Brookings Institution Press, Washington, D.C.

Ory, D. T. and P. L. Mokhtarian (2005). When is getting there half the fun? Modeling the liking for travel. *Transportation Research Part A: Policy and Practice,* **39,** 97-123.

Parker, D., T. Lajunen, and H. Summala (2002). Anger and aggression among drivers in three European countries. *Accident Analysis & Prevention,* **34,** 229-235.

Peirce, N. R., C. W. Johnson, and J. S. Hall (1993). *Citistates: how urban America can prosper in a competitive world,* Seven Locks Press, Washington, D.C.

Perez, B. G. (2003). A guide for HOT lane development. U.S. Department of Transportation, Federal Highway Administration, Washington, D.C.

Poole, R. W., Jr. (2004). Another Argument for Tolled Express Lanes. *Surface Transportation Innovations.*

Poole, R. W., Jr. and C. K. Orski (2000). HOT lanes: a better way to attack urban highway congestion. In *Regulation*, Vol. 23, pp. 15.

Pratt (R. H.) I. Associates (1977). *Traveler Response to Transportation System Changes: A Handbook for Transportation Planners,* Federal Highway Administration, Washington, D.C.

Pratt (R. H.) Consultant, I., Texas Transportation Institute, Cambridge Systematics, Inc., Parsons Brinckerhoff Quade & Douglas, Inc., SG Associates, Inc. and McCollom Management Consulting, Inc. (2000). Traveler Response to Transportation System Changes: Interim Handbook. In: *TCRP Web Document 12 (Project B-12): Contractor's Interim Handbook*, Vol. 2004 Transportation Research Board, Washington, D.C.

Pratt, R. H. (1981). *Traveler Response to Transportation System Changes,* Federal Highway Administration, Office of Highway Planning, Urban Planning Division, Washington, D.C.

Pratt, R. H. (2003). Traveler Response to Transportation System Changes: An Interim Introduction to the Handbook. *Research Results Digest (Transit Cooperative Research Program).*

Pucher, J. R. and J. L. Renne (2003). Socioeconomics of Urban Travel: Evidence from the 2001 NHTS. *Transportation Quarterly,* **57,** 49-77.

Rathbone, D. B. and J. C. Huckabee (1999). *Controlling Road Rage: A Literature Review and Pilot Study.* The InterTrans Group, pp. 1-40.

Rose, M. H. (1990). *Interstate Express Highway Politics, 1941-1989,* University of Tennessee Press, Knoxville.

Rose, M. H. and B. E. Seely (1990). Getting the Interstate system built: road engineers and the implementation of public policy, 1955-1985. *Journal of Policy History,* **2,** 23-55.

Ross, R. R. and D. H. Antonowicz (2004). *Antisocial drivers: prosocial driver training for prevention and rehabilitation.* Charles C Thomas, Springfield, Ill.

Rusk, D. (1993). *Cities without suburbs.* Woodrow Wilson Center Press, Distributed by the Johns Hopkins University Press, Washington, D.C.; Baltimore, Md.

Safirova, E., K. Gillingham, W. Harrington, and P. Nelson (2003). Are HOT lanes a hot deal? The potential consequences of converting HOV to HOT lanes in northern Virginia. In *Urban Complexities Issue Brief 03-03* Resources for the Future, Washington, D.C.

San Diego Association of Governments (2005). Interstate 15 managed lanes. Vol. 2005 San Diego.

Sarkar, S., A. Martineau, M. Emami, M. Khatib and K. Wallace (2000). Aggressive Driving and Road Rage Behaviors on Freeways in San Diego, California: Spatial and Temporal

Analyses of Observed and Reported Variations. *Transportation Research Record,* **1724,** pp 7-13.

Schwartz, G. T. (1976). Urban freeways and the Interstate System. *Southern California Law Review,* **49,** 406-513.

Seely, B. E. (1987). *Building the American Highway System: Engineers as Policy Makers,* Temple University Press, Philadelphia, Pa.

Sharkin, B. S. (2004). Road rage: risk factors, assessment, and intervention strategies. *Journal of Counseling and Development,* **v82,** 191-198.

Shinar, D. and R. Compton (2004). Aggressive Driving: an observational study of driver, vehicle, and situational variables. *Accident Analysis and Prevention,* **36,** 429-437.

Small, K. A., Winston, C. and C. A. Evans (1989). *Road work: a new highway pricing and investment policy.* Brookings Institution, Washington, D.C.

Small, K. A., C. Winston and J. Yan (2002). Uncovering the Distribution of Motorists' Preferences for Travel Time and Reliability: Implications for Road Pricing. In: *University of California Transportation Center Papers*, Vol. 2004.

Smart, R. G. and R. E. Mann (2002). Is Road Rage A Serious Traffic Problem? *Traffic Injury Prevention,* **3,** 183-189.

Smart, R. G., G. Stoduto, R. E. Mann and E. Adlaf (2004). Road Rage Experience and Behavior: Vehicle, Exposure, and Driver Factors. *Traffic Injury Prevention,* **5,** 343-348.

Stopher, P. R. (2004). Reducing road congestion: a reality check. *Transport Policy,* **11,** 117-131.

Sullivan, E. (2002). State Route 91 value-priced express lanes: updated observations. *Transportation Research Record,* **1812,** 37-42.

Surowiecki, J. (2004). *The wisdom of crowds: why the many are smarter than the few and how collective wisdom shapes business, economies, societies, and nations.* Doubleday, New York.

Swope, C. (2005). The fast lane. In *Governing*, pp. 24-30.

Taylor, B. D. (2000). When finance leads planning: urban planning, highway planning, and metropolitan freeways. *Journal of Planning Education and Research,* **20,** 196-214.

Taylor, B. D. (2002). Rethinking Traffic Congestion. *Access,* 8-16.

Taylor, B. D. (2004). The politics of congestion mitigation. *Transport Policy,* **11,** 299-302.

Tiebout, C. (1956). A Pure Theory of Local Expenditures. *Journal of Political Economy,* **64,** 416-424.

Tierney, J. (2004). The Autonomist Manifesto (Or, How I Learned to Stop Worrying and Love the Road). *New York Times Magazine* (Sept. 26).

Transportation Research Board, H. S. C. (2001). Inventory of Operational Characteristics of Selected Freeway/Expressway HOV Facilities. Vol. 2004.

U. S. Bureau of Public Roads (1939). Toll roads and free roads: message from the President of the United States.

U.S. Department of Transportation, F. H. A. (2002). *Regional transportation operations collaboration and coordination: a primer for working together to improve transportation safety, reliability and security.* Washington, D.C.

UPI NewsTrack (2004). Infant burned in road rage incident. (May 14, 2004).

Wachs, M. (1996). The Evolution of Transportation Policy in Los Angeles. In: *The City: Los Angeles and Urban Theory at the End of the Twentieth Century* (A. J. Scott and E. W. Soja, eds.), pp. 106-159, University of California Press, Los Angeles, Calif.

Wachs, M. (2002). Fighting traffic congestion with information technology. *Issues in Science and Technology,* **19,** 43.

Ward, E. E. (2003). A Systems Thinking Perspective on the Transportation and Regional Growth Study. In: *Transportation and Regional Growth Study,* University of Minnesota, Center for Transportation Studies, Minneapolis, MN.

Weiner, E. (1997). *Urban Transportation Planning In the United States: An Historical Overview,* U.S. Department of Transportation, Washington, D.C.

Weingroff, R. F. (2000). The genie in the bottle: the interstate system and urban problems, 1939-1957. *Public Roads,* **64.**

Wells-Parker, E., J. Ceminsky, V. Hallberg, R. W. Snow, G. Dunaway, S. Guiling, M. Williams and B. Anderson (2002). An exploratory study of the relationship between road rage and crash experience in a representative sample of US drivers. *Accident Analysis & Prevention,* **34,** 271-278.

NOTES

[1] Other strategies for addressing congestion, namely land use policies and efforts to improve accessibility, are also discussed in Chapter 5

[2] The author thanks Ann Forsyth for the reference to self psychology.

[3] For the purposes of this paper, land-use based TDM strategies are excluded because they are generally covered in Chapter 6.

[4] Alan Pisarsky, personal communication (November 5, 2004).

[5] The study's definition of reliability was the 80th percentile of the travel time savings minus the median time savings, which peaked at 3 minutes at about 8 a.m.

[6] The author thanks an anonymous reviewer for this observation.

[7] It is telling of the sensitivity of such activities that the nomenclature has evolved considerably. Initially, the term "new regional organizations" was adopted, but that aroused concern from existing organizations. Subsequently the term "regional operating organizations" was adopted, but that gave rise to concern that it implied the existence or creation of a formal regional organization.

Access to Destinations
D.M. Levinson and K.J. Krizek (editors)

CHAPTER 4

PLACE-BASED VERSUS PEOPLE-BASED ACCESSIBILITY

Harvey J. Miller, University of Utah

INTRODUCTION

Accessibility is a multi-faceted concept that ultimately centers on an individual ability to conduct activities within a given environment (Weibull, 1980). Accessibility is a fundamentally spatial concept: it is predicated on the ability to be present at some location where an activity such as shopping, education, health care, recreation, socializing or public events occur. Although traditional accessibility theory assumes physical presence, information and communication technologies (ICTs) afford the ability to be *telepresent* and participate without a physical presence at the activity location.

Traditional methods for measuring accessibility examine the spatial separation between some postulated key location in individuals' lives (typically, home or workplaces) and other locations where required or desired activities occur (Kwan and Weber, 2003). These are *place-based accessibility measures* since they are functions of locations rather than people. Place-based accessibility measurement is a sensible approach: homes and workplaces are critical to most individuals and often serve as bases for travel, communication and participation in activities. However, while still viable and useful, place-based accessibility measures are increasingly incomplete.

A place-based approach to accessibility by itself is no longer viable in a world where transportation and ICTs are drastically changing the relationships among place, space and person. Efficient transportation creates *space-time convergence*: some locations become more proximal with respect to required travel times, creating an intricate "warping" of space (Janelle 1969, 2004). ICTs essentially eliminate space, at least for some types of interactions and activities. These technologies in turn shape lives by changing the number and types of activities individuals can experience as well as their distribution in space and time. They also shape cities by altering a fundamental reason for urban settlement, namely, accessibility to people, activities and opportunities. This is creating a complex geography of accessibility that is only partially explained by simple distance relationships between places (Kwan and Weber, 2003).

Using place as a surrogate for people leads to tractable measures with data requirements that correspond with traditional place-based data collection methods such as census and surveys. However, the continuing development and deployment of geographic information systems (GIS) and other geospatial technologies are greatly enhancing the ability to collect and analyze data relevant to accessibility analysis. In particular, these technologies allow finer-grained depictions of travel, activity and interaction possibilities that can be tied directly to people rather than place.

This chapter argues that traditional, place-based measures of accessibility should be enhanced and complemented with people-based measures that are more sensitive to individual activity patterns and accessibility in space and time. The next section reviews place-based measures, including major methods and implementation issues. The following section highlights the relationships between new technologies and accessibility, both with respect to the impact of transportation and ICTs on activity participation and accessibility as well as the capabilities for collecting and analyzing fine-grained spatio-temporal data that can support people-based approaches. The fourth section discusses people-based accessibility; this includes time geography as the major theoretical framework, as well as existing and evolving approaches to people-based accessibility that use time geography and geospatial technologies as a basis. The final section concludes with some brief comments.

PLACE-BASED ACCESSIBILITY

Measuring place-based accessibility

Place-based accessibility measures focus on the spatial separation between key locations such as home and work and potential activity locations such as employment centers, retailing, health care facilities, recreation sites and so on. Major types of place-based accessibility measures include *distance, topological, attraction-accessibility* and *benefits* measures (Handy and Niemeier, 1997; Miller, 1999a; Pirei, 1979).

Distance measures

These are the simplest type of accessibility indicators. Distance measures view accessibility as exclusively a function of the spatial separation between two places. Assuming that overcoming physical separation is onerous to individuals, greater separation implies lower accessibility. It is possible to make simple nominal and cardinal statements about accessibility using distance measures (e.g., "Location A is within five miles of a hospital."). We can also compare the distances among location pairs in ratio measures of relative accessibility (e.g.,

"Location A is twice as far from the nearest park as Location B."). Absolute differences in distances among location pairs are also meaningful ("A is ten miles from the nearest hospital while B is only five miles away."). It is also possible to aggregate and summarize distance measures to accommodate multiple destinations (e.g., the average distance to recreation centers) or social cohorts (e.g., the average distance to the nearest supermarket from low income neighborhoods).

"Distance" is a general concept in these measures: although Euclidean distance is simple and straightforward, other distance metrics (such as Manhattan or empirically-estimated metrics; see Love, Morris and Wesolowsky, (1988)) and network-based shortest path lengths and travel times are possible and appropriate depending on the context.

An extension of distance measures are *cumulative opportunity measures*. These are cardinality measures that count the number of relevant destinations or *opportunities* within a fixed distance of an origin. As with distance measures, one can also make absolute or relative comparisons using cumulative opportunity counts. Cumulative opportunity measures are easy to compute using the buffer operation in most geographic information system (GIS) software; these can be overlaid with other social, economic or infrastructure data.

Topological measures

These measures examine the degree and pattern of connectivity of nodes within a network. A central component of topological measures is the *connectivity matrix*: each element records the presence or absence of a direct link between the node corresponding to the matrix row and the node corresponding to the column. We can also use measures of the cost of traversing the link, with very large values (conceptually equivalent to infinity) corresponding to the absence of a direct link. Using this matrix as a foundation, it is possible to compute various measures including power functions that show the number of direct and indirect links between origin-destination pairs or the *Shimbel* index showing the number of links in the shortest paths (Garrison, 1960; Taaffe, *et al.*, 1996).

Topological measures are useful for analyzing transportation problems and policy. They can be used as a measure of public or airline transportation access (e.g., the number of stops between an origin and destination). They can be extended to ICTs such as the Internet (see Grubesic and O'Kelly, 2002; O'Kelly and Grubesic, 2002). Topological accessibility measures are seeing renewed interest because of the widely discussed "small world" phenomenon in social and other networks where sparse patterns of local and global connectivity can lead to extensive interconnections in spatial systems (Watts, 1999).

Attraction-accessibility measures

A problem shared by both distance and topological measures is that both treat all destinations as equivalent. In reality, destinations differ with respect to characteristics that influence their attractiveness to individuals. Attraction-accessibility measures postulate a tradeoff between the utility of a destination and its required travel cost relative to a given origin. These types of measures have their foundation in spatial interaction and spatial choice theories. Spatial interaction theory addresses aggregate flows of people, material or information between origins and destinations based on their characteristics and the degree of spatial separation between origin-destination pairs. Spatial choice theory addresses individual-level preferences and choices among locations using utility functions that often incorporate destination attractiveness and spatial separation between an origin and destinations. Both theories are consistent: one can derive the spatial interaction model from a spatial choice foundation (see Fotheringham and O'Kelly, 1989).

Weibull (1976, 1980) develops a general and rigorous framework for attraction-accessibility measures that specifies the appropriate way that paired distance and attractiveness measurements can be meaningfully combined in accessibility measures. Historically, the most popular attraction accessibility measure is the *Hansen measure* (Hansen, 1959). This is the special case of the Weibull framework where paired distance and attractiveness measures combine additively:

$$A_i = \sum_j O_j f\left(C_{ij}\right) \tag{1}$$

where A_i is the accessibility of location i, O_j is the attractiveness of destination j and C_{ij} is the travel cost from i to j (typically measured in distance or time). $f(\ )$ is a distance impedance function such as the inverse power function $\left(f\left(C_{ij}\right) = C_{ij}^{-\beta}\right)$ or the negative exponential function $\left(f\left(C_{ij}\right) = \exp\left(-\beta C_{ij}\right)\right)$. Since the Hansen measure assumes that more opportunities are always better, it is a type of *additive attraction-accessibility measure* (also see Erlander, 1977; Erlander and Stewart, 1978; Geertman and van Eck, 1995). This is only one possibility. For example, if we postulate that only the best opportunity is relevant, than we can measure only the best attraction/distance pair to derive a *maxitive attraction-accessibility measures*. Another possibility is to use random utility principles to derive the expected maximum utility of a choice situation (see benefit measures discussed below). The result is a *transform-additive attraction accessibility measure*. The appropriateness of the alternative derivations depends on the behavioral assumptions appropriate for the given application context (see Weibull, 1976, 1980).

Despite their historic popularity, attraction-accessibility measures have some significant weaknesses. These measures assume that the ordering of alternatives is irrelevant to the individual; this is clearly not the case when individuals have less than complete knowledge and must acquire information through a search process. Attraction-accessibility measures also deny the possibility of a hierarchical decision process where individuals mentally cluster individual choices into aggregates (e.g., making a choice between downtown versus suburban shopping malls prior to choosing individual stores). Finally, attraction-accessibility measures can be difficult to interpret. For example, researchers often interpret the Hansen measure as a gauge of "potential interaction"; however, it is unclear exactly what this means beyond simple ordinal relationships (e.g., "A has more potential interaction than B.") (Miller, 1999a).

Benefit measures

Benefit measures draw from the random utility framework and the microeconomic theory of consumer surplus. These measures equate accessibility with the benefits provided to an individual from a spatial choice situation. Benefit measures are closely related to attraction-accessibility measures: one can derive these measures within the Weibull framework (Miller, 1999a).

We can form an accessibility benefits measure using the random utility choice framework where individuals' utilities can only be observed to a stochastic residual   If we assume that the unobserved utilities have identical but independent distributions and scale the observed utilities appropriately, we can derive the expected maximum utility measure from the nested logit choice model (Ben-Akiva and Lerman, 1979, 1985):

$$A_i = \ln \sum_{j \in \mathbf{K}_i} \exp\left(u_{ij}\right) \tag{2}$$

where u_{ij} is the observed utility of destination j for person i, and $\mathbf{K}_i$ is the choice set for person i. Although this does not appear to be a place-based measure, in practice the person i is linked to a location and the utilities $\left\{u_{ij}\right\}$ typically include origin-destination travel costs measures as disutilities.

There is a close correspondence between the expected maximum utility of a choice situation and the concept of *consumer surplus* in microeconomic theory (Ben-Akiva and Lerman, 1979, 1985; Jara-Diaz and Friesz, 1982; Small, 1992; Williams, 1976, 1977; Williams and Senior, 1978). Consumer surplus measures the benefits to an individual from the prevailing price in a market. The log-sum expected maximum utility is consistent with two major consumer surplus measures, namely, *Marshallian surplus* (an individual's willingness to pay above the prevailing market price) and the *compensating variation* (the income transfer required to maintain the same utility level given a small change in market price) (Jara-Diaz and Farah, 1988).

Although consistent with consumer surplus, there are some problems with this interpretation of the log-sum expected maximum utility. First, it is unclear that it can be interpreted as a "willingness to pay" unless strict (and somewhat unrealistic) requirements are met with respect to the individual's utility of money (McFadden, 1998). Without a common metric such as money, it difficult to absolute comparisons among differences in accessibility. However, it is still possible to make relative comparisons with benefit measures. Second, interpreting the measure as consumer surplus requires assuming that demand cross-elasticities are constant among all choice-pairs; this is inconsistent with the theoretically-correct demand shifts in transportation markets. Since transportation modes typically have different degrees of substitutability, cross-elasticities are not constant (Jara-Diaz and Friesz, 1982).

Implementation issues

Place-based accessibility measures share several implementation issues that are common among most types of place-based analysis such as spatial analysis and travel demand modeling. These include *the degree and type of disaggregation, specification of origins and destinations, measuring destination attractiveness* and *estimating travel impedance* (Handy and Niemeier, 1997).

Aggregation and disaggregation

Accessibility analysis often uses spatial zoning systems based on census data or traffic analysis zones due to data availability and consistency with other exercises such as travel demand forecasts. However, spatial aggregation can affect the results of an accessibility analysis. A well-established result from the spatial analysis literature is the *modifiable areal unit problem* (MAUP). If the spatial zoning system is artificial (or "modifiable") than the results of analysis based on that system are arbitrary: the results can vary simply by changing the zoning system (see Cressie, 1996; Fotheringham and Wong, 1991; Openshaw and Taylor, 1979). There are two dimensions to this problem, namely, *scale* and *zoning*. Scale refers to the level of spatial aggregation and zoning refers to the spatial partitioning given the level of aggregation. There is no unequivocal way to solve the MAUP; methods for mitigating its effects include: i) performing a sensitivity analysis; ii) designing optimal spatial units; iii) using better interzonal distance measures such as true average distances between zones rather than the centroid-to-centroid surrogate (Miller, 1999b). A general rule of thumb is that greater spatial disaggregation is better (Handy and Niemeier, 1997), although only complete disaggregation to the atomic units of analyses will eliminate MAUP effects.

A related issue is whether the individual or household is the atomic unit of analysis. Accessibility is an individual-level phenomenon, but ignoring within-household organization and decision-making can miss critical aspects of accessibility, including task specialization among members, joint activity participation and competition for transportation and other resources within households (Hanson and Schwab, 1987; Janelle and Goodchild, 1983; Jones, 1989; Kwan, 1999).

Other aggregation dimensions include socio-economic cohorts and trip purpose Accessibility measures can also be aggregated or disaggregated according to socio-economic status, life cycle, gender, culture and so on, supporting comparative analysis. Trip purpose can be highly general, disaggregated into generic types (work versus non-work), as well as disaggregated within the generic types (employment categories, types of retail outlet and so on). These aggregation decisions relate to data availability as well as particular problem or policy being analyzed (Handy and Niemeier, 1997).

Specifying origins and destinations

Place-based accessibility measures require specification of trip origins and destinations (Handy and Niemeier, 1997). A common trip origin is the individual's or household's place of residence. This has obvious relevance to daily lives, particularly for accessibility to employment opportunities and household maintenance-related services (such as grocery stores). However, the relative decline of home-based trips for some types of travel purposes mean that this assumption must be used with caution. In particular, multipurpose and multistop trips raise complex issues since accessibility is dictated by intermediate stops on the trip as well as the trip's ultimate origin. Southworth (1983) develops circuit-based accessibility measures that combine the attraction-accessibility approach with the topological approach to capture complex multi-stop/multipurpose interdependencies.

The selection of destinations is closely related to disaggregation by trip purpose. The relevant set of potential destinations is a function of trip purpose, and therefore trip purpose disaggregation should dictate the destination choice set. A more subtle problem is the spatial boundary on the destination choice set. All potential destinations in a study area may not be reasonable due to the distance or travel time required. Although large travel times or distances discount a destination in attraction-accessibility and benefit measures, including unrealistic destinations will still overestimate accessibility (Handy and Niemeier, 1997). One can constrain these choices using a maximum travel time or distance as a delimiter of the destination choice set, although this can be arbitrary unless it is based on first principles or time-use data. Limited operating hours also means that some destinations are not available to an individual depending on the trip timing; this is not often considered in accessibility analysis (Weber and Kwan, 2002).

Destination attractiveness

Measuring destination attractiveness is critical but difficult task in attraction-accessibility and benefits measures. The simplest measure is binary: this indicates only the existence of an opportunity. An extension of binary measures is simple counts for aggregate destinations such as shopping centers malls or large medical facilities (e.g., the number of stores, the number of doctors on staff). An often used surrogate for attractiveness in the retail choice literature is the square footage of a destination; this corresponds (sometimes only roughly) to the number and variety of opportunities at the destination (e.g., the range of goods available at a grocery store). More generally, the attractiveness of destination can encompass many dimensions including cost, quality of service, convenience, perceived social status, familiarity and so on. Collapsing multiple dimensions into a real number that represents attractiveness requires detailed stated or revealed preference data combined with sometimes elaborate (and therefore time and resource consuming) scaling and estimation techniques (see Fotheringham and O'Kelly, 1989, Chapter 6).

Travel impedance

Attraction-accessibility and benefits measures also require estimation of travel impedance. This requires hypothesizing an impedance functional form (e.g., power versus exponential) calibrating its associated parameter using stated or revealed preference data. This should occur in conjunction with scaling and estimating the parameters associated with destination attractiveness since these effects are closely intertwined. Estimating travel impedance associated with public transit and multimodal trips is particularly complex since this can involve access time (such as walking to a light rail stop), wait time, line-haul time, transfer time and egress time; all can have different perceived burden to individuals. There is also perceived risks and consequences associated with missing transfers, levels of comfort, security, and so on; a similar problem to the multidimensional measurement problem associated with destination attractiveness.

NEW TECHNOLOGIES AND ACCESSIBILITY ANALYSIS

While places are still important, the context for accessibility and its analysis is changing. Transportation and information/communication technologies are greatly changing the relationships among place, space and individual. GIS and other geospatial technologies are greatly enhancing the ability to collect data relevant to accessibility analysis. This section reviews these technological changes and their effects on accessibility and its analysis.

Transportation and Information/Communication Technologies

Transportation and technologies exist to alter relationships between time and space. Transportation technologies increase the efficiency of trading time for space in physical movement. Perhaps the most profound experience of humanity over the past few centuries has been *space-time convergence* or the collapse of space with respect to travel time due to transportation technologies. Innovations such as the stagecoach, clipper ship, railroad, automobile and airplane allow less time to be traded per unit space, freeing this scarce resource for other activities including travel itself (Janelle, 1969, 2004).

ICTs facilitate an even more dramatic collapse of space with respect to time, allowing essentially instantaneous exchange of information subject to limitations such as the speed of light, bandwidth capacity and network congestion. ICTs can fragment traditional geographies: they allow some activities to be disconnected from space (Couclelis and Getis, 2000). For example, there is no longer an unequivocal location associated with the activity "work": one can work at an office, at home, at a coffee shop, in a park, and even in a moving automobile, subject to the ability to afford the equipment and high-speed Internet services required as well as employment that allows virtual rather than physical interaction.

The collapse of distance with respect to time has profound impacts. First, as geographer Waldo Tobler points out, although the world is shrinking, it is also shriveling: relative differences in transportation costs and times are increasing. This can occur due to simple geometry: for example, since even uniform travel time improvements have greater effects on distant places than proximal places, urban fringes will typically benefit more than city centers from reductions in transport cost (Janelle, 2004). It also relates to varying transportation resources at the individual level, as well as uneven public and private sector investments at the neighborhood, regional, national and international levels. This is creating a complex and uneven geography of travel and accessibility (Kwan and Weber, 2003). Space-time convergence is also an accelerant: it allows more activities to be conducted in more locations and times. This can create complex interpersonal organizational problems since more activities have to be co-located and scheduled.

The relationships between transportation, ICTs and accessibility are complex and not easily explained by the simple "Death of Distance" argument that dominated the early literature on these topics. ICTs and transportation can serve as *substitutes* for each other: for example, higher ICT usage may reduce transportation demand if this allows the substitution of in-home (or at-work) activities for those that previously required travel. ICTs can also *complement* transportation: for example, higher ICT usage may lead to more transportation demand if individuals can acquire information about modes, routes and destination options that were previously unknown. ICTs and transportation can also *modify* the demands for both by changing the location and timing of travel or communication activities without a net increase or decrease in demand. ICTs can also affect the supply of transportation, and vice-versa; for example, ICTs can allow more efficient use of existing transportation facilities, increasing capacity without the need for additional infrastructure. This can have consequent longer-term impacts on their relative demands (see Mokhtarian, 1990; Salomon, 1986).

Krizek and Johnson (2003) classify potential transportation-ICT relationships based on the underlying activity and the potential relationships between transportation and ICTs with respect to the given activity. They classify activities as fixed or flexible (that is, difficult or easy to relocate or re-schedule, respectively; see the section on "time geography" below), with fixed activities further subdivided into subsistence and maintenance activities. Table 1 provides their typology with examples.

Table 1. Relationships between transportation/ICTs and activities (Krizek and Johnson, 2003)

| Activity type | Relationship between transportation and ICTs | | |
	Substitute	Complement	Modify
Fixed Subsistence activities	Telecommuting and homeschooling	Arranging and booking work-related travel	Work-related communication while at home, traveling or participating in other activities
Maintenance	Online banking and shopping	Online shopping guides (to "bricks and mortar" shops)	Completing errands involving travel while telecommuting from home
Flexible activities	On-demand video; Purchasing concert or film tickets online	Online restaurant, nightlife and travel guides	Better coordination and increased flexibility in communicating and meeting with friends and family.

The fundamental lesson is that accessibility has increasingly complex and in some cases fractured relationship with places. In many parts of the world, individuals are increasingly more mobile: they are in fixed locations (particularly their homes) for smaller proportions of their daily and weekly cycles. Distance from a single, key location or mathematical functions of this simple relation no longer describe adequately the context for accessibility for many people: trip origins are spread over space and time and are linked in complex ways. There is also no longer a one-to-one relationship between activities and locations. This makes accessibility inferences from place alone increasingly dangerous. Individual and household characteristics can have much stronger influences on accessibility than place (Weber and Kwan, 2003). Different people in the same home or workplace can have strikingly different activity, travel and interaction patterns (Kwan, 1998). However, this does not mean that place be ignored: this can result in an "individualistic fallacy" where interaction and synergy among individuals at particular locations are not captured (National Research Council, 2002).

Location-aware technologies

At the same that technologies are changing human activity in space and time, they are also increasing researchers' abilities to collect and analyze activity data (Miller, 2003). Traditional methods for collecting travel and activity data collection have well-known problems. Asking people to report activities, locations and times in real-time using traditional diary methods is bothersome and therefore fraught with omissions (Brog *et al.*, 1982; Golledge and Zhou, 2001;

Purvis, 1990). People also have trouble recalling activities that comprise a "typical" day or week (Golledge and Zhou, 2001).

Location-aware technologies (LAT) are devices that can capture geo-spatial coordinates at high levels of temporal resolution. The global positioning system (GPS) is a well-known example, but other possibilities include triangulation methods that piggyback on wireless communication. LAT continue to shrink in size and cost; in the near-future they will be easily deployed and even disposable. LATs can collect high resolution and high accuracy data on travel and activities in space and time. The patterning and clustering of these paths can illuminate sharp differences among individuals and social groups with respect to the limited environments or *activity spaces* occupied during a daily or weekly cycle. (Golledge and Stimson, 1997). These can reflect both differences in accessibility as well as denote the spatio-temporal constraints on accessibility.

In addition to their scientific benefits, LATs can also provide unprecedented levels of privacy invasion. Space-time paths and activity spaces are in a sense of type of "signature" that can reveal much about a person's life (indeed, this is why they are so useful to social research). *Locational privacy* is an emerging issue in geographic information science, and new techniques are being developed such as *locational masking* that protect location privacy by introducing known and controlled error into the data (see Armstrong, 2002; Armstrong *et al.*, 1999).

PEOPLE-BASED ACCESSIBILITY

The increasingly complex relationship between people, place, space and activities suggests that the place-based approach to accessibility should be enhanced with measures that are directly tied to the individual in space and time. This section reviews the emerging people-based approach to accessibility analysis. This includes *time geography* as the major theoretical framework for conceptualizing people-based accessibility, as well as attempts to resolve weaknesses of classical time geography for accessibility analysis through the use of the geospatial technologies as well as the underlying geographic information science that informs these technologies.

It is important to note that a people-based approach to accessibility does not negate the traditional place-based perspective. Rather, a people-based perspective complements the traditional approach by focusing on the individual in space and time, including their use of places. It is a flexible approach that can accommodate accessibility of people to people as well as people to places. In contrast, a place-based perspective can only accommodate accessibility among places; this is a reason why complex mobility patterns and ICTs cannot be handled using the traditional approach.

Time geography

Time geography (Hägerstrand, 1970) examines the simple question: "how does participating in an activity at a given place and time limit a person's abilities to participate in activities at other places and times?" This simple question has potent implications for conceptualizing and measuring accessibility.

Time geography views activities as different with respect to pliability in space and time. *Fixed activities* are relatively difficult to re-locate and re-schedule. Examples include work and many types of personal and familial activities such as sleep and child care. *Flexible activities* are relatively easy to re-locate and re-schedule; these include shopping, recreation and socializing. The existence of fixed activities means that individuals have a limited *time budget* to allocate among flexible activities. Individuals must also allocate time for travel among activity locations. The combination of space-time *anchor points* (fixed activity locations and durations), the time budget and the ability to trade time for space using transportation technologies determines an individual's accessibility to resources and opportunities that exist in relatively few places for limited durations (Miller, 2004).

Two entities are fundamental to time geography. The *space-time path* traces the individual's physical movement in space with respect to time. This highlights the constraining effects of a person's need to be at different locations at different times as well as the role of transportation in mitigating these constraints. Figure 1 illustrates a space-time path among several space-time *stations* or locations for conducting activities The path slope illustrates the efficiency of the given travel mode: e.g., a steeper slope indicates more time required per unit space in movement. Space-time paths can reveal a great deal about varying activity patterns and spaces among different social groups. Methods for deriving this information from the voluminous path data sets generated by LATs include visualization (Huisman and Forer, 1998; Kwan, 2000; van der Knaap, 1997) and data mining (Arentze *et al.*, 2000; Joh, Arentze and Timmermans, 2001).

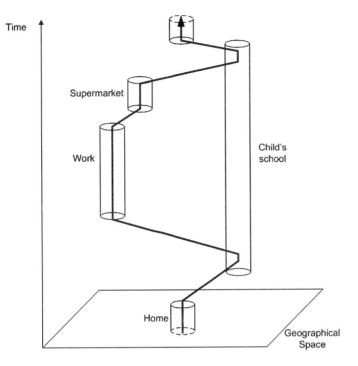

Figure. A space-time path and stations

The *space-time prism* delimits the possible locations for the space-time path. Figure 2 illustrates a prism with a coincident anchors; more generally, the anchors may be different locations, or the second anchor may be undefined (see Burns 1979). Fixed activities anchor a space-time prism since (by definition) these allow only one spatial possibility during their duration. Associated with the anchors are the earliest time the person can leave that first fixed activity and the latest time they can arrive at the second fixed activity (t_i and t_j, respectively). The time interval $t_{ij} = (t_j - t_i)$ is the individual's time budget for travel and activity participation. The person must stop at some location to conduct an activity that will require at least a time units. Finally, the person can move with an average maximum velocity v. The interior of the prism is the *potential path space*: this shows the points in space and time that the person could occupy during this episode. The projection of the potential path space to geo-space provides the *potential path area*: all spatial locations that the person could occupy during the time interval (Miller, 2004).

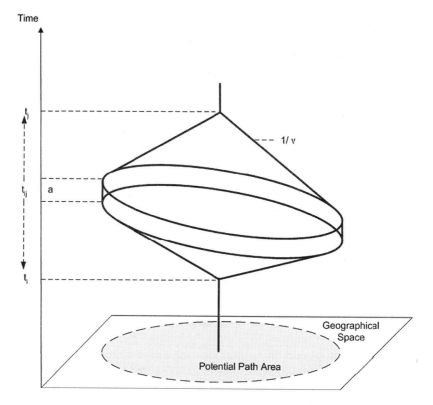

Figure 2. A space-time prism (Miller 2004).

The path and prism are foundations for measuring individual accessibility. The number of paths allowed by a given fixed and flexible activity schedule is a surrogate for accessibility (Lenntorp, 1976, 1978). The prism can also be used to delimit potential activity locations. For example, a person cannot participate in an activity unless its station intersects the potential path space to a sufficient degree. Also, a person cannot participate in an activity unless its location falls within the potential path area (ignoring the temporal duration of the activity; see Kwan, 1998). We can also form accessibility measures that are a function of the prism size, or the activities and participation times allowed by a prism (Burns, 1979; Miller, 1999a).

Problems with classical time geography in accessibility analysis

There are several problems with the classical space-time prism as an accessibility measure. One problem is the assumption of a uniform travel velocity over space and time. This is highly unrealistic. Most physical movement is constrained into corridors by the built environment, including buildings, sidewalks, streets and highways, as well as the modes used to traverse these corridors (including, walking, bicycling, driving, riding public transit or combinations of these). In addition, the increasingly saturated nature of many transportation networks means that travel times within these conduits can be highly variable over time due to congestion.

A second problem with the classic prism is it a conceptual entity that cannot support the high-resolution measurement and corresponding analytical possibilities allowed by LATs as well as GIS and other geo-spatial technologies. Until recently, there were no analytical statements of basic time geographic entities other than the informal, geometric descriptions of Burns (1979), Lenntorp (1976, 1978) and Hornsby and Egenhofer (2002). "Analytical" in this case refers to the ability to make statements about time geographic properties (such as the prism extent) to arbitrary levels of spatio-temporal resolution. Burns (1979) and Lenntorp (1976, 1978) only provide verbal descriptions of constructible geometric objects that can support summary measures such as the prism volume given the strict assumptions discussed above. Hornsby and Egenhofer (2002) provide more rigorous definitions of some basic time geographic including the path and prism, but they define these entities through simultaneous inequalities; this is cumbersome for high-resolution measurement and analysis since the entities are defined implicitly (Miller, 2005a).

A third problem with the path and prisms is their physical basis. Although time geography recognizes the possibility of virtual interaction (see Hägerstrand, 1970), these modes are downplayed relative to physical movement and presence. Some progress is being made in extending time geography to virtual interaction, including graphical techniques (e.g., Adams, 2000) as well as behavioral frameworks (e.g., Kwan, 2001). However, these extensions are not analytical or well-integrated into time geography.

Advances in geospatial technologies and the underlying geographic information science can allow more realistic and useful time geographic analysis of accessibility. The next subsection discusses some of the emerging approaches to measuring people-based accessibility within network, multidimensional and virtual spaces.

Measuring people-based accessibility

Network spaces

GIS and related geo-spatial technologies allow researchers to form more realistic space-time prisms. Geospatial technologies such as LATs, mobile GIS, remote sensing and Intelligent Transportation Systems (ITS) allow collection of high resolution data on transportation infrastructure and conditions at increasingly lower costs. Social data are readily available in digital form. GIS supports the integration, maintenance and visualization of these data, some built-in analytical functionality and the ability to link to customized or user-built analytical software. This is a powerful platform for accessibility analysis.

A *network time prism* (NTP) is a space-time prism defined using the topology, travel distances and velocities dictated by transportation infrastructure and conditions. A basic NTP product is the *potential path tree* (PPT): this shows all network nodes that can be reached by an individual based on space-time anchors (also defined at nodes), a time budget and static arc travel velocities. An extension of the PPT, the *potential network area* (PNA) resolves to arbitrary locations within the network: it shows the accessible locations in the network to the sub-arc level. Figure 3 illustrates a PPT and PNA (Miller, 2004).

NTPs are not limited to single-mode networks. We can also define the NTP within multimodal networks since we can represent any mode using networks, distances and time, including waiting and transfer times. (Note that, unlike attraction-accessibility and benefits measures, we do not need to calibrate parameters associated with the differing perceived costs of these time modes: time geography is concerned with constraints on choice rather than choice per se.) O'Sullivan, Morrison and Shearer (2000) use off-the shelf GIS software to estimate prisms based on a multimodal public transit network and walking to/from transit stops.

The NTP can also be integrated with other georeferenced socio-economic or behavioral data. Kwan and Hong (1998) integrate cognitive constraints into a NTP through a GIS overlay with georeferenced survey data indicating individuals' locational preferences and knowledge about the environment.

We can also extend the NTP to capture time-varying conditions such as congestion. A *dynamic network prism* (DNP) is a space-time prism within a network with time-varying velocities. It is tractable to calculate DNP products such as the dynamic PPT using discrete-time dynamics such as those generated from a discrete time dynamic flow model or instantaneous ITS data captures at discrete intervals of time (see Wu and Miller, 2000). Calculating NTP products within networks with continuous-time dynamics is an open research issue.

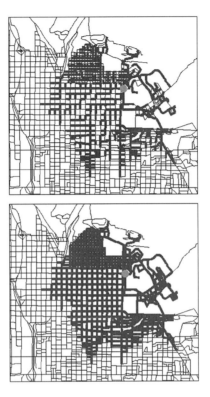

Figure 3. A network potential path tree (top) and potential path area (bottom) (Miller 2004).

Multidimensional spaces

It is possible to both generalize time geography as well as improve its analytical foundation through temporal disaggregation. Analytical statements of time geographic entities and relations are possible for given moment in time since these instantaneous entities and relationships are simple geometric objects and relations. These analytical definitions are also general to multi-dimensional spaces, although we are only concerned about one, two and three-dimensional space directly in time geography.

For example, the spatial extent the space-time prism at any given instant in time t is the intersection of simple and compact spatial sets as illustrated in Figure 4. At any instant t, the space-time prism defined by anchors $\mathbf{x}_i, \mathbf{x}_j$ with required presence at times t_i, t_j (respectively) and maximum travel velocity v is:

$$Z_{ij}(t) = \left\{ \mathbf{x} \mid f_i(t) \cap p_j(t) \cap g_{ij} \right\} \tag{3}$$

where:

$$f_i(t) = \left\{ \mathbf{x} \middle| \|\mathbf{x} - \mathbf{x}_i\| \le (t - t_i)v \right\} \tag{4}$$

$$p_j(t) = \left\{ \mathbf{x} \middle| \|\mathbf{x}_j - \mathbf{x}\| \le (t_j - t)v \right\} \tag{5}$$

$$g_{ij} = \left\{ \mathbf{x} \middle| \|\mathbf{x} - \mathbf{x}_i\| + \|\mathbf{x}_j - \mathbf{x}\| \le (t_j - t_i - a)v \right\} \tag{6}$$

$f_i(t)$ is the *future disc*: the locations that can be reached by time t when leaving from $\mathbf{x}_i$ at time t_i $p_j(t)$ is the *past disc*: these are the locations that can reach $\mathbf{x}_j$ by the remaining time $t_j - t$. We refer to these sets as "discs" since they are compact spatial sets consisting of all locations within a fixed distance of a point. g_{ij} is the *geo-ellipse*: it constrains the prism locations to account for any stationary activity time a during the time interval. It is equivalent to the *potential path area* of classical time geography. We refer to this as an ellipse since consists of all locations within a fixed distance of two locations. These sets are simple geometric forms: the future and past discs are lines in one spatial dimension, circles in two dimensions and spheres in three dimensions. The geo-ellipse is a line, ellipse and ellipsoid in one, two and three spatial dimensions respectively. It is also possible to simplify the calculations by solving for the time boundaries within the interval to determine subintervals when a set is encompassed by other sets and can be ignored (Miller, 2005a).

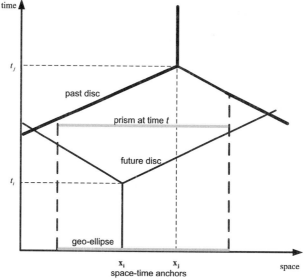

Figure 4. Analytical definition of the space-time prism

The multidimensional analytical framework for time geography can support high-resolution measurement through LATs. The instantaneous solutions can match the high-density but discrete time locational tracking performed by these technologies. It also can be used to analyze imperfect measurement (including intentional error to provide locational privacy) and the propagation of uncertainty since the simple geometry allows analytical solutions to error propagation analysis. The geometric solutions also provide functional requirements for software implementation of time geographic queries (Miller, 2005a).

Virtual spaces

Time geography can encompass ICTs and virtual interaction by extending the measurement theory discussed above. Define a *portal* as a type of space-time station where actors can access appropriate communication services. A portal consists of three components: i) a point *source* for ICT access; ii) a *range* for ICT access, indicating the maximum distance from the source that an actor can access the service; and, iii) an ordered list of operating spans when the portal is available. Portals correspond to real-world entities such as wired telephony and Internet connections, wireless access points and cellular telephone base stations. Figure 5 illustrates two portals and a space-time path. An individual can access a communication service only if his or her paths or prism (reflecting actual and potential movement) intersect with the service footprint of an appropriate portal (Miller, 2005b).

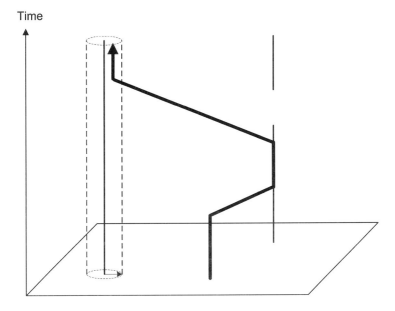

Figure 5. Two portals and a space-time path (Miller, 2005b)

The places and times when individuals access their respective portals combined with the spatio-temporal constraints imposed by the communication mode circumscribes virtual interaction among individuals. A communication mode can require either physical presence or telepresence and either synchronous or asynchronous communication. *Synchronous presence* (SP) corresponds to face-to-face (F2F) interaction. *Synchronous telepresence* (ST) requires only coincidence in time; this includes telephones, text messaging and television. *Asynchronous presence* (AP) requires coincidence in space but not time; this includes hospital charts. *Asynchronous telepresence* (AT) does not require coincidence in space and time: this mode includes printed media, email, text messages and webpages (Janelle, 1995, 2004). Table 2 summarizes these communication modes.

Temporal	Spatial Physical presence	Telepresence
Synchronous	**SP** Face to face (F2F)	**ST** Telephone Instant messaging Television Radio Teleconferencing
Asynchronous	**AP** Refrigerator notes Hospital charts	**AT** Mail Email Fax machines Printed media Webpages

Table 2. Spatial and temporal constraints on communication (based on Harvey and Macnab, 2000; Janelle, 1995, 2004; Miller, 2005b).

Message windows are intervals of time corresponding to potential or actual communication events. A *general message window* is a temporal interval when an individual can access a portal. A *strict message window* is a temporal interval corresponding to an actual message. Figure 6 illustrates general and strict message windows. Since communication can be asymmetric, we also distinguish between *send* and *receive* message windows. Table 3 indicates the types of questions we can ask about the potential for virtual interaction using these windows.

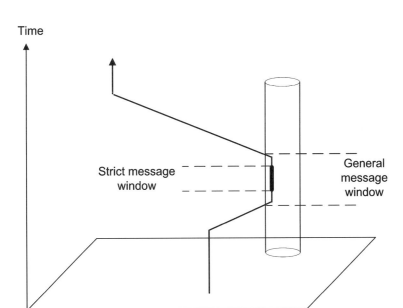

Figure 6. General and strict message windows (Miller, 2005b).

Send	Receive	
	General	**Strict**
General	General spatio-temporal bounds on virtual interaction	Who could have sent a specific message?
Strict	Who could receive a specific message?	Process theory: Actual message transmission

Table 3. Types of interactions between message windows (Miller 2005b)

Send and receive windows must occur at the same portal for presence modes, although these windows may not necessarily be the same portal for telepresence modes. We can solve for the temporal constraints on message windows using the well-known Allen time predicates that encompass all possible relationships between two intervals of time (Allen, 1984). For example, consider the case of a strict send window and a general receive window. The temporal constraints that dictate which general windows could receive a sent message are:

Synchronous $\qquad t_i^s \geq t_k^r \wedge t_j^s \leq t_l^r$ (7)

Asynchronous $\qquad t_j^s < t_l^r$ (8)

where $t^s = \left[t_i^s, t_j^s \right]$ and $t^r = \left[t_k^r, t_l^r \right]$ are the time intervals corresponding to the send and receive windows respectively, and $\wedge$ is the logical predicate AND. (7 states that the general window must be at least as large as the strict window, i.e., a person must interact with a receiving portal the entire time during synchronous communication. (8 states that the general window must begin anytime after the strict window ends, i.e., a person must interact with a receiving portal after the message is sent. We can also solve for side conditions dictating the required lengths of each window in the asynchronous case:

$$
\begin{aligned}
t_l^r - t_k^r \geq t_j^s - t_i^s \,, \quad t_j^s \leq t_k^r \\
t_l^r - t_j^s \geq t_j^s - t_i^s \,, \quad t_j^s > t_k^r
\end{aligned}
\tag{9}
$$

The first condition and second conditions correspond to no temporal overlap and temporal overlap between the windows, respectively. Similar results for all possible pairings of general and strict message windows for both synchronous and asynchronous modes can be found in Miller (2005b).

CONCLUSION

Places such as home and work are important, and therefore place-based accessibility analysis will always have its applications. However, the continuing development and impact of transportation and ICTs in many parts of the world are creating complex relationships between person, place, space and activity. People are more mobile, both physically and virtually, and activity patterns are consequently more complex and dispersed in space and time. This implies that accessibility to activities is also more convoluted than can be explained by a place-based perspective alone. People-based measures examine accessibility from the perspective of a mobile and active person in space and time. These measures can complement place-based analysis by providing a more intricate view of accessibility relative to individuals' busy and convoluted lives.

Using time geography as a theoretical base and geospatial technologies as media, researchers are developing and applying people-based measures to the problem of accessibility analysis. Preliminary empirical results indicate that people-based measures provide striking different portrayals of accessibility than place-based measures (Kwan, 1998). However, additional research and development is required to provide practical people-based accessibility measures for researchers and practitioners.

With respect to theory and modeling, a critical research question is the development of accessibility measures that can exploit high-resolution locational tracking technologies. The time geographic measurement theory and its virtual extension discussed in this paper only provide a foundation: they allow the researcher to infer basic time geographic entities and relationships from measurement of physical activities and virtual interaction. Building these entities and relationships into practical accessibility measures is still an open research question. Also critical are aggregate, summary measures: individual, people-based accessibility measurement can generate a morass of detail, and there is a risk that synoptic patterns and relationships can be lost in this detail.

While useful, people-based measures are potentially intrusive. There is a need to balance the enhanced view of individual activities in space and time with individuals' rights to privacy. One possible strategy is to introduce controlled error into measured space-time activities and paths in a manner that provides some degree of privacy without harming accessibility and other analyses. This strategy extends the geographic masking approach of Armstrong, Rushton and Zimmerman (1999) to include mobile entities. Specific methods and their effects on measured activities and accessibility is an open research question.

REFERENCES

Adams, P. C. (2000). Application of a CAD-based accessibility model. In: *Information, Place and Cyberspace: Issues in Accessibility*, (D. G. Janelle and D. C. Hodge, eds.), pp. 217-239. Springer, Berlin.

Allen, J. F. (1984). Towards a general theory of action and time. *Artificial Intelligence*, **23**, 123-154.

Arentze, T. A., F. Hofman, H. van Mourik, H. J. P. Timmermans and G. Wets (2000). Using decision tree induction systems for modeling space-time behaviour. *Geographical Analysis,* **32,** 330-350.

Armstrong, M.P. (2002). Geographic information technologies and their potentially erosive effects on personal privacy. *Studies in the Social Sciences,* **27**, 19-28.

Armstrong, M.P., G. Rushton and D. L. Zimmerman (1999). Geographically masking health data to preserve confidentiality. *Statistics in Medicine*, **18**, 497-525.

Ben-Akiva, M. and S. R. Lerman (1979). Disaggregate travel and mobility-choice models and measures of accessibility. In: *Behavioural Travel Modelling* (D. A. Hensher and P. R. Stopher eds.), pp. 654-679. Croom-Helm, London.

Ben-Akiva, M. and S. Lerman (1985). *Discrete Choice Analysis: Theory and Application to Travel Demand*, MIT Press, Cambridge, MA.

Brog, W., E. Erl, A. H. Meyburg, and M. J. Wermuth (1982). Problems of non-reported trips in surveys of nonhome activity patterns. *Transportations Research Record* **891**, 1-5.

Burns, L. D. (1979). *Transportation, Temporal and Spatial Components of Accessibility*. Lexington Books, Lexington, MA

Couclelis, H. and A. Getis (2000). Conceptualizing and measuring accessibility within physical and virtual spaces. In: *Information, Place and Cyberspace: Issues in Accessibility*, (D. G. Janelle and D. C. Hodge, eds.), pp. 15-20.. Springer, Berlin.

Cressie, N. (1996). Change of support and the modifiable areal unit problem. *Geographical Systems,* **3**, 159-180.

Erlander, S. (1977). Accessibility, entropy and the distribution and assignment of traffic. *Transportation Research*, **11**, 149-153.

Erlander, S. and N. F. Stewart (1978). Interactivity, accessibility and cost in trip distribution. *Transportation Research*, **12**, 291-293.

Fotheringham, A. S. and M. E. O'Kelly (1989). *Spatial Interaction Models: Formulations and Applications.* Kluwer Academic Dordrecht, The Netherlands.

Fotheringham, A. S. and D. W. S. Wong (1991). The modifiable areal unit problem in multivariate statistical analysis. *Environment and Planning A*, **23**, 1025-1044.

Garrison, W.L. (1960) Connectivity of the interstate highway system. *Papers and Proceedings of the Regional Science Association*, **6**, 121-137

Geertman, S. C. M. and J. R. R. van Eck (1995). GIS and models of accessibility potential: An application in planning. *International Journal of Geographical Information Systems*, **9**, 67-80.

Golledge, R. G. and R. J. Stimson (1997). *Spatial Behavior: A Geographic Perspective.* Guilford: New York.

Golledge, R. G. and J. Zhou (2001). *GPS Tracking of Daily Activities*, final report, UCTC Grant # DTRS99-G-0009, Department of Geography and Research Unit on Spatial Cognition and Choice, University of California at Santa Barbara.

Grubesic, T. H. and M. E. O'Kelly (2002). Using points of presence to measure city accessibility to the commercial Internet. *Professional Geographer*, **54**, 257-278.

HägerstrandT. (1970) What about people in regional science? *Papers of the Regional Science Association*, **24**, 7-21.

Handy, S. L. and D. A. Niemeier (1997). Measuring accessibility: An exploration of issues and alternatives. *Environment and Planning A*, **29**, 1175-1194

Hansen, W. G. (1959). How accessibility shapes land use. *Journal of the American Institute of Planners*, **25**, 73-76

Hanson, S. and M. Schwab (1987). Accessibility and urban travel. *Environment and Planning A*, **19**, 735-748.

Harvey, A. and P. A. Macnab (2000). Who's up? Global interpersonal temporal accessibility," In: *Information, Place and Cyberspace: Issues in Accessibility* (D. G. Janelle and D. C. Hodge, eds.), pp. 147-170. Springer, Berlin.

Hornsby, K. and M. J. Egenhofer (2002). Modelling moving objects over multiple granularities. *Annals of Mathematics and Artificial Intelligence*, **36**, 177-194.

Huisman, O. and P. Forer (1998). Computational agents and urban life spaces: A preliminary realisation of the time-geography of student lifestyles.   *Proceedings, GeoComputation 98*, http://www.geocomputation.org/.

Janelle, D. G. (1969). Spatial organization: A model and concept. *Annals of the Association of American Geographers*, **59**, 348-364

Janelle, D. G. (1995). Metropolitan expansion, telecommuting and transportation. In: *The Geography of Urban Transportation*, 2ed. (S. Hanson, ed.), pp. 407-434, Guilford, New York.

Janelle, D. G. (2004). The impact of information technologies. In: *The Geography of Urban Transportation*, 3ed. (S. Hanson and G. Giuliano eds.), pp. 86-112, Guilford, New York.

Janelle, D. G. and M. F. Goodchild (1983). Transportation indicators of space-time autonomy. *Urban Geography,* **4**, 317-337.

Jara-Diaz, S. R. and M. Farah (1987). Transport demand and users' benefits with fixed income: The goods/leisure trade off revisited. *Transportation Research B*, **21B**, 165-170.

Jara-Diaz, S. R. and T. L. Friesz (1982). Measuring the benefits derived from a transportation investment. *Transportation Research B*, **16B**, 57-77

Joh, C.-H., T. A. Arentze and H. J. P. Timmermans (2001). Multidimensional sequence alignment methods for activity-travel pattern analysis: A comparison of dynamic programming and genetic algorithms. *Geographical Analysis*, **33**, 247-270.

Jones, P. (1989). Household organization and travel behaviour. In: *Gender, Transport and Employment: The Impact of Travel Constraints*. (M. Grieco, L. Pickup and R. Whipp eds.), pp. 46-74, Gower, Oxford Studies in Transport, Aldershot, UK.

Krizek, K. J. and A. Johnson (2003). Mapping the terrain of information and communication technology (ICT) and household travel. *Proceedings, 82nd annual meeting of the Transportation Research Board*, CD-ROM.

Kwan, M.-P. (1998). Space-time and integral measures of accessibility: A comparative analysis using a point-based framework. *Geographical Analysis*, **30**, 191-216.

Kwan, M.-P. (1999). Gender and individual access to urban opportunities: A study using space-time measures. *Professional Geographer*, **51**, 210-227.

Kwan, M-P. (2000). Interactive geovisualization of activity-travel patterns using three-dimensional geographical information systems: A methodological exploration with a large data set. *Transportation Research C: Emerging Technologies*, **8**, 185-203.

Kwan, M.-P. (2001). Cyberspatial cognition and individual access to information: The behavioral foundation of cybergeography. *Environment and Planning B: Planning and Design*, **28**, 21-37.

Kwan, M.-P. and X.-D. Hong (1998). Network-based constraints-oriented choice set formation using GIS. *Geographical Systems*, **5**, 139-162.

Kwan, M.-P. and J. Weber (2003). Individual accessibility revisited: Implications for geographical analysis in the twenty first century. *Geographical Analysis*, **35**, 341-353.

Lenntorp, B. (1976). *Paths in Time-Space Environments: A Time Geographic Study of Movement Possibilities of Individuals*. Gleerup, Lund Studies in Geography Series B: Human Geography, Lund, Sweden.

Lenntorp, B. (1978). A time geographic simulation model of individual activity. In: *Timing Space and Spacing Time: Human Activity and Time Geography* (T. Carlstein, D. Parkes and N. Thrift eds.), pp. 162-180, Edward Arnold, London.

Love, R. F., J. G. Morris and G. O. Wesolowsky (1988). *Facilities Location: Models and Methods*. North-Holland, New York.

McFadden, D. (1998). Measuring willingness-to-pay for transportation improvements. In: *Theoretical Foundations of Travel Choice Modelling* (T. Gärling, T. Laitila, and K. Westin eds.), pp. 339-364, Elsevier Science, Amsterdam.

Miller, H. J. (1999a). Measuring space-time accessibility benefits within transportation networks: Basic theory and computational methods. *Geographical Analysis*, **31**, 187-212.

Miller, H. J. (1999b) Potential contributions of spatial analysis to geographic information systems for transportation (GIS-T). *Geographical Analysis*, **31**, 373-399

Miller, H. J. (2003) What about people in geographic information science? *Computers, Environment and Urban Systems*, **27**, 447-453.

Miller, H. J. (2004). Activities in space and time. In: *Handbook of Transport 5: Transport Geography and Spatial Systems* (D.A. Hensher, K. J. Button, K. E. Haynes and P. R. Stopher, eds.), pp. 647-660, Elsevier Science, Amsterdam.

Miller, H. J. (2005a). A measurement theory for time geography. *Geographical Analysis*, **37**, 17-45.

Miller, H. J. (2005b). Necessary space-time conditions for human interaction. *Environment and Planning B: Planning and Design*, in press.

Mokhtarian, P. L. (1990). A typology of the relationships between telecommunication and transportation. *Transportation Research A,* **24A**, 231-242.

National Research Council (2002) *Community and Livability: Data Needs for Improved Decision Making – Transportation and Quality of Life*, National Academy Press. Washington, DC.

O'Kelly, M. E. and T. H. Grubesic (2002) Backbone topology, access, and the commercial Internet. *Environment and Planning B: Planning and Design*, **29**, 533-552.

Openshaw, S. and P. J. Taylor (1979). A million or so correlation coefficients: Three experiments on the modifiable areal unit problem. In: *Statistical Applications in the Spatial Sciences* (N. Wrigley ed.)*,* pp. 127-144, Pion, London.

O'Sullivan, D., A. Morrison and J. Shearer (2000). Using desktop GIS for the investigation of accessibility by public transport: An isochrone approach," *International Journal of Geographical Information Science*, **14**, 85-104

Pirie, G. H. (1979). Measuring accessibility: A review and proposal. *Environment and Planning A*, **11**, 299-312

Purvis, C. L. (1990) Survey of travel surveys II. *Transportation Research Record*, **1271**, 23-32.

Salomon, I. (1986). Telecommunications and travel relationships: A review. *Transportation Research A*, **20A**, 223-238.

Small, K. A. (1992) *Urban Transportation Economics*. Harwood Academic, Chur, Switzerland.

Southworth, F. (1983). Circuit-based indices of locational accessibility. *Environment and Planning B: Planning and Design*, **10,** 249-260.

Taaffe, E. J., H. L. Gauthier and M. E. O'Kelly (1996) *The Geography of Transportation*, 2ed., Prentice-Hall,Upper Saddle River, N.J

van der Knaap, W. G. M. (1997) Analysis of time-space activity patterns in tourist recreation complexes: A GIS-oriented methodology. In: *Activity-based Approaches to Travel Analysis* (D. F. Ettema and H. J. P. Timmermans, eds.), pp. 283-311, Elsevier, Amsterdam,

Watts, D. J. (1999) *Small Worlds.* Princeton University Press, Princeton, N. J.

Weber, J. and Kwan, M.-P. (2002). Bringing time back in: A study on the influence of travel time variations and facility opening hours on individual accessibility. *Professional Geographer*, **54**, 226-240.

Weber, J. and Kwan, M.-P. (2003). Evaluating the effects of geographic contexts on individual accessibility: A multilevel approach. *Urban Geography*, **24**, 647-671.

Weibull, J. (1976). An axiomatic approach to the measurement of accessibility. *Regional Science and Urban Economics*, **6**, 357-379

Weibull, J. W. (1980). On the numerical measurement of accessibility. *Environment and Planning A*, **12**, 53-67.

Williams, H. C. W. L. (1976). Travel demand models, duality relations and user benefit analysis. *Journal of Regional Science*, **16**, 147-166

Williams, H. C. W. L. (1977). On the formation of travel demand models and economic evaluation measures of user benefit. *Environment and Planning A*, **9**, 285-344.

Williams, H. C. W. L. and M. L. Senior (1978). Accessibility, spatial interaction and the spatial benefit analysis of land use transportation plans. In: *Spatial Interaction Theory and Planning Models* (A. Karlqvist, L. Lundqvist, F. Snickars and J. W. Weibull, eds.), pp. 253-287, North-Holland, Amsterdam.

Wu, Y.-H. and H. J. Miller (2001). Computational tools for measuring space-time accessibility within dynamic flow transportation networks. *Journal of Transportation and Statistics*, **4** (2/3), 1-14

Access to Destinations
D.M. Levinson and K.J. Krizek (editors)

CHAPTER 5

THE TRANSPORTATION-LAND USE POLICY CONNECTION

Gerrit-Jan Knaap, University of Maryland
Yan Song, University of North Carolina, Chapel Hill

INTRODUCTION

The discovery of the transportation-land use connection is not new. As Cervero (1991) notes, little has changed since the 1954 publication of *Urban Traffic: A Function of Land Use* (Mitchell and Rapkin 1954). But the discovery certainly is older than that. The Robber Barons of the 19th Century, for example, probably had some sense of the relationship when they accepted land as payment for constructing the transcontinental railroad. So, most likely, did the Dutch when they purchased Manhattan Island from Native Americans for 24 dollars. In more recent times, recognition of the transportation-land use connection rose following the completion of the interstate highway system and the rise of urban sprawl. And as public interest in growth management revived following the 1980 recession, a consensus grew that it was impossible to pave our way out of congestion. As a result, the transportation-land use connection became a central theme of the movement that became known as Smart Growth. At approximately the same time, transportation planning models—long based on a four-step model that accepted land use as a given—began to incorporate feedbacks from transportation to land use. LUTRAC, the pioneering effort in Portland, Oregon, to defeat the construction of a circumferential highway is perhaps the seminal application of such models.

In this paper, we explore the transportation-land use *policy* connection. More specifically, we consider the question: can land use policy be used to alter transportation behavior? The answer is of some importance. If the answer is yes, then there is hope that land use policies can be designed and implemented that will bring some relief to the congestion and complex transportation problems facing US metropolitan areas. This is the underlying assumption behind most smart growth policy reforms. If the answer is no, then land use policy may still be important, but is not likely to play an important role in resolving transportation issues.

We proceed as follows. First we offer a schematic that identifies necessary conditions for land use policy to play a role in addressing transportation issues. Specially, we argue that for land use policy to play an effective role, three conditions must hold. First, land use must be able to alter transportation behavior. Second, transportation infrastructure must not fully determine land use. Third, the condition on which we consider most extensively, land use policy must significantly and constructively affect land use. After presenting the schematic, we consider the evidence on each of these conditions. Based on our review of the evidence, we conclude that land use policy can play an effective role in addressing transportation issues, but that the role is likely to be small, often counter productive, and most effective at the neighborhood scale.

A FRAMEWORK FOR ANALYSIS

In this paper we focus not on the transportation-land use connection, but on the transportation-land use *policy* connection. Thus land use in our analysis plays an intermediary role. That is, for land use policy to affect transportation behavior, not only must land use affect transportation behavior, but land use policy must affect land use. And for land use policy to affect land use, land use must not be fully determined by transportation infrastructure. Our logic is illustrated in Figure 1 below. In Figure 1, transportation infrastructure and land use policy in some combination affects land use, which in turn affects transportation behavior. Clearly this schematic ignores many important factors, and fails to consider the feedback between land use and transportation that we have now come to recognize. Still it provides structure for our argument and helps isolate key issues.

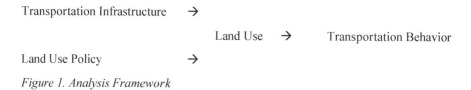

Figure 1. Analysis Framework

The land use-transportation behavior connection

For land use policy to affect transportation behavior, land use must affect transportation behavior. Fortunately, this relationship has been the subject of considerable research over the last two decades, motivated in large by the quest to lower automobile travel through urban design. The research has been greatly facilitated by the rapid rise in GIS technology and by new, innovative approaches of capturing travel behavior (such as global positioning recording instruments, accelerometers, and digital travel diaries). As a result, we are now able to capture features of the landscape and patterns of travel in ways unimaginable just a few years ago. The literature on this subject has grown large, diffuse, and difficult to synthesize—even though the subject is now the focus of several good review articles and books. Yet this voluminous literature features more argumentation than consensus. For our purposes this is not all bad, since it provides us the opportunity to frame the evidence in a manner that supports our argument.

As Crane (1999) demonstrates, much of the research on the influence of land use on transportation behavior focuses on the relationship between outcome measures of transportation behavior (total travel, trip generation rates, car ownership, mode choice, and length of journey to work) and measures of urban form (density of population and employment, land use mix, street and sidewalk circulation patterns, and jobs-housing balance) each measured at the trip origin, trip destination, and points along the trip route. Some more recent work, however, focuses on travel tours and trip chains (Wallace *et al* 2000; Krizek 2003). Again, the primary interest is in finding ways to reduce travel in cars. After a careful review of the evidence, Crane offers a conclusion quite similar to the proposition offered by Cervero (2002): high density, a mixture of land uses, more open circulation patterns, and pedestrian-friendly environments are all associated with less travel by car. Similar conclusions are drawn by Frank *et al.* (2003), who also highlight the importance of thresholds and interactive effects.

But as Crane cautions, these results offer very little of use for designing land use policy. Not only is most of the research plagued with problems of endogeneity—that is, those who choose to live in high density, mixed use, pedestrian friendly environments are probably predisposed to less automobility—but most of the work also fails to address both the benefits and opportunity costs of high intensity urban environments. Perhaps many would trade pedestrian accessibility for large houses and private open space; but many more would not. Further, nearly all the research is based on data and analyses at the neighborhood level. From this it is difficult to extrapolate to cities or entire metropolitan levels.

A few papers have, however, examined the difference between local and regional accessibility. Handy (1992 1993) examined the effects of local and regional accessibility on transportation behavior in 550 neighborhoods in the San Francisco metropolitan area. In the first of these studies, Handy conducted a case study of four neighborhoods: two had good local accessibility (i.e., neotraditional designs) and two had good regional accessibility (i.e., had good access to regional retail centers). She found that residents did make more walking trips in the neighborhood with good local accessibility, but that these trips did not seem to replace trips to regional shopping centers. In the second study, Handy found that travel distances, but not the number of trips increased as both local and regional accessibility fell. That is, people made shorter trips when destinations—both local and regional—were closer by, but the number of trips remained relatively constant. Further, she found, local and regional trips were partial substitutes; this increasing local accessibility is most effective at reducing automobile travel in places with low regional accessibility.

Our review of research in this area leads us to the following conclusions. First, there does appear to be a trade off between high intensity urban living and automobility. Thus it appears possible to design neighborhoods in ways that reduce automobile travel. Still the effects are marginal, complicated by trip chaining, and may lead to greater concentration of congestion and air quality degradation. Further, there are many who would choose not to live in such environments even if forced to bear the full social costs of driving a car. Thus there indeed seems to be a land use-transportation behavior connection and we know how this relationship works at the neighborhood scale. We know much less, however, about how this operates at a regional scale, or how to manage land use so as best to serve the interests of those who would choose intense urban living over automobility as well as those who would choose otherwise.

THE TRANSPORTATION INFRASTRUCTURE-LAND USE CONNECTION

For land use policy to serve as an effective means of shaping transportation behavior, land use cannot be fully determined by transportation infrastructure. At one level the proposition seems easily dismissed. Land use is determined in part by history, land ownership patterns, topology, culture and many other factors besides transportation infrastructure. But the question here is more subtle and perhaps best illustrated by example. If investments in light rail and light rail stations lead by themselves to complementary land uses around light rail stations, then there is little need for land use policy to accomplish the same. Thus the question is: do investments in all forms of transportation infrastructure lead to appropriate configurations of land uses in the absence of intervening land use policy.

Like the effects of land use on transportation behavior, the effects of transportation infrastructure on land use are complex and contentious. To oversimplify, the literature has focused primarily on two questions: do investments in roads and highways lead to development patterns widely characterized as sprawl and do investments in public transit lead to transit-oriented development? Once again, despite volumes of research, the evidence on both questions is decidedly mixed, and leaves plenty of room for interpretations that support underlying predispositions.

The proposition that the extension of highways leads to urban decentralization and low-density development patterns is strongly supported by economic theory and common sense. According to economic theory, land rent gradients, and thus urban structure, are largely determined by the trade off between accessibility and transportation costs. Further, highway extension lowers transportation costs, flattens land rent gradients, and causes urban expansion. Common sense suggests that development will take place where roads provide access. Almost no one disputes these general propositions. The disputes centers on issues of causality, elasticity, and significance.

Economic theory and common sense notwithstanding, Guiliano (1989) argues that the effects of highway investments on land use have significantly diminished. Specifically, she argues, "transport cost is a much less important factor than location theory predicts." Her argument is supported by a study by a team of researchers at the Transportation Center at the

University of Illinois at Chicago (1998). They found that the decentralization of the Chicago metropolitan area began long before the construction of the metropolitan highway system and would have occurred even without the highways. On the other hand, Boarnet and Houghwout (2000, p. 12), following a detailed review of the literature conclude that: "the evidence suggests that highways influence land prices, population, and employment changes near the [highway] project, and that land use effects are at the expense of losses elsewhere." After a more recent review, Handy (2005) concludes:

> it is reasonable to conclude that new highway building will enable or encourage additional sprawl to some degree, although to exactly what degree is uncertain and depends on local conditions. However, the converse of this proposition is probably not true: not building more highways will probably not slow the rate of sprawl, at least not much.

Research on the effects of transit on land use patterns has focused almost entirely on rail transit and is somewhat less ambiguous. Again, economic theory and common sense strongly suggests that increased accessibility around rail stations should cause increases in property values and stimulate high-density development. Some empirical research provides supportive evidence of these effects (Huang 1996). But significantly more evidence suggests that these effects are small and perhaps inconsequential without supportive land use policies. Cervero, for example concludes, "LRT can be an important, though unlikely a sufficient, factor in changing land use" (1984, p. 46). Handy (2005) concludes:

> The evidence thus supports the proposition that investments in light rail transit will increase densities – but only under the right conditions. These studies point to several important lessons about the conditions under which the proposition will hold: a region that is experiencing significant growth, a system that adds significantly to the accessibility of the locations it serves, station locations in areas where the surrounding land uses are conducive to development, and public sector involvement in the form of supportive land use policies and capital investments. Without these conditions, increased densities are unlikely. With these conditions, increased densities are not assured but they are possible.

The evidence on the effects of highways and transit on land use leads us to the following conclusions. First, it is clear that both highways and transit can have land use impacts. As economic theory suggest, highways can contribute to urban decentralization and transit stations can lead to nodes of high density, mixed use development. Both effects, however, are conditional. Urban decentralization seems to occur even without highway construction (Mieszkowski and Mills 1993) and, perhaps, can be mitigated with offsetting land use policies. Transit oriented development, meanwhile, rarely seems to occur without supportive land use policies.

To return to the question of interest, therefore, transportation infrastructure does not appear to fully determine land use. This leaves plenty of room for land use policy to play a role in shaping land use and, perhaps, transportation behavior.

THE LAND USE POLICY-LAND USE CONNECTION

We now turn to the focal question of this paper: does land use policy affect land use? Again, this question would appear on its face to be self evident. Certainly land use policy affects land use. Without government policies that specify the rights of land owners, determine the location of transportation infrastructure, and specify the conditions on which development is allowed to occur, no form of urban development could take place. But again, the question must be considered for its contextual and substantive implications. Can land use policy, as it is formulated and implemented in the United States, shape land uses in ways that significantly alter transportation behavior? We submit that the answer to this question is far from obvious. We address this question at two levels—first in the context of governance frameworks or policy regimes, then by focusing on specific types of land use policies.

Policy Regimes

The notion of policy regimes is a bit ephemeral, especially in the domain of land use. By policy regimes we mean the larger institutional context and statutory framework in which land use policies are imposed. We include in this the definition of property rights, the state statutory framework that govern local land use policy, and the culture and practice of local land use planning. The evidence on which we draw is far from definitive but helps to illustrate larger institutional issues.

Interesting insights on the effects of institutional context on land use patterns, are provide by Bertaud and his colleagues at the World Bank (2004). Bertaud and his colleagues have examined patterns of urban development in metropolitan areas around the world and considered the impact of land use regimes on urban structure. Some of the results of that work are illustrated in Figures 2–4.

Figure 2 illustrates the pattern of urban development in Paris, a city largely constructed in period when land use decisions were shaped by land markets. As shown in the Figure, the density of development in Paris displays the well-known pattern of exponential decay caused, according to economic theory, by declining land prices and corresponding capital-land substitution. Figure 3 illustrates the pattern of urban development in Brasilia, Moscow, and Johannesburg, cities largely constructed during periods in which land use decisions were made without land markets. As shown in Figure 3, the density of urban development in these cities displays quite different patterns than Paris which, according to Bertaud, reflects the absence of market discipline on the relationship between accessibility and land use.

Figure 4 displays the pattern of development in North American, European, and Asian cities constructed under the influence of market forces. As shown in Figure 4, Asian cities, where transportation was dominated by walking and biking, display the steepest density gradients; European cities, where transportation was dominated by transit, have less steep gradients; while North American cities, where transportation was and is dominated by the car (especially in Atlanta) have the flattest density gradient. Bertaud uses these diagrams to make two points. First, he argues, when allowed to do so, market forces impose a discipline on development patterns that reflect the trade-off between accessibility and urban intensity. Where accessibility is high, land prices are high, and urban densities are high. Second, where market forces dominate, development patterns are largely shaped by the dominant mode of transportation. These arguments suggest that land use policy regimes can and do make a significant difference.

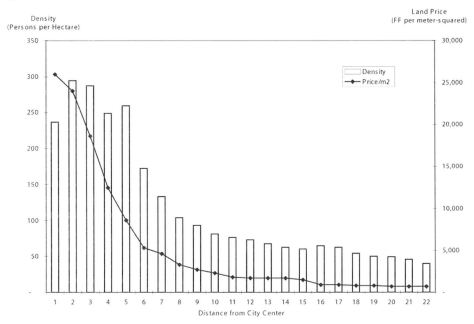

Figure 2. Density Profile for City of Paris (Bertaud, 2004)

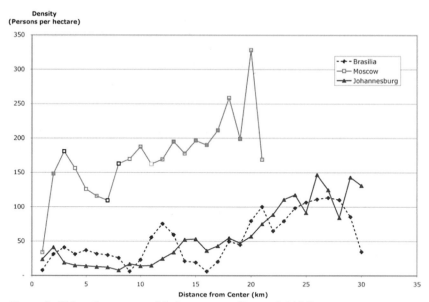

Figure 3. Urban Structure and Infrastructure (Bertaud, 2004)

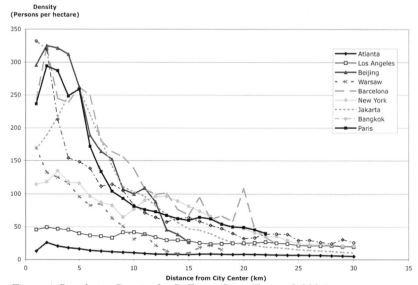

Figure 4. Population Density for Different Cities (Bertaud, 2004)

The work by Rolf Pendall on policy regimes provides additional insights. In a series of papers, Pendall (1999, 2000, and 2001) identifies how land use policy regimes differ across the country and shows how differences in land use regimes lead to differences in land use and demographic composition. Based on a nationwide survey of local government planning practices, and using cluster analysis, Pendall identifies eight distinct policy regimes in the United States: Big City, Growth Management, Mitigated Growth Control, Suburban Growth Control, Suburban Sprawl Control, Laissez Faire, Modified Exclusion, Exclusionary Zoning. Then using analysis of variance, Pendall examined the relationship between policy regimes and development density. He finds that between 1982 and 1997, development under Growth Management regimes became significantly denser than under Laissez Faire regimes. Further, Pendall finds that development under Exclusionary Zoning regimes resulted in lower overall densities and, perhaps, the exclusion of racial minorities. This work suggests that land use policies can facilitate more favorable transportation-land use connections, but can also lead to low density development and exclusion.

These works provide illustrative, if not compelling, evidence that land use regimes can significantly affect land use. The work by Bertaud suggests that land use regimes dominated by market forces creates a pattern of land uses that reflects the influence of accessibility; further, where such accessibility facilitates high-density, such developments will be forthcoming. The work by Pendall suggests that land use regimes dominated by exclusionary interests can lead to low density development and exclusion. Judging by these two sources alone, it appears that policies regimes can both strengthen and weaken the transportation-land use connection at the metropolitan scale.

Policy Instruments

Research on the effects of specific policy instruments on land use is more voluminous and thus more ambiguous. Policy instruments vary widely in type and strength of implementation. Here we focus on three types of land use instruments: land use regulations, land use plans, and development incentives.

Regulations.

The literature on land use regulations is large and dominated by economists. Here we consider three types of regulations: zoning, urban growth boundaries, and subdivision regulations. The literature on <u>zoning</u> is by far the most developed. Most of the research, especially by economists, explores the effects of zoning on land and housing prices. If zoning affect prices, economists infer, then zoning affects land use and land allocation. With few exceptions, economists find that zoning does affect land and housing prices. Many economists, however, have argued that zoning follows rather than constrains the market. That is, zoning regulations tend to mimic market forces, thus the effects of zoning on land values may simply reflect the misspecification of an endogenous effect. Still the overriding conclusion of this body of research is that zoning at best does little to alter land use and when it does, it tends to limit development density and do more harm than good.

With support from the U.S. Department of Housing and Urban Development, a team of researchers at the National Center for Smart Growth are currently exploring the effects of zoning on housing density and single family-multifamily mix. Toward this end, the research team has collected zoning data in GIS format from seven metropolitan areas. The team will then compare actual densities with zoning densities to examine whether zoning constrains housing density and mix. At present the work is incomplete, but a quick examination of the relationship between zoning density and actual density in the Portland metropolitan area provide some insights.

Preliminary results of analyses in the Portland metropolitan area suggest that the difference between zoned density and actual density (or available development capacity) is greatest in the city center and in the subcenters of the metropolitan area. Substantial development capacity also exists at Max (light rail) station areas and along major transportation corridors. This clearly reflects Portland's attempts to concentrate development in urban centers and transportation corridors as articulated in its 2040 plan. But it just as clearly it illustrates the difficulty of promoting density using land use policy. Though it is still early in the life of the 2040 plan, it is clear that zoning parcels for high density does not immediately or necessarily lead to higher density development—even when growth is contained within an UGB.

Research on urban growth boundaries (UGBs) is much less voluminous and dominated by discussion on the effects of Portland's UGB. Long ago Knaap (1985) examined the effects of Portland's UGB on land values on the Oregon side of the metropolitan area. Though the UGB had been in place for only a few years, he found that land values were higher inside than outside the UGB. Corroborating evidence was soon offered by Nelson (1986) in his examination of the UGB in Salem. Since then a debate has raged as to whether Portland's UGB has adversely affected housing affordability or served as an effective tool for managing urban growth. Recent papers by Knaap 2001 and Pendall *et al.*, however, suggests that Portland's UGB has actually had little effect direct effect on land prices or development densities. Knaap shows that housing prices and densities in Portland do not differ substantially from other western metropolitan areas. Further, he argues, because the UGB must always contain the capacity to accommodate 20 years of growth, the UGB has been effective at framing a regional planning effort but has had little effect on the quantity or density of Portland's growth.[1]

Most of the research on the effects of subdivision regulations is qualitative. Many claim that subdivision regulations impede the development of mixed use, pedestrian friendly neighborhoods. Based on these claims there have been developed several model subdivision regulations and development codes with intent to overcome this impedance. To explore this question, Talen and Knaap (2003) collected zoning codes and regulations from a large sample of cities and counties in Illinois and concluded that such regulations imposed more restrictive requirements on set backs, parking spaces, lot sizes, and street widths than were necessary compared to widely accepted contemporary standards. But the extent to which such standards actually alter development patterns remains largely unexplored.

In a recently published paper Song and Knaap (2004), computed several measures of urban form for Washington County, Oregon, and examined how they changed over time. Though the paper did not specifically focus on subdivision regulations, they noted how local subdivision regulations were shaped by the 2040 plan, the plan prepared by Metro, Portland's regional government. Changes over time in these measures for the entire Portland metropolitan area are illustrated in Figure 5. Figure 5 illustrates the urban form of the "median" TAZ developed in each decade, where the age of the TAZ is determined by the median "year built" attribute of the housing stock and the median TAZ is the TAZ for which the various measures of urban form come closest to the median value of the entire metropolitan area. (For more detail on methods, see Appendix A) As shown, internal connectivity, illustrated by the ratio of red dots (cul-de-sacs) to total dots, was high in the 1940s, fell until the 1970s, and started rising again in 1980. External connectivity, illustrated by the length of the line segments around the edge of the neighborhood, exhibits a similar trend. Single family lot sizes rose from 1940 to 1970 then fell continuously to reach an all time low after the year 2000. As depicted by the mixture of the color of the lots, land use mix has continuously fallen over the same period. A combination of improved proximity to commercial uses and internal street network connectivity has brought increased pedestrian accessibility since 1990. Based on these results Song and Knaap concluded that changes in land use policies in general, and in subdivisions regulations in particular, urban form in the Portland metropolitan area had improved at the neighborhood scale but not at the regional scale.

Land Use Plans.

The effect of land use plans on land use is also largely unexplored and inconclusive. The extant literature on the influence of plans largely focuses on the content of plans or on narrow measures of implementation. Alterman and Hill (1978) compared land uses and densities in a plan with eventual development densities and found approximately 66 percent congruence. Talen (1996) examined plans for parks with the eventual size and locations of parks in Dallas, Texas, 20 years later and found a reasonable level of consistency. Connerly and Muller (1993) examined the number of times a plan was used in decision making but did not examine how plans affected land use. Others have looked at whether the land market responds to the information contained in plans. Ding, Hopkins and Knaap (2002), for example, examined whether plans for light rail stations in the Portland metropolitan area affected land values before the light rail system began operation. Though the magnitude was small, Ding *et al.* found that land values did increase within a half mile of the planned station area. Similar results are reported by Ferguson, Goldberg, and Mark (1988), Gatzlaff and Smith (1993), and McDonald and Osuji (1995). These results, we argue, offer evidence that markets respond to the information content in plans and that land values and development patterns would subsequently be affected.

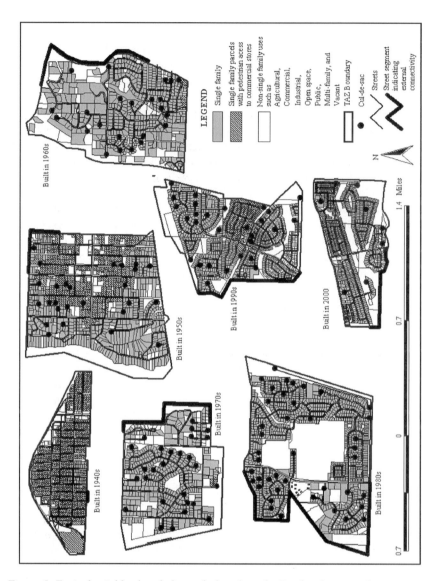

Figure 5. Typical neighborhoods by each decade in the Portland metropolitan area

Incentives.

Research on the effects of incentives on land use has grown in recent years. The use of incentives in place of regulations is an approach pioneered by the state of Maryland. Thus on this subject we focus our comments on policies in Maryland by researchers at the University of Maryland.

Cohen and Pruess (2002) examined the efficacy of Montgomery County's well-known transferable development rights program. Under this program development rights in Montgomery County's agricultural reserve could be sold or transferred to areas within the existing urban envelope. Cohen and Pruess found the price of development rights falling over time, the supply of receiving areas diminishing, and the extent to which the programs preserves farmland in doubt. Further, because the program failed to target the most the most fertile soils as sending areas, and failed to provide adequate and timely infrastructure in receiving areas, the popularity of the program has fallen significantly.

Sohn and Howland (forthcoming) examined the effects of Maryland's Priority Funding Areas on investments in sewer infrastructure from 1997 to 2002. According to Maryland's Smart Growth Statutes, passed in 1997, the state will only invest in urban infrastructure inside PFAs. They found that of the total amount invested in sewer infrastructure by counties, 25 percent was invested on sewer infrastructure outside PFAs. But of the total amount invested by the state, 29 percent was invested on sewer infrastructure outside PFAs. Most of these investments were used to repair nonperforming septic systems. Still, these findings suggest that even the state is having difficulty conforming to smart growth incentives.

Sohn and Knaap (2002) examined the effects of Maryland's job creation tax credit program (JCTC), which, since 1997, provides greater credits for job creation inside than outside priority funding areas. Using data on job growth in Maryland from 1996 to 2000 and spatial econometric techniques, Sohn and Knaap found that job growth was greater inside than outside PFA's, holding other things constant, but only for jobs in the service sector. Based on these results, they concluded that Maryland's JCTC program can help to concentrate job growth in PFA's but that the contribution of the JCTC program toward such concentration is likely to be small.

The most thorough analysis of the effects of PFAs was recently completed by Shen *et al.* (2005). Using Landsat data Shen *et al.* examined changes in land use before and after the passage of Maryland's smart growth legislation. With these data they examined changes in urban development patterns both inside and outside of PFAs. They found that the likelihood of urban development was higher inside rather than outside PFAs both before and after they were drawn, but that the density of urban development has fallen since 1973, and fell even more rapidly after 1997.

Shen *et al.* also conducted a statistical analysis of the determinants of urban growth in four Maryland counties. In each county, they created grids and identified whether each grid was largely urban or rural. Each grid cell then served as observation in several regression models. In these models, the probability of urban conversion was specified as a probit function of economic, fiscal, and policy variables. The results were plausible and robust across several specifications and sub-regions. In general, the probability of conversion was higher for land closer to the city center, closer to other urban grids, and inside PFA's. The probability of development was also lower for parcels in rural legacy areas and for areas outside PFAs. This was true, however, both before and after 1997, so it is not possible to say whether the likelihood of development inside PFAs increased after 1997. The results generally showed that the probability of development was greatest in the suburban areas of the state but lower where local governments had stronger regulatory systems.

So what can be surmised about Maryland's experiment with smart growth? In short: it has accomplished about what could be expected. It is likely that Maryland's use of the state budget has had and could continue to have marginal effects on residential location decisions and urban containment, where the extent of the margin is determined by the availability of state funds and the commitment of the administration in power. But these effects are likely to remain small and overwhelmed by powerful and pervasive market forces and local controls.

SUMMARY

The above brief and highly selective review of the literature on the effects of land use policy on land use, like the other sets of literature, is highly mixed and subject to interpretation. The literature on land use policy suggests that policy regimes differ significantly with potentially significant effects on land use patterns. Somewhat ironically, though, the studies reviewed here suggest that regimes that rely on market forces are better able to coordinate land use with transportation and that more invasive regulatory regimes in the US can lead to more exclusion and urban sprawl. The literature on specific regulatory instruments provides corroborative evidence. Zoning has been shown to affect land prices and land allocation but the result has generally been lower densities and a limited ability to increase densities. Urban growth boundaries can provide an effective framework of land use planning but also have limited ability to increase urban densities. Subdivision regulations, on the other hand, can affect street network patterns at the neighborhood scale but not at the regional scale. The literature on the effects of incentives and plans, is no more encouraging. While markets appear to react to plans, the reaction is typically small. And based on analyses of smart growth programs in Maryland, incentives have minor effects on the location of urban development at best.

CONCLUSIONS

In this paper we have offered a brief examination of the transportation-land use policy connection. By selectively drawing on three large bodies of literature we have come to the following conclusions:

- Land use can affect transportation behavior, but the evidence is more compelling on how land use affects transportation behavior at the neighborhood scale than at the regional scale;

- Transportation infrastructure can affect land use but the effects are often small without accommodating, or countervailing, land use policies

- Land use regimes and regulations can affect land use, but many land use regulations are much more effective at limiting development than increasing densities.

So, to return to the central question: can we use land use policy to shape transportation behavior? Perhaps. At the regional scale we have limited knowledge about how to design transportation and land use policies that meet the needs of those who prefer to live in urban environments without the car and those who prefer to live in suburban environments and with the car. Further, land use institutions in the US are at present—and probably for the foreseeable future—ill suited to design and implement policies at that scale. Thus, success at the regional scale will require regional institutions with the capacity to design integrated land use and transportation plans and the regulatory capacity to assure their implementation.

At the local scale, the obstacles are different though no less formidable. We know what kinds of urban environments lead to less automobile use for those who choose to live in them. We also know that we should remove existing policies that preclude the development of such urban environments. The simple removal of regulatory constraints, however, is unlikely to produce adequate results. Further, plans and incentives alone are unlikely to produce adequate results. What we appear to need at the local level is strong regulatory policies, public-private partnerships, or direct public investments in transit-oriented development. At the local level the problems are less technical than political. We have the knowledge and the tools. Whether we will generate the political will remains to be seen.

REFERENCES

Alterman, R. and M. Hill (1978). Implementation of urban land use plans. *Journal of American Planning Association* **44(3)**, 274-86.

American Planning Association (1998). *The Principles of Smart Development*. PAS report #479, Chicago, IL: APA.

Bertaud, A. (2004). *The Spatial Organization of Cities: Deliberate Outcome or Unforeseen Consequence*, http://alain-bertaud.com/images/AB_The_spatial_organization_of_cities_Version_3.pdf.

Boarnet, M. G. and A. Houghwout (2000). *Do Highways Matter? Evidence and Policy Implications of Highways' Influence on Metropolitan Development*. Brookings Institution, Washington, DC.

Cervero, R. (1984). Light rail transit and urban development. *Journal of the American Planning Association,* **50(2)**, 133-147.

Cervero, R. (1991). Congestion relief: The land use alternatives. *Journal of Planning, Education and Research*, **10(2)**, 119-130.

Cervero, R. (2002). Induced travel demand: Research design, Empirical evidence and normative policies. *Journal of Planning Literature,* **17**, 3-20.

Cohen, J. and I. Preuss (2002). *An Analysis of Social Equity Issues in the Montgomery County (MD) Transfer of Development Rights Program*, University of Maryland, College Park, MD. Available at:
http://www.smartgrowth.umd.edu/research/pdf/TDRequity.text.pdf

Connerly, C.E. and N.A. Muller (1993). Evaluating housing elements in growth management comprehensive plans. In: *Growth Management: The Planning Challenge of the 1990s,* (J. Stein, ed.) 185-199. Sage, Newbury Park, CA.

Crane, R. (1999). The Impacts of Urban Form on Travel: An Interpretive Review. *Journal of Planning Literature,* **15(1)**, 3-23.

Ding, C., G. J. Knaap and L. Hopkins (2002). Does planning matter? The effects of light rail plans on land values in station areas. *Journal of Planning, Education and Research* **21(1)**, 32-39.

Fergusan, B. G., M. A. Goldberg, and J. Mark (1988). The pre-service impacts of the Vancouver Advanced Light Rail Transit System on single family property values. In: *Real Estate Market Analysis: Methods and Applications,* (J. M. Clapp and S. D. Messner, eds.). Praeger, New York.

Frank L. D., P. O. Engelke and T. L. Schmid (2003). *Health and Community Design: The Impact of the Built Environment on Physical Activity.* Washington, DC: Island Press.

Gatzlaff, D. H. and M. T. Smith (1993). "The impact of the Miami Metrorail on the value of residences near station locations," *Land Economics,* **69(1)**, 54-66.

Guiliano, G. (1989). New directions for understanding transportation and land use. *Environment and Planning A,* **21**, 145-159.

Handy, S. (2005). Smart growth and the transportation-land use connection: what does the research tell us. *International Regional Science Review,* **28**, 146-167.

Handy, S. (1992). Regional versus local accessibility: Neotraditional Development and its implications for non-work travel, *Built Environment,* **18(4)**, 253-267.

Handy, S. (1993). Regional versus local accessibility: Implications for non-work travel, *Transportation Research Record,* **1400**, 58-66.

Huang, H. (1996). The land-use impacts of urban rail transit systems. *Journal of Planning Literature,* **11(1)**, 17-30.

Knaap, G. J. (1985). The price effects of urban growth boundaries in metropolitan Portland, Oregon. *Land Economics,* **61(1)**, 26-35.

Knaap, G. J. (2001). The urban growth boundary in Metropolitan Portland, Oregon: Research, rhetoric, and reality. *American Planning Association,* PAS memo.

Krizek, K. (2003). Neighborhood services, trip purpose, and tour-based travel. *Transportation,* **30**, 387-410.

McDonald, J. F. and C. I. Osuji, (1995). The effect of anticipated transportation improvements on residential land values. *Regional Science and Urban Economics,* **25**, 261-278.

Mieszkowski, P. and E. Mills (1993). The causes of metropolitan suburbanization. *Journal of Economic Perspectives,* **7(3)**, 135-147.

Mitchel, R. and C. Rapkin (1954). *Urban Traffic: A Function of Land Use*. Columbia University Press, New York.

Nelson, A. C. (1986). Using land markets to evaluate urban containment programs. *Journal of the American Planning Association*, **52(2)**, 156-71.

Pendall, R. (1999). Do land-use controls cause sprawl? *Environment and Planning B: Planning and Design*, **26**, 555-571.

Pendall, R. (2000). Local land-use regulation and the chain of exclusion. *Journal of the American Planning Association*, **66**, 125-142.

Pendall, R. (2001). Municipal plans, state mandates, and property rights: Lessons from Maine. *Journal of Planning Education and Research* **21(2)**, 154-165.

Pendall, R, J. Martin, and W. Fulton (2002). *Holding the Line: Urban Containment in the United States.* Brookings Institute, Washington, DC.

Shen, Q, J. Liao and F. Zhang (2005). *Changing Urban Growth Patterns in a Pro-Smart Growth State: The Case of Maryland, 1973-2000*. National Center for Smart Growth, College Park, MD.

Sohn, J. and M. Howland (Forthcoming). Has Maryland's priority funding areas initiative constrained the expansion of water and sewer investments? *Land Use Policy*, in press.

Sohn, J. and G. Knaap (2002). *Does the job creation tax credit program in Maryland induce spatial employment growth or redistribution?* Presented at the 49th Annual North American Meeting of the Regional Science Association International.

Song, Y. and G. J. Knaap (2004). Measuring urban form: is Portland winning the battle against urban sprawl? *Journal of the American Planning Association*, **70(2)**, 210-225.

Southworth, M. (1997). Walkable suburbs? An evaluation of neotraditional communities at the urban edge. *Journal of American Planning Association*, **63**, 28-44.

Talen, E. (1996). Do plans get implemented? A review of evaluation in planning. *Journal of Planning Literature*, **10(3)**, 248-259.

Talen, E. and G. J. Knaap (2003). The implementation of smart growth principles: an empirical study of land use regulation in Illinois. *Journal of Planning Education and Research*, **22(4)**, 345-359.

Transportation Center, University of Illinois at Chicago (1998). *Highways and Urban Decentralization.* University of Illinois at Chicago, Chicago, IL.

NOTES

[1] Subsequent work by Song and Knaap (2002) suggests, however, that the UGB and other policy instruments have had a measurable effect on Portland's urban form.

Access to Destinations
D.M. Levinson and K.J. Krizek (editors)
© 2005 Elsevier Ltd. All rights reserved.

CHAPTER 6

PERSPECTIVES ON ACCESSIBILITY AND TRAVEL

Kevin J. Krizek, University of Minnesota

LAND USE AND TRANSPORTATION – ACCESSIBILITY

Urban form, whether it is compact, multi-nodal, or sprawling, impacts the type and cost of transportation systems needed to serve residents of a metropolitan area. On the other hand, the type and location of major transportation facilities greatly influences urban form (Kelly, 1994). Almost a half-century's worth of study on the link between two provides a solid foundation to understand some inherent interactions between land use and transportation. These interactions manifest themselves in two forms:

1. The influence of urban form on transportation systems, travel demand, and urban travel behavior; and

2. The influence of transportation systems and transportation investments on metropolitan urban form.

The two phenomena share a common heritage; however each asks different questions, and they often relate to different scales of analysis. This chapter endeavors to describe the issues that emanate from the former question—that is, what do we know about the manner in which land use patterns affect household travel. In doing so, this chapter discusses how the relationship between urban form and transportation has historically been conceptualized and summarize some of the existing research. It then turns to describing how this history relates to new and pressing research questions that provide the impetus for studying more in depth matters related to accessibility.

Travel is generally considered to be a derived demand. Notwithstanding instances of joyriding and excess travel (Moktarian and Salomon, 1998), the act of travel occurs because someone wants to do something somewhere else. This idea was articulated by Robert Mitchell and Chester Rapkin who suggested that the amount and nature of movement is derived from the amount and nature of activities (Mitchell and Rapkin, 1954). Thus, an individual's location vis-à-vis the distribution of potential activity sites is an important determinant of travel behavior or location decisions.

Efforts to quantify the relationship between movement and activities have been a longstanding component of transportation planning and modeling (e.g., providing services so that residents can freely travel to valued destinations). Examining travel that is derived from activity sites requires merging two phenomena: (1) land use patterns (where are people traveling?), and (2) the transportation system (how are they getting there?).

The label of accessibility is often used to describe this relationship in its most basic sense. As has been well described elsewhere (Handy and Niemeier, 1997), measures of accessibility represent the spatial distribution of potential destinations, the ease of reaching each destination, and the magnitude, quality, and character of the activities found there. Any reputable interpretation of the concept requires defining at least two basic tenets:

1. The pattern of potential activities—their quantity, quality, and variety—commonly referred to as the attractiveness of a place. This could include the spatial distribution of people, of socioeconomic opportunities (especially jobs), or retail opportunities.

2. The connectivity between activities as provided by the transportation system. This is commonly referred to as the resistance or impedance function measured in terms of distance or time by particular modes.

Handy and Niemeier (1997) describe how measures of accessibility generally have historically subscribed to one of three categories: gravity based measures, cumulative opportunity measures, and behavioral measures. Even some of the more advanced measures presented in Chapter 4 of this volume by Harvey Miller, comprise a variation of those briefly described below.

Gravity

As most commonly conceived, the roots of accessibility measures are derived from the gravity model (Reilly, 1953; Huff, 1963). An early application to traffic modeling can be traced to Casey (1955) in which the laws of gravity where used to forecast parking and access requirements for shopping malls. Similar applications are firmly grounded in the widely used Urban Transportation Planning System (UTPS) software and its descendants, where the gravity model formula of accessibility is commonly employed in trip-distribution modeling. To the chagrin of transportation modelers, UTPS has undergone very little change over the past forty or so years.

Hansen's (1959) article is commonly referenced as the first application of accessibility within a context directly applicable to land use modeling. It also forms the basis of most of the accessibility indices employed in this volume. Hansen presents a hypothetical model showing how differences in accessibility—constructing an express highway—could be used as the basis for a residential land use model. In this context and others (Patton and Clark, 1970), highways provide accessibility that is used to explain residential (or other) locations.

In these applications, accessibility measures weight opportunities (e.g., the quantity of an activity as measured by employment) by impedance (e.g., a function of travel time or cost). Many measures alter the impedance function by using the composite utility of travel to destinations because it more accurately approximates travel times. Under this framework, accessibility is typically described by the following equation:

$$A_i = \sum_{j=1}^{n} O_j f(C_{ij})$$

where A_i = accessibility from a zone (i) to the considered type of opportunities (j)
O_j = opportunities of the considered type in zone j (e.g., employment, shopping, etc.)
C_{ij} = generalized (or real) time or cost from i to j
$f(C_{ij})$ = Impedance function (exponential or power functions are most often used)

Cumulative opportunity

Cumulative-opportunity measures of accessibility index the number of opportunities that can be reached from an origin of interest within specified travel distances or times. This approach was commonly used in early studies (Wachs and Kumagi, 1973). Because cumulative-opportunity measures fail to discount measures of opportunity over distance (all potential destinations within a threshold are weighted equally), the impedance function in effect equals one if the opportunity is within the travel time limit and zero otherwise. For this reason, cumulative-opportunity measures are considered a specific form of the gravity-based measure.

Behavioral

A final approach to measuring accessibility introduced by Ben-Akiva and Lerman (1978) and Koenig (1980) aims to predict the probability that an individual makes a particular choice, depending on the utility of that choice relative to the utility of all other choices. Using a logit model, individuals are assumed to assign a utility to each choice in a specified choice set. If they select the alternative that maximizes their utility, accessibility can be defined as the denominator of the multi-nomial logit model, also known as the logsum. This measure has been the most frequent application of accessibility used in integrated land use-transportation models.

What has become increasingly clear over the decades of wrestling with applications of accessibility is that the term *accessibility* is used in countless contexts and in countless ways.

For example, the concept was initially used as a framework for highway routing or other capital improvement projects. Throughout the 1970s, many studies advanced interpretation of the term, applying measures in different ways. They ranged from defining accessibility as a simple evaluation tool for social services to highly mathematical econometric formulas for predicting travel behavior (e.g., mode split, trip frequency). Notwithstanding considerable progress during this period, several fundamental questions continue to pervade the most appropriate strategies for specification. For example:

- At what rate should opportunities be discounted (impedance)?

- To what degree should measures be sensitive to zone sizes and zonal configurations?

- From where should accessibility be measured?

- What is the most appropriate type of utility function?

Concerns echoed in this volume (particularly in Chapter 4 by Harvey Miller) and elsewhere in the literature, clearly suggest that what constitutes a "best" measure is far from clear.

TRAVEL AND REGIONAL ACCESSIBILITY, A BRIEF REVIEW OF EVIDENCE

The majority of available research centers around questions considering travel behavior relative to the intensity of an urban core, population density, employment centers or other macro-scale urban form measures. To the extent that conclusions from dozens of studies can be easily synthesized, a thorough review of this research (Handy, 1992a) suggests the following. Higher densities generally decrease the number of trips taken, the percentage of the trips taken by automobile and the average trip speed. As jobs decentralize to suburbs, trips tend to be shorter, but more often by automobile. Although if jobs greatly exceed housing, trips are longer but more often involve ridesharing. Relationships between spatial structure of a region and travel patterns are relatively unclear. For the most part, there is evidence that the polycentric form leads to fewer trips, shorter trips, and less total energy than dispersed or monocentric forms.

A finding that has caught the eye of the land use-transportation community is that, under some conditions, dispersed urban form leads to shorter travel time than monocentric form (Gordon, Kumar, and Richardson, 1989; Levinson, 1998) This is because of the increasing migration of employment and services to more suburban communities, which reduce travel distances to common destinations. The corollary to this, however, is that dispersed urban forms also have considerably higher rates of trips by automobile because the density of such development does not reach levels that make auto travel unattractive or transit attractive. Thus, while employment and services are migrating to outlying suburban areas to complement residential areas, the actual style of development may be doing little to impact non-auto related travel.

Regional versus neighborhood accessibility

The majority of early research examining land use and transportation focused on relatively macro-scale units of analysis. These ranged from entire cities to census tracts or transportation analysis zones. The rising importance of development at the neighborhood scale, however, prompts increasing attention on the need to differentiate between multiple scales of analysis.

Take, for instance, the need to consider both neighborhood and regional accessibility. The importance of differentiating between each is that both may influence the amount a household travels. The character of the particular neighborhood in which the household lives is important (neighborhood accessibility) as is the position of the neighborhood in the larger region (regional accessibility). Each may affect travel behavior differently and each speaks to slightly different policy initiatives.

Compare a mall, complete with a wide array of services, largely isolated, and containing large parking lots, with a neighborhood complete with corner stores within one-half mile (800 m) of many residences. Using a single measure of urban form, these two development patterns may have the same level of accessibility. Yet the implications for travel patterns may be substantially different. Other reasons for capturing varying scales of accessibility include:

- The regional context of the neighborhood may provide more opportunities that may mean more overall travel, or

- The characteristics of the regional structure may simply dwarf any variations in the local, neighborhood structure (e.g., a neighborhood immediately adjacent to a downtown core would likely generate different travel than that same neighborhood if it were placed on the urban fringe 30 miles (50 km) from any economic activity).

To better differentiate and understand the relative impact of these two phenomena, Handy used the terms regional and local accessibility and measured each using different criteria (Handy, 1993). Regional accessibility was defined by the regional structure of a metropolitan area based on variables such as location and type and of activity that affect shopping behavior. Local accessibility, on the other hand, was primarily determined by nearby activity (where nearby is used to refer to the neighborhood unit, approximately between one-half (800 m) to one mile (1.6 km) in residential areas). Areas with higher local accessibility would be oriented to convenience goods, such as supermarkets and drug stores, and located in small centers.

Gur (1971), however, is credited with recognizing the need to distinguish between scales of accessibility and the resulting travel implications. He contended that the rate of trips being generated depended on at least two factors based on consumer theory. The first factor was the ease and worth of travel to destinations far away; the easier the travel to various opportunities there, the more trips are going to be made to those distant destinations. The second factor was the availability of trips close by; the more opportunities that are available close by, the greater the likelihood that an activity which may require a long trip is substituted for by an activity close by. It was shown that these factors have a significant effect on travel demand, and that an increase in the ease and worth of making a trip to distant destinations was positively related to trip generation.

Cervero and Gorham (1995) provide additional evidence for the importance of considering two different scales of accessibility. They found little variation between measures of mode-split in Los Angeles for transit versus automobile neighborhoods. They contend one reason for the finding is because the Los Angeles region is so expansive and laced by over 500 miles (800 km) of freeways. In some contexts, the form of the region as a whole may have at least as great a role in influencing modal choice as neighborhood design, if not greater. The important point is that the effect of each phenomenon—regional access versus neighborhood access—should be disentangled; put otherwise, the relative magnitude of each is an empirical question with pressing policy significance.

Differentiating measures between the two levels of accessibility is a messy process. Many policy initiatives speak to increasing accessibility on both scales—regional and local; and, while the two scales are intricately related, each calls for different policies at different scales. For example, regional land use-transportation policies may speak to issues of urban growth boundaries, increasing densification, and diversifying the geographical distribution of employment centers. It is not likely that such regional policies prescribe development regulations for specific neighborhoods. Neighborhood accessibility policy initiatives speak more to issues of mixing uses on a parcel or neighborhood scale, site design, and more directly, facilitating circulation patterns that enhance walking, cycling and transit use. While increasing regional access helps to create neighborhood access, it is not essential.

Moudon and Hess (2000) show how patterns of development may exhibit relatively high regional access in terms of density and mixing of uses but still not be conducive to pedestrian travel. Neighborhood access is an intricate composite of elements that includes, but is not limited to compact development, mixing land uses, mixing development types, sidewalks, and landscape provisions.

This concept is graphically depicted in Figure 1, showing how the relatively amorphous notions of regional access and neighborhood access are not clear divisions. While each can be independently measured on respective scales, each lies on a continuum that is subject to different perspectives, different travel purposes, and/or different travel modes. Regional accessibility is defined primarily by the characteristics of the transportation analysis zone relative to the region at large. Neighborhood access, on the other hand, is defined primarily by the urban form characteristics within a quarter-mile (400 m) of each household. Such geographic boundaries are by no means cut and dry.

A neighborhood could score high on regional access and low on local access, denoted by area "A" below. Alternatively, a neighborhood could score high on neighborhood access and low on regional access, denoted by area "C." This latter neighborhood would likely be more pedestrian friendly than the former, though, it may still lack in terms of regional access. Only those neighborhoods that score high on both (denoted by area "B") represent the most accessible neighborhoods in the study area. For this reason, policy initiatives focusing on neighborhood accessibility tend to be more important in terms of creating communities considered to be more pedestrian-friendly.

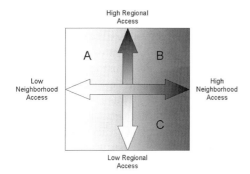

Figure 1. Graphical depiction showing the continuous nature of accessibility across dimensions of low and high as well as local and regional.

Travel and neighborhood accessibility, examining the evidence

The potential of either neighborhood or regional accessibility to moderate travel is a topic of intense policy interest recently. In response, it has been the subject of between 50 and 100 empirical studies. Any single review cannot do justice to the many issues, approaches, findings, and shortcomings involved in synthesizing the wealth of these studies. At least two bibliographies cover the literature in annotated form (Handy, 1992a; Ocken, 1993). A handful of literature reviews are also available (Handy, 1996a; Pickrell, 1996; Crane, 2000). As mentioned in Ewing and Cervero (2001), the reader may wonder whether another literature survey can add much value.

For this reason, the review offered in this chapter does not examine in detail existing literature related to urban form and travel. The reader is urged to consult Handy (1996a) and Crane (2000) as both reviews focus on the different approaches used in past studies, explaining their techniques, strengths, and weaknesses. Instead, this section serves two purposes. The first is to provide the reader with a more thorough understanding of both the complexity and disparity of existing research. The second is to clearly articulate issues—shortcomings if you prefer—of past research.

Given the complex array of issues at stake in such a research endeavor, any number of data, research approaches, and analysis strategies could be employed. Consequently, any review of such research could be organized in a variety of ways. For example, Crane (2000) lists the different strategies in which such studies could be organized (Table 1).

Table 1. Taxonomy of ways to classify studies related to accessibility and travel.

Travel Outcome Measures	Urban Form and Land Use Measures	Methods of Analysis	Other Distinctions and Issues
• Total distance traveled • Trip generation • Vehicle trip generation • Time spent traveling • Car ownership • Mode of travel • Congestion • Commute length • Other commute measures (e.g., speed, time)	• Density • Land use pattern • Land use mixing • Traffic calming • Circulation pattern • Jobs/housing balance • Pedestrian features (e.g., sidewalks, perceived safety, visual amenities) • Composite indices	• Simulation • Description of observed travel behavior in different settings (e.g., commute length in large vs. small cities) • Multivariate statistical analysis of observed behavior	• Land use and urban design features as the trip origin vs. the destination vs. the entire route • Composition of trip chains and tours (e.g., use of commute home to buy groceries • Use of aggregate vs. individual level traveler data and aggregate vs. site specific urban form data.
Table adapted from Crane, 2000.			

The first category relates to the travel (dependent) variables being analyzed. Depending on data availability, most studies separately examine one dimension of travel (e.g., trip generation for work versus non-work travel). Doing so reduces the extent to which different studies can be compared because they often analyze different phenomena.

A second category separates studies according to the independent variable. For example, Ewing and Cervero (2001) discuss analyses according to the findings with respect to at least four different dimensions of the built form: land use patterns, transportation network, urban design features or composite indices. While such a review may help better understand the relative effect of each element, it has at least two principal shortcomings. First, many such studies examine more than one dimension of urban form in concert with other dimensions. Second, some studies use a single measure (e.g., street pattern) to represent multiple dimensions of accessibility. Thus, assessing the independent effect of one variable without fully considering the range of other variables fails to do justice to the specific dimension in question. It speaks more to the limitations of singling out the individual effect of one element of the built environment as opposed to attempting to fully capture the myriad dimensions of access.

The third category groups studies that use similar methods of analysis. Research approaches can be broken down according to simulation studies, aggregate analysis, disaggregate analysis, choice models, and activity-based analysis (Handy, 1996a). But even within each grouping, there remains considerable variation. For example, studies with similar methods of analysis may still analyze different dependent variables; alternatively, they may employ different analysis techniques (e.g., regression models versus discrete choice models).

The fourth category reviews even additional strategies. Confounding issues stem from varying units of analysis (e.g., disaggregate versus aggregate) or measuring only trips from certain origins and/or destinations.

Despite such disparities in data, methods, or approaches, it is nonetheless helpful to shed light on some of the findings from an active line of research. Doing so provides a better appreciation for the range of issues discussed, travel behavior variables used, urban form measures employed, and general pattern of results.

Early work primarily used matched pair analysis and aggregate statistics to examine travel outcomes in neighborhoods with varying degrees of access. Crudely simplifying this stream of research suggests the following:

- less distance traveled in neighborhoods with higher density and better transit access (Holtzclaw, 1994).

- fewer vehicle and more pedestrian and transit trips in neighborhoods that are more pedestrian-friendly (1000 Friends of Oregon, 1993).

- fewer total trips and slightly higher ratios of transit use and pedestrian activity in traditional neighborhoods versus standard suburban neighborhoods (McNally and Kulkarni, 1997).

- higher percentages of transit use for commuting in some transit neighborhoods relative to automobile neighborhoods (Cervero and Gorham, 1995).

- two-thirds more vehicle hours of travel per person for households in sprawling type suburbs versus comparable households in a traditional city (Ewing *et al.*, 1994).

- more pedestrian activity in mixed use centers with site design features that include sidewalks and street crossings (Hess *et al.*, 1999).

In later work more disaggregate approaches analyze the travel behavior of individual households within neighborhoods to better understand travel choices and relationships with access features. These studies use analysis of variance, regression or logit models to compare the relative influence of different urban form characteristics relative to socio-demographic characteristics. Again, simplifying the results suggests:

- more walking to shopping and potentially less driving to shopping for residents in some traditional neighborhoods (Handy, 1996b; Handy, 1996c).

- fewer vehicle hours of travel for residents in neighborhoods with higher accessibility (Ewing, 1995).

- higher percentages for transit and non-motorized trips for residents closer to the bus or rail and in higher density neighborhoods (Kitamura *et al.*, 1997).

- reduced trip rates and more non-auto travel for individuals living in neighborhoods with higher density, land use mix, and better pedestrian orientation (Cervero and Kockelman, 1997).

- walking to transit stations is more likely where retail uses predominate around stations (Loutzenheiser, 1997).

- higher trip frequency in areas of high accessibility to jobs or high accessibility to households (Sun *et al.*, 1998).

- a reduced number of non-work auto trips in ZIP code areas with higher retail employment densities (Boarnet and Greenwald, 2000).

- higher transit passenger distance in areas with fewer jobs and grocery stores within one kilometer (Pushkar *et al.*, 2000).

- more walking and transit use—and less vehicle distance and less frequent auto trips— in areas with higher composite indices (Lawton, 1997).

- more likely use of non-auto modes in areas with greater mixing of commercial-residential uses (only in middle suburbs), auto use is less likely in areas (Pushkar, Hollingworth *et al.*, 2000).

Each of the above studies shows that urban form appears to influence travel in hypothesized (and expected) directions. However, R-squared values rarely exceed 0.40 in such work, suggesting that there remain many unexplained factors that influence travel. Crane (2000) asserts that basic relationships between urban form and travel, to date, have not been analyzed within a behavioral framework considering basic tenets such as the cost (in terms of time or convenience) of each trip. This echoes of the results found in recent work (Boarnet and Sarmiento, 1998; Crane and Crepeau, 1998). These studies offer skepticism about urban form's potential to moderate travel demand, especially with respect to vehicle trip generation. As a result, Crane goes so far as to contend that "not much can be said about the relationship between travel preferences and urban form" (Crane, 2000).

SHORTCOMINGS OF PREVIOUS RESEARCH

Throughout any of the issues or studies previously described, however, there remain at least three overarching matters. The following section of this chapter describes each complexity, explaining how it contributes to the shortcomings of past research. The focus of this discussion primarily serves to better position the research and approaches offered throughout this volume.

The influence of household preferences versus urban form features

Ever since the earliest studies examining the influence of the neighborhood scale (Handy, 1992c; 1000 Friends of Oregon, 1993), it has been posited that access does nothing more than attract residents with certain attitudes, values, or demographic and socio-economic attributes.

Recent thought, however, challenges the conventional wisdom of research that points to correlations between urban form and travel as evidence that changing land use patterns can impact travel. Residents may be selecting residential locations in part to match their travel preferences; therefore, differences in travel between households who live in neighborhoods with different designs should not be credited to urban design alone. Differences could be attributed to the broader *preferences* that triggered the choice to locate in a given neighborhood.

If there is a self-selection bias at work, policies designed to induce travel changes through land uses may not have the expected or desired effect—or, their impact may be marginal. For example, many contend that managing travel demand through land use policy is limited by the size of a "niche" market of households who *prefer* to locate in neighborhoods that are pedestrian-friendly. They state that aiming to use urban design tools to induce unwilling auto-oriented households to drive less may be futile. The corollary asserts that the success of access in mitigating travel may be based on the relatively small market of households who currently live in transit-oriented neighborhoods and/or those who will bring their non-auto using behavior with them to newer neighborhoods. In short, transit-oriented households would likely remain transit-using; auto-oriented households would likely remain auto-using. The success of access, it is hypothesized, may be largely based on the size of this "niche" market. Therefore, the research should be conceptualized somewhat differently: can urban form be relied upon as a means of reducing travel once residential sorting (self-selection) is considered?

A handful of recent efforts (described below) attempt to better disentangle the two relative effects of urban form versus preferences. Doing so helps better understand the relative magnitude of the independent effect of urban design on travel, an effect that may become marginalized once preferences are accounted for.

Prevedouros (1992) measured personality characteristics and analyzed their association with choice of residential neighborhood type. Kitamura *et al.* (1997) attempted to untangle whether attitudinal factors (typically not included in urban-form/travel studies) are more (or less) correlated with a household's travel characteristics than with land use characteristics. They did so by first estimating "base" regression models to confirm that neighborhood characteristics—in particular residential density, public transit accessibility, distance to nearest park, and presence of sidewalks—are significantly associated with trip generation by mode and modal split. They then used attitudinal information from 39 different questions to assess their subjects' attitudes toward various aspects of urban life. The questions were divided into eight groups—private automobile, ridesharing, public transit, urban transportation, time, environment, housing and economy—and factor analysis was used to reduce the dimensionality. The eight factors where then introduced into the base models. Although the neighborhood descriptors introduced into the best models improved the statistical fit of the models, the contributions of the attitude factors were in general greater than those of the neighborhood descriptors in the best model. The authors therefore concluded that attitudes are more strongly, and perhaps more directly, associated with travel than are land use characteristics.

Using the same dataset within a system of structural equations, Bagley and Mokhtarian (2000) examined relationships between urban form and travel, incorporating attitudinal, lifestyle, and demographic variables. In terms of both direct and total effects, they concluded that attitudinal and lifestyle variables had the greatest impact on travel demand among all the explanatory variables.

A different approach offered by Boarnet and Sarmiento (1998) and Boarnet and Greenwald (2000) used instrumental variables representing residential location decisions to validate the theory that households choose their residential locations based in part on their desired travel behavior. They added measures to an existing base model to determine the extent to which pre-existing attitudes determine travel. Instead of measures of the attitudes themselves, they used four non-transport neighborhood amenities as instrumental variables: percentage population that is black, percentage population Hispanic, percentage housing stock built before 1940 and percentage housing stock built before 1960. These demographic and housing stock variables are likely to be correlated with land use patterns measuring residential location. Their findings argue for the importance of controlling for residential choice.

Incomplete account of total travel

Travel behavior is often measured using a single dimension such as trip frequency or travel distance. Simplifying the dependent variable in this way does not do justice to possible tradeoffs between either of the variables. For example, Handy's work (1996b; 1996c) provides empirical evidence of Crane's assertion (1996a) that open and gridded circulation patterns make for shorter trip distances and may even stimulate trip taking. He argues that residents with higher neighborhood access may shop more often and drive longer distances overall. Using regression or logit models on a limited number of dependent variables only speaks to one piece of the puzzle and is unable to shed light on possible tradeoffs between trip frequency and travel distance.

In addition, most studies analyze individual trips independently. Doing so often implicitly assumes that each trip originates from home. In reality, both of these assumptions defy how decisions related to travel behavior are made or acted upon. Many individuals link multiple purpose trips together; many start trips from locations other than the home. Transportation modeling applications have attempted to mitigate such problems by specifying trips by different types: home-based-work, home-based-other, and non-home-based. This strategy attempts to better address the issue of different types of trips originating from different locations. Using such coding may no longer implicitly assume each trip purpose originates from home.

While specifying trip type better addresses the origin part of the trip, it still fails to consider linked travel. Analyzing individual trips masks sequential and multi-purpose travel because the nature of many trips is often a function of the preceding trip. A vehicle trip to the dry cleaner may not be because a car was required for this trip, but because the dry cleaner trip was done on the way to the grocer, a trip that required a car in the first place.

It is just as important to examine multiple trip purposes—both work and non-work. Commute data is often analyzed because it is readily available and has long been considered the lion's share of metropolitan travel flow; non-work trips are analyzed because they represent trip types most directly influenced by levels of neighborhood access. But if a commuter stops for a cup of coffee and then proceeds to work, the first leg is often classified as home-based-shopping and the second leg may be classified as non-home-based. The primary purpose of the travel—work—is lost in the classification process. Thus, any analysis that separates work from non-work trips can say very little about sequential trip-making that combines both. Over two decades ago, Hanson (1980) stressed the importance of jointly analyzing work and non-work travel because separating trips by type fails to capture linked and multi-purpose travel behavior that exists.

The bottom line is that examining only individual trips instead of the larger pattern of linked trips fails to work with the basic forces that generate and influence travel and may provide an incomplete account of the travel behavior picture.

The manner in which urban form is operationalized

The manner in which the land use dimension (i.e., the opportunities) is measured presents at least four separate, but related challenges. They include (a) using relatively large units of analysis, (b) delineating neighborhoods using artificial boundaries, (c) being able to distinguish between regional and local accessibility, and (d) observing little variation within study areas. Each challenge is briefly described below.

The majority of past research aggregates urban form information to census tracts, ZIP code areas, or transportation analysis zones (TAZs). These units often do little justice to the central aim; they can be quite large—almost two miles (3.2 km) wide—and contain over 1,000 households. The ecological fallacy lies in that urban form characteristics that are measured constitute summary data that may or may not apply to individual households who reside in a particular zone. Research in the Central Puget Sound (Moudon and Hess, 2000) identified almost one hundred concentrations of multifamily housing within one mile (1.6 km) of retail centers and/or schools. By aggregating measures of commercial intensity, each zone reveals the same measure, although each development pattern is likely to affect travel behavior differently (see left side of Figure 2). Because census tracts or TAZs average these types of concentrations with adjacent lower-density development, it is difficult to associate many neighborhood-scale aspects with travel demand.

Furthermore, census tracks or TAZs are often delineated by artificial boundaries (e.g., main arterial streets) that bear little resemblance to the neighborhood-scale phenomenon being studied in terms of their size or shape. Consider the example graphic displayed on the right side of Figure 2. The first example shows two households living on opposite sides of the street from one another but the same distance to a corner grocery store. Using TAZ geography, household A is linked with TAZ #1 and household B is linked with TAZ #2. The second example shows how a four-way intersection with retail activity on all four corners divides this retail center into different TAZs. Such division dilutes the measure of commercial intensity in any single zone. In terms of affecting travel behavior, however, the commercial intensity of all four corners should be grouped together.

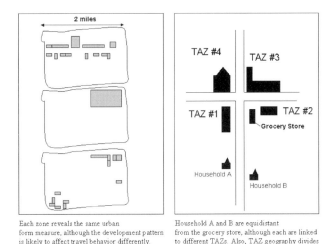

Each zone reveals the same urban form measure, although the development pattern is likely to affect travel behavior differently.

Household A and B are equidistant from the grocery store, although each are linked to different TAZs. Also, TAZ geography divides the retail center into different zones.

Figure 2. Confounding issues caused by relatively large units of analysis.

Third, as described in the section above (Regional versus neighborhood accessibility), it is also important to distinguish the effects of urban form at the neighborhood scale from those at the regional scale. Because a household's travel may be influenced by both the immediate locale— the character of the particular neighborhood in which the household lives—and the position of the neighborhood in the larger region, it is important to be able to identify the relative contribution of each. The regional context of a neighborhood, too often neglected in previous research on the subject, may provide more opportunities that mean more travel. Or, the regional structure may simply dwarf any variations in the local, neighborhood structure. In a similar manner, it would be ideal to examine correlations between urban form not only surrounding the residential location, but also an individual's work location and features of the corridor along the main axis from one's residence to their work location. After all, all three are likely to play out in some manner.

A final challenge relates to being able to measure study areas that exhibit enough variety in different types of accessibility for which there is likely to be a difference in travel mode. For example, Orange County, California—the context for a fair amount of work on this subject (McNally and Kulkarni, 1997; Boarnet and Sarmiento, 1998; Crane and Crepeau, 1998)—is "hardly a place of great land use diversity" (Cervero, 1996b). The thresholds at which one may expect detectable differences in mode split or vehicle miles of travel may not have been reached. Given the uniformity of most residential development in the U.S., urban forms that exhibit expected thresholds are difficult to locate. Therefore, it is important to carefully consider the different urban forms within a study area to ensure that adequate variety is met.

ROUNDUP AND ADDITIONAL PERSPECTIVES

Research efforts to document relationships between accessibility and travel behavior enjoy a rich history that dates back several decades. Using accessibility as the key concept that bonds land use together with transportation, recent work often delves into more specific issues of consideration. The bulk of such efforts have focused on examining the prospects of access to moderate travel. But within this emerging research agenda, many of the results are still difficult to compare. One research effort may extend methodologies to measure travel; another may make notable contributions to measure a variant of an accessibility measure. Still other studies try to sort out the inherent endogeneity problem associated with self-selection. The variety of studies and variety of approaches reveal even greater complexities than may have been initially realized. Understandably, available evidence exists in either direction and is very messy. Echoing the sentiments expressed by both Handy (1996a) and Crane (1999), sorting out the link between urban form and travel is difficult.

Conceptual framework for how accessibility affects travel

Within this discussion, it is important to clarify the many ways in which patterns of urban form influence travel behavior. Doing so helps to provide a clearer discussion and interpretation of the research and results that follow. Random choice utility theory provides a means to incorporate many of the important tenets of this research in a single framework.

According to choice theory, individuals (or households) assign a utility (or value) to each choice within a set of alternatives (Ben-Akiva and Lerman, 1979). Decision makers select the choice that maximizes their utility. For purposes of this application, utility is loosely defined as the likelihood that households will engage in behavior that includes heightened levels of walking and transit use and less automobile travel.

The utility of a particular choice can be derived in a number of ways. Generally speaking, it can be partitioned into two sub-functions: one depends on measures of accessibility that can be directly observed, such as the physical attributes of the land use-transportation system; another sub-function depends on aspects of utility less easily observed, including the preferences or tastes that households embody with respect to the overall feel of the neighborhood in which they live or how they travel. Thus, utility could be expressed as:

$$U_{ni} = V_{ni} + \varepsilon_{ni} \quad \text{where:}$$

U_{ni} = the true utility of choice i for household n
V_{ni} = the observable utility of choice i for household n
ε_{ni} = the unobserved utility of choice i for individual n.

Land use patterns inform the utility equation presented above in a variety of ways. Considering a relatively short-term time horizon (less than a year), at least three ways stand out as presented in *Figure 3*. The first two affect the observable utility; the last affects the unobserved utility. (Figure 3 presents a simplified depiction of the relationship between urban form and travel. It is conceivable that household travel behavior may influence urban form patterns, suggesting a bi-directional relationship. However, this effect would likely be seen over a time horizon of multiple years.)

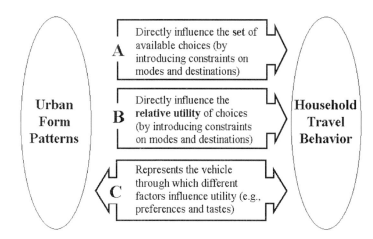

Figure 3. Relatively short-term influences of urban form patterns on travel behavior.

In the first case (A), urban form plays an important role in directly influencing the travel possibilities that exist. The nature of the urban form may create constraints (or opportunities) that define the choice set. Consider the following overly simplistic scenario (*Table 2*). The Jones household lives in an area with high neighborhood access; the Smith household lives in a neighborhood with only *relatively* high access. Consequently, each household has a wide of range of travel alternatives that are feasible. It is possible for them to walk to the corner store to buy a pint of milk. They may not take advantage of the full range of travel options; nonetheless, each of them is a feasible alternative. In contrast, the Barnes household lives in a neighborhood with low access where it is impossible (or at least extremely difficult) to walk to the corner store to buy a pint of milk (the corner would be very far away). In this case, the choice set—as defined by the land use pattern—consists of one feasible travel mode, the automobile.

Table 2. Land use patterns defining the choice set of travel options.

Household	Accessibility	Example picture	Choice set of feasible travel alternatives	Attractive travel alternatives
Barnes household	Low access		Automobile	Automobile
Smith household	Relatively high access		Automobile, bus, walking, cycling	Automobile, bus
Jones household	High access		Automobile, bus, walking, cycling	Automobile, bus, walking, cycling

In the second case (B) presented in *Figure 3*, the patterns of urban form may directly affect travel behavior by introducing constraints (or opportunities) that in turn influence the utility derived from each choice. The Smith household may have a variety of travel choices that are feasible. However, the urban form characteristics may influence the relative utility (or attractiveness) of each alternative. Even if the corner store is within walking distance, the walk may be along a six lane arterial without sidewalk. While cycling is technically an option, there may be no shoulders on the roadway and no sidewalk. In this case, only the automobile and the bus would emerge as attractive modes. The overall utility of choosing each of these modes for the milk trip is directly influenced by the urban form characteristics.

In cases A and B in Figure 3, the influence of urban form can be observed by measuring the attributes of the land use-transportation system. In one respect, urban form defines the choice set; in another, it influences the relative utilities of each choice. A third case (C) considers the unobservable utility of each choice or the residual term, ε_{ni}. In this case, the unobservable utility of traveling in less auto-dependent ways would be reflected by the decision maker's preferences or tastes towards the type of neighborhood and/or how they wish to travel. These preferences or tastes are manifested jointly by urban form and travel behavior characteristics, thus the dual causality as indicated in case C. For example, if a household wants to be able to ride transit, they are then most likely going to select an urban form combination that allows them to value this overall utility. The combined urban form-travel relationship would represent the vehicle through which such preferences would be expressed.

Such an explanation—albeit somewhat simplified in application—allows the reader to better understand the conditions under which accessibility affects travel. It introduces a single conceptual framework that helps to more clearly articulate the research approaches and results that follow.

The review offered as part of this chapter endeavored to respond to a variety of calls from the community of researchers and academicians as well as the community professional planners and policy makers. It provides a context in which to approach the range of research applications that follow. Many of the following papers call for a better understanding of how urban form and travel relate. Considering the line of research focused on in this review, for example, Cervero (1996b), Boarnet and Sarmiento (1998) and Frank (1994b) all call for more a behavioral analysis of how urban form may affect trip chaining. Other than McCormack (1997), little to no research explicitly examines how accessibility affects trip chaining. Hess (1996) describes how conventional units of analysis (e.g., census tracts) fall short of the task of measuring pedestrian-scale activity and therefore calls for more precise measures of urban form. The emergence of increasingly available disaggregate urban form and travel data combined with improving technological capacity provide a means to account for more disaggregate approaches.

This review also directly responds to the needs of the community of professional and practicing planners by providing them with better information on which to base land use-transportation policy. Understanding the nature of the relationships that are posited in many research applications have direct implications for integrated land use-transportation models and policy practice. Also, more explicitly addressing and tackling the role that household preferences have in influencing integrated residential location and travel decisions helps comment on the prospects of using accessibility to moderate the demand for travel. This review endeavors to provide the reader with an understanding of important components to such questions in a more informed and robust manner where past review efforts have fallen short.

REFERENCES

1000 Friends of Oregon (1993). *Making the Land Use Transportation Air Quality Connection, Vol. 4A: The Pedestrian Environment.* LUTRAQ, with Cambridge Systematics, Inc., Calthorpe Associates, and Parsons Brinkerhoff Quade and Douglas, Portland, OR.

Bagley, M. N., P. L. Mokhtarian, *et al.* (2000). *A methodology for the disaggregate, multidimensional measurement of residential neighborhood type.* U.C.-Davis, working paper.

Ben-Akiva, M. and J. L. Bowman (1998). Integration of an activity-based model system and a residential location model. Urban Studies, **35(7)**, 1131-1153.

Ben-Akiva, M. and S. Lerman (1979). Disaggregate travel and mobility choice models and measures of accessibility. In: *Behavioural Travel Modelling.* (D. A. Hensher and P. R. Stopher, eds.), pp. 654-679. Croom Helm, London.

Boarnet, M. G. and M. Greenwald, J. (2000). Land Use, Urban Design, and Non-Work Travel: Reproducing Other Urban Areas' Empirical Test Results in Portland, Oregon. *Transportation Research Record,* **1722**, 27-37.

Boarnet, M. G. and S. Sarmiento (1998). Can Land-Use Policy Really Affect Travel Behavior? A Study of the Link Between Non-work Travel and Land-Use Characteristics. *Urban Studies*, **35(7)**, 1155-1169.

Casey, H. J. (1955). The Law of Retail Gravitation Applied to Traffic Engineering. *Traffic Quarterly,* **9(1)**, 23-25.

Cervero, R. (1996b). *Urban Design Issues Related to Transportation Modes, Designs and Services for Neo-Traditional Developments.* Urban Design, Telecommuting and Travel Forecasting Conference, Williamsburg, VA.

Cervero, R. and R. Gorham (1995). Commuting in Transit Versus Automobile Neighborhoods. *Journal of the American Planning Association,* **61(2),** 210-225.

Cervero, R. and K. Kockelman (1997). Travel Demand and the Three Ds: Density, Diversity, and Design. *Transportation Research, Part D*, **2(2),** 199-219.

Crane, R. (1996a). Cars and Drivers in the New Suburbs: Linking Access to Travel in Neotraditional Planning. *Journal of the American Planning Association,* **62(1),** 51-65.

Crane, R. (1999). The Impacts of Urban Form on Travel: A Critical Review. Lincoln Institute of Land Policy: 38.

Crane, R. (2000). The Influence of Urban Form on Travel: An Interpretative Review." *Journal of Planning Literature*, **15(1),** 3-23.

Crane, R. and R. Crepeau (1998). Does neighborhood design influence travel?: A behavioral analysis of travel diary and GIS data. *Transportation Research Part D-Transport and Environment,* **3(4)**, 225-238.

Ewing, R. (1995). Beyond Density, Mode Choice, and Single Purpose Trips. *Transportation Quarterly,* **49(4),** 15-24.

Ewing, R. and R. Cervero (2001). *Travel and the Built Environment: Synthesis.* Transportation Research Board, Washington, D.C.

Ewing, R., P. Haliyur, *et al.* (1994). Getting Around a Traditional City, a Suburban Planned Unit Development, and Everything in Between. *Transportation Research Record,* **1466**, 53-62.

Frank, L. D. and G. Pivo (1994b). *Relationships between land use and travel behavior in the Puget Sound Region,* Washington State Transportation Center. Seattle.

Gordon, P., A. Kumar and H. W. Richardson (1989). Congestion, Changing Metropolitan Structure, and City Size in the United States. *International Regional Science Review,* **12(1),** 45-56.

Gur, Y. (1971). *An Accessibility Sensitive Trip Generation Model.* The Chicago Area Transportation Study, Chicago.

Handy, S. L. (1992a). How Land Use Patterns Affect Travel Patterns: A Bibliography, CPL Bibliography.

Handy, S. L. (1992c). Regional Versus Local Accessibility: Variations in Suburban Form and the Effects on Non-Work Travel. City and Regional Planning. Berkeley, University of California, Berkeley.

Handy, S. L. (1993). Regional Versus Local Accessibility: Implications for Nonwork Travel. *Transportation Research Record,* **1400,** 58-66.

Handy, S. L. (1996a). Methodologies for Exploring the Link Between Urban Form and Travel Behavior. *Transportation Research D,* **1(2),** 151-165.

Handy, S. L. (1996b). Understanding the Link Between Urban Form and Nonwork Travel Behavior. *Journal of Planning Education and Research,* **15(3),** 183-198.

Handy, S. L. (1996c). Urban Form and Pedestrian Choices: Study of Austin Neighborhoods. *Transportation Research Record,* **1552,** 135-144.

Handy, S. L. and D. A. Niemeier (1997). Measuring Accessibility: an exploration of issues and alternatives. *Environment and Planning A,* **29,** 1175-1194.

Hansen, W. (1959). How Accessibility Shapes Land Use. *Journal of the American Institute of Planners,* **25,** 73-76.

Hanson, S. (1980). The Importance of the Multi-purpose Journey to Work in Urban Travel Behavior. *Transportation,* **9,** 229-248.

Hess, P. M. (1996). Studying Pedestrian Activity in Small Suburban Clusters: Issues of Land Use, Development Patterns and Scale. *Urban Design and Planning.* Seattle, WA, University of Washington.

Hess, P. M., A. V. Moudon, *et al.* (1999). Site Design and Pedestrian Travel. *Transportation Research Record,* **1674,** 9-19.

Hess, P. M., A. V. Moudon, *et al.* (2000). *Measuring land use patterns for transportation research.* Transportation Research Board, Washington D.C.

Holtzclaw, J. (1994). *Using Residential Patterns and Transit to Decrease Auto Dependence and Costs.* Natural Resources Defense Council, San Francisco,

Huff, D. L. (1963). A probabilistic analysis of shopping centre trade areas. *Land Economics,* **39,** 81-90.

Kelly, E. D. (1994). The Transportation Land-Use Link. *Journal of Planning Literature,* **9(2),** 128-145.

Kitamura, R., P. L. Mokhtarian, *et al.* (1997). A Micro-Analysis of Land Use and Travel in Five Neighborhoods in the San Francisco Bay Area. *Transportation,* **24,** 125-158.

Levinson, D. (1998). Accessibility and the Journey to Work. *Journal of Transport Geography,* **6(1),** 11-21.

Lawton, T. K. (1997). *The Urban Environment Effects and a Discussion of Travel Time Budget.* Portland Transportation Summit.

Loutzenheiser, D. (1997). Pedestrian Access to Transit: Model of Walk Trips and Their Design and Urban Form Determinants Around Bay Area Rapid Transit Stations. *Transportation Research Board,* **1604,** 40-49.

McCormack, E. D. (1997). A Chained-Based Exploration of Work Travel by Residents of Mixed Land Use Neighborhoods. *Geography,* Seattle, University of Washington.

McNally, M. G. and A. Kulkarni (1997). Assessment of Influence of Land Use-Transportation System on Travel Behavior. *Transportation Research Record,* **1607,** 105-115.

Mitchell, R. B. and C. Rapkin (1954). *Urban Traffic: A Function of Land Use,* New York, Columbia University Press.

Moktarian, P. L. and I. Salomon (1998). Travel for the fun of it. *Access,* **15,** 26-31.

Moudon, A. V. and P. M. Hess (2000). Suburban Clusters: The Nucleation of Multifamily Housing in Suburban Areas of the Central Puget Sound. *Journal of the American Planning Association,* **66(3),** 243-264.

Ocken, R. (1993). *Site Design and Travel Behavior: a Bibliography.* 1000 Friends of Oregon, Portland.

Patton, T. and N. Clark (1970). Towards an accessibility model for residential development. Conference title: Analysis of Urban Development (Volume 5), University of Melbourne. Proceedings of the Tewksbury Symposium. (N. Clark, ed.), pp. 266-288

Pickrell, D. (1996). Transportation and Land Use. In: In: *Essays in transportation economics and policy: a handbook in honor of John R. Meyer* (J. R. Meyer, J. A. Gómez-Ibáñez, W. B. Tye, and C. Winston, eds.), pp. 403-435. Brookings Institution Press, Washington DC.

Prevedouros, P. (1992). Associations of Personality Characteristics with Transport Behavior and Residence Location Decisions. *Transportation Research A,* **26,** 381-391.

Pushkar, A. O., B. J. Hollingworth, *et al.* (2000). *A Mulitvariate Regression Model for Estimated Greenhouse Gas Emissions from Alternative Neighborhood Designs.* Transportation Research Board, Washington, DC.

Reilly, W. S. (1953). *The Law of Retail Gravitation.* Pillsbury, New York,

Sun, A., C. G. Wilmot, *et al.* (1998). Household travel, household characteristics, and land use. *Transportation Research Record,* **1617,** 10-17.

Wachs, M. and T. G. Kumagi (1973). Physical Accessibility as a Social Indicator. *Socio-Economic Planning Sciences,* **7,** 437-456.

Access to Destinations
D.M. Levinson and K.J. Krizek (editors)

CHAPTER 7

PLANNING FOR ACCESSIBILITY: IN THEORY AND IN PRACTICE

Susan Handy, University of California, Davis

INTRODUCTION

Pick up a transportation plan for a major metropolitan area in the US and you are likely to find improved mobility and accessibility highlighted as goals. The 2020 Regional Transportation Plan for the Austin region, for example, stated that "The primary goal of the CAMPO 2020 Plan is to provide an acceptable level of mobility and accessibility for the region's residents with the least detrimental effects." The 2020 Regional Transportation Plan for the Chicago region aimed to "provide an integrated and coordinated transportation system that maximizes accessibility and includes a variety of mobility options that serve the needs of residents and businesses in the region." Such statements are likely influenced by the Transportation Equity Act for the 21st Century, commonly known as TEA-21, which established seven "planning factors" for consideration in the planning process, including "increase the accessibility and mobility options available to people and for freight."

What exactly these plans mean by "mobility" and "accessibility" isn't clear, nor whether the agencies writing these plans themselves have a clear sense of what they mean in using these terms. If so, they may have missed an important opportunity to clarify their objectives and direct their planning efforts more effectively. This paper takes the position that mobility and accessibility are distinct concepts with vastly different implications for planning. First, the paper looks at these concepts in theory, articulating a distinction between mobility and accessibility and outlining the possible implications of planning for mobility versus planning for accessibility. Second, the paper looks at the current use of these concepts in practice, by examining a sample of regional transportation plans. This exploration yields a mixed picture: although these plans continue to reflect a traditional concern with mobility, they also show many indications of a concern with planning for accessibility, even if they don't label their efforts as such.

IN THEORY

The American Heritage Dictionary Fourth Edition defines "mobility" as "the quality or state of being mobile" and "mobile" as "capable of moving or of being moved readily from place to place" (Picket et al. 2000). The Oxford English Dictionary defines "mobility" as the "ability to move or to be moved... facility of movement" (OED 2002). In the context of transportation planning, mobility has been defined as the potential for movement, the ability to get from one place to another, an ability to move around (Hansen 1959; Handy 1994). Traditional level-of-service measures used in transportation planning are measures of mobility; higher volume-to-capacity ratios mean slower travel times, less ease of movement, and thus lower mobility. Mobility is sometimes also measured by actual movement, either numbers of trips made or total kilometers traveled. Actual movement is not necessarily an accurate measure of the potential for movement, however. First, potential movement can exceed actual movement, for example, if individuals choose to drive less than they could. Second, increases in actual movement can mean decreases in potential movement, as is the case when roads are congested.

Accessibility has been harder for planners to both define and to measure. The American Heritage Dictionary Fourth Edition defines "accessibility" as "easily approached or entered" (Picket et al. 2000). The Oxford English Dictionary defines "accessibility" as "the quality of being accessible, or of admitting approach" (OED 2002). Accessibility was perhaps more clearly defined for the planning context by Hansen (1959) as "the potential for interaction." Accessibility can be thought of as an ability to get what one needs, if necessary by getting to the places where those needs can be met. In most cases, measures of accessibility include both an impedance factor, reflecting the time or cost of reaching a destination, and an attractiveness factor, reflecting the qualities of the potential destinations. Researchers have used many different forms of accessibility measures and have raised many important issues about these measures (Handy and Niemeier 1996). Simple "cumulative-opportunities" measures, which count the number of destinations of interest within a certain time or distance of the origin point, seem to be coming into greater use in transportation planning, as discussed below. Choice is an important element of accessibility: more choices in either destinations or modes of travel mean greater accessibility by most definitions.

Part of the confusion in the use of these terms may stem from the relationship between them. Mobility, the potential for movement, is related to the impedance component of accessibility, in other words, how difficult it is to reach a destination. Policies to increase mobility will generally increase accessibility as well by making it easier to reach destinations. But it is possible to have good accessibility with poor mobility. For example, a community with severe congestion but where residents live within a short distance of all needed and desired destinations has poor mobility but may still have good accessibility. If destinations are close, the travel times will still be reasonably short, even if travel speeds are relatively low. In this case, accessibility is not dependent on good mobility (though it does depend on having some mobility by one mode or another). It is also possible to have good mobility but poor accessibility. For example, a community with ample roads and low levels of congestion but with relatively few destinations for shopping or other activities or with undesirable or inadequate destinations has good mobility but poor accessibility. *Good* mobility is neither a sufficient nor a necessary condition for *good* accessibility.

Planning for mobility, then, means making it easier to get around and, in practice, has largely meant making it easier to drive around. To plan for mobility is to focus on the means without direct concern for the ends: can people move around with relative ease? The traditional emphasis on road building in the U.S. is consistent with a planning-for-mobility perspective in that the aim is to accommodate growing levels of travel and increase the potential for movement. The planning process traditionally started with a projection of future traffic volumes that was followed by a determination of the capacity needed to accommodate those volumes at acceptable levels-of-service. The focus was on the performance of the system. This approach can lead to a vicious circle, however. When it's easier to drive around, people are more likely to drive around. Higher levels of driving put pressure on policy makers to accommodate more driving, and the cycle begins again (Figure 1).

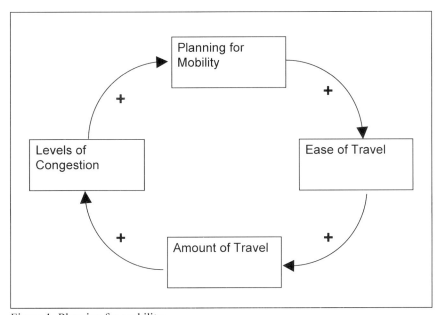

Figure 1. Planning for mobility

Although planning for mobility can be compatible with planning for accessibility, the traditional focus on mobility in transportation planning in the U.S. has over time helped to decrease accessibility. As a result of this emphasis, accessibility in the U.S. is largely mobility-dependent, and mobility in the U.S. is largely car-dependent. In the suburban areas of metropolitan regions, transit service is relatively sparse and destinations are generally beyond walking distance, leaving residents with no option but to drive. The result is a lower level of accessibility, at least for those who need or would like to travel by modes other than the automobile. But even for those residents who prefer to drive, accessibility is threatened. As traffic levels invariably increase in these areas, getting around by car becomes harder, and accessibility ultimately declines (Handy 1993).

Planning for accessibility, in contrast, means making it easier to get where you need to go. To plan for accessibility is to focus on the ends rather than the means and to focus on the traveler rather than the system: do people have access to the activities that they need or want to participate in?   Transit services that focus on linking specific groups of users to their desired destinations, such as reverse commute programs and other client transportation services, are an example of planning for accessibility. Land use policies designed to bring destinations within walking distance of residential areas are another example of planning for accessibility. Efforts like these reduce the need to drive, although they don't necessarily reduce actual driving. Either way, they help to break the planning-for-mobility cycle (Figure 2). Although such efforts are not new, they have largely played a secondary role to the primary focus on planning for mobility. As congestion levels continue to worsen and funding for capacity-expanding road projects dwindles relative to needs, however, interest in a planning-for-accessibility approach seems to be increasing.

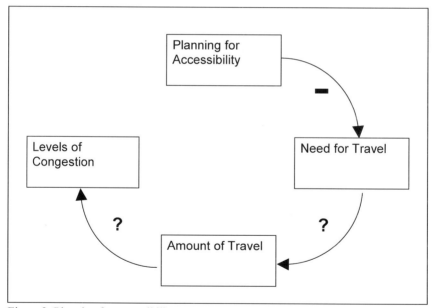

Figure 2. Planning for accessibility.

IN PRACTICE

To what degree do current transportation plans continue to emphasize planning for mobility and to what degree do they now reflect planning for accessibility? For most plans, it's hard to tell. Not only do most plans not clearly define these concepts, they may implicitly address accessibility or mobility without explicitly saying so. The orientation of a plan, whether a mobility orientation or an accessibility orientation, may be evident in several different places: the stated goals and objectives for the plan, the performance measures used to evaluate alternatives, and the kinds of investments and policies chosen.

As an initial step towards determining the degree to which regional transportation plans reflect an accessibility orientation, I reviewed four regional transportation plans from Northern California and their supporting documents (Figure 1, Table 1). This sample of plans, though too small to allow for generalization, nevertheless illustrates some of the different ways in which mobility and accessibility may appear in regional transportation plans. The four regions are increasingly connected from the standpoint of daily travel patterns, as the population commuting from San Joaquin and Merced Counties to the Bay Area and Sacramento and even from Sacramento to the Bay Area continues to grow. Significant variations in the orientations of the plans of these interconnected regions could thus be interesting. For each plan, I addressed three questions:

What do goals and objectives say about accessibility and mobility?

What kinds of performance measures are used to evaluate alternatives?

What kinds of strategies are included in the plan?

In addressing Question 1, I looked for direct as well as indirect mention of accessibility and/or mobility (Table 2). Any goals that used the terms "mobility" or "accessibility" were highlighted. Indirect evidence of mobility orientation included goals that focus on congestion reduction or improvement in travel times. Indirect evidence of accessibility orientation included goals that focus on increasing the ease of reaching specific destinations, reducing the need for travel, increasing travel choices, or addressing the needs of specific populations.

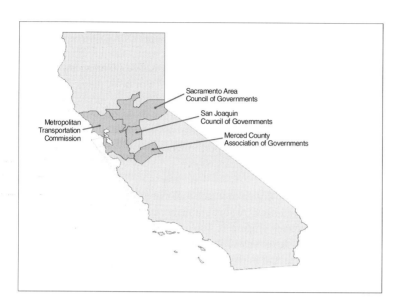

Figure 3. Regional plan areas.

TABLE 1 Sample of Regional Transportation Plans

MPO	2000 Population	Plan	Date
MTC - Metropolitan Transportation Commission	6.8 million	2001 Regional Transportation Plan for the San Francisco Bay Area	Amended November 2002; draft of updated plan available for review November 2004
SACOG - Sacramento Area Council of Governments	1.9 million	A Bold First Step for Mobility in the Sacramento Region: Metropolitan Transportation Plan for 2025	Adopted July 2002; updated plan due for adoption in June 2005
SJCOG - San Joaquin Council of Goverments	0.6 million	2004 San Joaquin Council of Governments' Regional Transportation Plan: Vision 2030	Adopted 2004
MCAG - Merced County Association of Governments	0.2 million	2004 Regional Transportation Plan - Merced County	Adopted August 2004

TABLE 2 Criteria for Distinguishing Mobility and Accessibility in Transportation Plans

	Goals	Measures	Strategies
Mobility	✓ Uses the term "mobility" ✓ Focuses on reducing congestion or improving travel times	✓ Traditional level-of-service measures ✓ Vehicle-miles travelled ✓ Average travel time ✓ Other measures of congestion	✓ Projects intended to increase road capacity to enable higher volumes of traffic ✓ ITS projects that aim to increase the capacity of the road system ✓ Transit projects that do not link key origins and destinations ✓ Bicycle and pedestrian projects that do not link key origins and destinations
Accessibility	✓ Uses the term "accessibility" ✓ Focuses on increasing the ease of reaching specific destinations ✓ Focuses on reducing the need for travel ✓ Focuses on increasing options for travel ✓ Focuses on meeting the needs of specific population groups	✓ Share of jobs or other destinations within specified travel times or distances ✓ Measures of travel options ✓ Measures that focus on the needs of specific population groups	✓ Land use strategies intended to reduce the need for travel and/or promote the viability of alternatives to driving ✓ Transit investments or services that clearly link key origins and destinations ✓ Bicycle and pedestrian projects that link origins and destinations

In addressing Question 2, I looked at the measures used to compare the performance of alternative plan scenarios. These scenarios generally include a no-build scenario as well as at least one and often more than one scenario consisting of a set of proposed improvements to the transportation system. As noted above, transportation plans were traditionally analyzed using the four-step travel demand model that produces measures of congestion. However, these models can be used to produce measures of accessibility as well, and plans may opt to evaluate scenarios using measures that are produced by means other than travel demand models. Measures of level of service, vehicle delay, congestion levels, and vehicle-miles-traveled were classified as mobility measures. Measures of the number or share of jobs or other activities or destinations within specified travel distances or times are categorized as accessibility measures, as are measures of travel choices and measures that focus on specific groups

Question 3 is the most challenging. To answer this question, I looked at the types of projects funded as well as other strategies included in the plan, and to the relative share of funding by project type (focusing on "Tier 1" funding, that is, funding that the MPOs believe is reasonably assured). Deciding what counts as a mobility-oriented project and what counts as an accessibility-oriented project is not always straightforward, however. As a starting point, I drew on previous work (Handy, 2002) in which I defined mobility-oriented strategies as including road expansion (either new roads or expansion of existing roads) and intelligent transportation systems (ITS) technologies designed to increase the effective capacity of the road system. Accessibility-oriented strategies included transit investments, investments in bicycle and pedestrian infrastructure, land use policies designed to increase the viability of these modes, and programs targeted to the needs of specific groups.

But these categories are overly simplified. Some road expansion projects can be accessibility-oriented, for example, if they aim to increase the ease of reaching particular destinations rather than to reduce congestion in general. Some transit projects can be mobility-oriented, for example, if by making use of available rights-of-way they do not efficiently serve important origins and destinations; trails projects designed for recreational purposes are often more mobility than accessibility oriented. In other words, projects must be categorized according to the goal towards which they aim. Thus, I defined mobility-oriented projects as road expansion and ITS projects intended to increase capacity or reduce congestion, as well as transit and bicycle/pedestrian projects that do not explicitly link origins and destinations; road maintenance projects were also included in this category. Accessibility-oriented projects include transit and bicycle/pedestrian projects that link origins and destinations, land use strategies designed to reduce the need for travel, strategies that focus on the needs of specific population groups, as well as road expansion intended to increase the ease of reaching specific destinations and ITS projects that focus on specific travel needs (Table 2). In reviewing the regional transportation plans, I attempted to discern these often subtle differences. Where these differences were not clear, I defaulted to the original categorization of strategies.

Bay Area

The San Francisco Bay Area is defined as the nine counties that border the San Francisco Bay. These nine counties encompass 100 cities, 7,000 square miles (18,000 km^2), and a total population of 6.8 million. By 2025, the population of the region is expected to reach 8.2 million. The Metropolitan Transportation Commission (MTC) adopted its current regional transportation plan in 2001. MTC is currently in the process of developing a new plan, and the draft was made available to the public in November 2004.

The 2001 plan lays out six broad policy goals: mobility, safety, equity, environment, economic vitality, and community vitality. Although mobility is not explicitly defined in the plan, the concept used in the plan appears consistent with the definition of mobility as offered in this paper. The aim of this goal is to "improve mobility of persons and freight." Specific objectives within this goal include improving travel time in congested corridors, but also improvements in transit, system reliability, and traveler information. Although accessibility is not specifically mentioned as a goal, it may be implicit in at least two of the other goals. The equity goal aims to "promote equity for system users," and the community vitality goal aims to "promote vital and livable communities." Within the equity goal, one objective is to provide equitable levels of transportation service for elderly, disabled, minority, and low-income persons. The community vitality goal includes encouragement of transit-oriented development and improvements in convenience and safety for bicyclists and pedestrians, objectives that are more oriented towards accessibility than mobility. In other words, accessibility is not an explicit goal of the plan, though several goals are consistent with an accessibility orientation.

Accessibility is used, however, a one of "a set of complementary measures" in evaluating the proposed transportation system. In fact, accessibility measures are used as performance measures for several of the goals in the plan (Table 3). Most notably, accessibility to jobs and shopping opportunities is used as a performance measure for mobility of people and freight. The *Performance Measures Report for the 2001 RTP* states that "Accessibility is a significant measure of mobility because transportation is rarely an end in itself; it is most often a means for getting to other activities" (pg. 9). Travel time between selected geographic origins and destinations could also be considered an accessibility measure rather than a mobility measure. Measures of accessibility are also used as performance measures for economic vitality, community vitality, and equity. In fact, accessibility measures dominate the list of performance measures, and traditional measures of mobility, such as levels of congestion, are not used. Although the stated goal is to improve mobility, the use of accessibility measures to evaluate the plan gives it an overriding accessibility orientation.

The accessibility measures used in the plan are variations of cumulative opportunities measures. For example, accessibility to jobs and shopping is defined as the share of all regional work and shopping opportunities that residents can reach within specified time intervals and is measured for driving, transit, and walking or biking (MTC 2001). Access of employers to the region's workforce is measured as the number of jobs that can be reached from home for employed residents within specified time intervals and is also measured for multiple modes. To measure performance with respect to equity, access to jobs is compared for communities with more than a specified share of minority or low-income residents and those with less. These measures are derived from the travel demand model.

The strategies outlined in the plan also suggest a shift in emphasis away from a mobility orientation toward an accessibility orientation. The website for the plan boasts that the plan provides "funding for dozens of congestion relief projects on Bay Area freeways" and sets aside "nearly $11 billion for new rail and bus projects that will improve mobility and enhance connectivity for residents throughout the region" (MTC, 2003). The first of these statements emphasizes mobility, but the latter while mentioning mobility suggests a concern with accessibility as well through the term "connectivity." On the mobility side, the plan includes widenings or other improvements to reduce congestion at key points in the freeway system and highlights maintenance of the existing network and improvements in system management, such as the Freeway Service Patrol (FSP) and call box network, and pavement management and traffic engineering technical assistance programs. These projects total about $3.9 billion out of a total of $8.6 billion of funding allocated in the plan (this total does not include funding already committed by law, voter mandates, or prior MTC actions).

TABLE 3 MTC Performance Measures by Goal

Goal	Performance Measures
Mobility of freight and people	Travel time Travel time between selected geographic origins and destinations Accessibility to jobs and shopping opportunities
Safety	No measures included
Economic vitality	Accessibility of region's work force to employers Economic efficiency of transportation investments
Community vitality	Population and employment within walking distancse of transit intermodal/rail stations Use of walking to access transit
Environment	Air quality and global warming - vehicle emissions
Equity	Comparison of changes for low-income and minority communities relative to other communities in: - Travel time - Accessibility to jobs - Transit travel time from target communities to major job centers

Source: MTC 2001

On the accessibility side, the plan emphasizes the expansion of regional transit and a regional bicycle master plan. Included in the Regional Transit Expansion Program are projects to extend BART from Fremont to San Jose and to build a Central Subway in San Francisco that would extend to Chinatown; these transit projects can be classified as accessibility oriented. The Regional Bicycle Master Plan aims to fill gaps in the regional network of bike routes and focuses bicycling as a means of transportation, reflecting an accessibility orientation. The plan also includes several innovative programs oriented toward accessibility: the Lifeline Transportation Program (described below), the Transportation for Livable Communities (TLC)/Housing Incentive Program (HIP) (also described below), the TransLink transit smart card, rideshare programs, and regional transit information and marketing programs. These programs total $4.8 billion, or about 56% of the total funding allocated in the plan.

A good example of an accessibility-oriented strategy is the Lifeline Transportation Program. This program focuses on the needs of low-income persons and consists of multiple components. The Low-Income Flexible Transportation Program (LIFT) includes improvements to existing fixed-route transit services as well as innovative alternatives where fixed route service is not viable, including demand-responsive van and taxi service and guaranteed ride home programs. The Lifeline Transportation Network, defined based on the region's public transit system, helps to target resources to spatial and temporal gaps in the network that affect low-income communities. A third element addresses transportation affordability and includes a pilot program to provide subsidized transit passes to students. These efforts help to increase mobility – by making it easier to get around – but their focus on meeting the needs of a specific segment of the population reflects an accessibility orientation.

The Transportation for Livable Communities (TLC) program and Housing Incentive Program (HIP) both aim to link land use planning and transportation planning and may help to increase transportation choices. The TLC program provides planning and capital grants for "small-scale transportation projects that enhance community vitality" (pg. 7). The HIP program "rewards cities for fostering compact housing with easy access to public transit lines" (pg. 7). The higher the density of the housing project, the higher the grant amount, and projects earn bonuses for affordable units. The guidelines for the HIP program call for cities to use the grants to fund TLC-type projects. As the website for the regional planning notes, "While focused on mobility investments, the Regional Transportation Plan also triples funding for MTC land use-oriented initiatives" (MTC 2003). By promoting development that facilitates transit and non-motorized modes, these programs represent a significant shift toward an accessibility orientation.

Sacramento

The Sacramento region comprises 6 counties and a population of 1.9 million as of 2000. The region is expected to grow rapidly, reaching a population of 2.8 million by 2025. The Sacramento Area Council of Governments (SACOG) most recently adopted a regional transportation plan in 2002. In October, 2004 SACOG adopted *The Final Interim Metropolitan Transportation Plan*, necessitated by a lapse in the conformity analysis for the 2002 plan. SACOG expects to adopt a new regional transportation plan in June 2005.

The current plan, the *MTP for 2025*, was subtitled "A bold first step for mobility." Although this title suggests a mobility orientation, the plan itself suggests at least a balance between mobility and accessibility. The "overarching" goal of the plan is "quality of life" (pg. 27). The plan does not define quality of life, noting that this concept means different things to different people; it thus puts focus on the needs of people rather than the needs of the system. It lists as its second goal "Access and Mobility" and aims to "Improve access to goods, jobs, services, housing, and other destinations" and to "Provide mobility for people and goods throughout the region, in a safe, affordable, efficient and convenient manner" (pg. 28). The Sacramento Plan is the only one of those reviewed here to define these terms: access is defined as "the ability to get somewhere" and mobility as "the ability to move easily and quickly to get there" (pg. 28); the plan notes that these concepts are interrelated. These definitions are generally consistent with those offered earlier in this paper, although "somewhere" as a destination is rather vague.

Consistent with a goal of enhancing both mobility and accessibility, the plan uses different performance measures for each. Mobility is measured by congestion levels, as determined by the travel demand forecasting model. The model produces measures of vehicle miles traveled, traffic congestion levels on roadways, transit travel times, and hours of delay in traffic jams. The plan also uses a congestion index that reflects time spent driving in congested conditions on a peak hour trip. These measures all reflect the ease of travel, the ability to move around. Accessibility is measured by "the ability to reach job centers in a reasonable period of time by auto and transit" (pg. 30) and also makes use of the travel time information produced by the travel demand model. The accessibility index reflects how many of the region's ten largest job areas can be reached from each residential community within 20 minutes driving time or 45 minutes riding transit (pg. 41). This index is an example of a cumulative opportunities measure.

Not only does this plan provide clear definitions of mobility and accessibility, with appropriate measures for each, the Environmental Impact Report (EIR) for the plan (as required by the California Environmental Quality Act) explicitly categorizes strategies in the plan as designed to address accessibility or mobility. The accessibility-oriented strategies as listed in the EIR include improvements to the transit system and bicycle and pedestrian projects. The proposed transit projects include commuter rail linking Sacramento to the Bay Area, extensions to the light rail system to serve key destinations such as the airport, and bus rapid transit in three commute corridors. Accessibility-oriented projects also include a number of "connector projects" – road improvements that connect specific communities to major freeways. In contrast, mobility-oriented projects are described as "projects aimed at reducing the most critical areas of congestion from a region-wide viewpoint" (pg. 27). Examples include carpool lanes on existing freeways, highway bypasses around outlying communities, and improved highway interchanges in urban areas. Intelligent Transportation Systems projects are also counted in the mobility category, as are transportation demand management strategies and local road improvements.

With respect to funding, the plan stresses that it "gives first priority to expanding the transit system" (pg. 5). The adopted plan includes $7.6 billion in funding for public transit, plus an additional $1 billion for "regional programs," including "clean air" projects, regional bicycle and pedestrian projects, community design plans and projects "to support smart growth," transportation demand management, and landscaping and other enhancements. The plan also allocates $281 million for local bicycle and pedestrian projects. The connector projects total $944 million, bringing the total for accessibility-oriented projects to $9.8 billion. The total for road, highway, and bridge projects other than the connector projects is $6.7 billion, with an additional $5.8 billion allocated for road maintenance, bringing the total for roads up to $12.5 billion. Although mobility-oriented projects thus garner a larger share of the funding, accessibility-oriented projects are well supported.

San Joaquin County

San Joaquin County is located directly south of the Sacramento region and to the east of the San Francisco Bay Area. Cities in the region, including Stockton and Tracy, are growing rapidly, and many residents commute to the Bay Area. With a 2000 population of just over half a million, the county is projected to nearly double in population by 2025, to just under 1 million. The San Joaquin Council of Governments (SJCOG) adopted a regional transportation plan, called *Vision 2030*, in 2004.

The 2004 plan continues the goals, performance measures, and policies of the previous plan. The plan lays out the general goals of increasing mobility, accessibility, safety, reliability, preservation of the transportation system, and the environment. More specific goals, along with corresponding objectives and policies, are outlined for the overall system and for specific modes. The overarching goal is to "design a transportation system that will enhance the quality of life in San Joaquin County" (pg. 2-1). Tied to this goal is a policy to "design a transportation system that will meet the travel needs of both citizens and businesses." This theme is echoed in the goals specific to the road system, where the need to develop a road and street system complementary to other modes of transportation is also noted. Objectives related to the expansion of the road system appear only within the goods movement category, where the plan recognizes the need for adequate road capacity to handle growing levels of freight traffic. The terms "access" and "mobility" appear in the specific goals only with respect to aviation, where the objective is to "improve ground access to the regional airports, with emphasis on alternatives to the single occupancy vehicle" (pg. 2-8). Although accessibility is not explicitly emphasized in the specific goals, they reflect a strong accessibility orientation.

The performance measures are grouped by a somewhat different list of goals: mobility, accessibility, cost-effectiveness, sustainability/system maintenance, environmental quality, environmental justice, and safety (pg. 2-13). In this list, the mobility goal is stated this way: "Transportation system should meet public need for improved access and for safe comfortable convenient and economical movement of goods and people." While the goal mixes accessibility and mobility concerns, the measures of mobility do not: daily VMT, daily vehicle hours of delay, and peak lane miles. The accessibility goal is defined as a transportation system that ensures "the ease with which transportation options are available" (pg. 2-13), thus emphasizing the importance of choice to accessibility. However, accessibility is measured by peak hour freeway travel speed, person trips by mode, farebox recovery ratios, transit frequency and timeliness, and numbers and distance of bus stops. Only the last two measures might be considered accessibility measures by conventional definitions.

With respect to strategies, the plan promotes "a 'balanced' multi-modal transportation system" and calls for "an increased investment in alternative transportation modes, while accommodating a necessary amount of new highway capacity" (pg. 4-1). For road and highway projects, the plan emphasizes the need for efficient movement of people and freight and focuses on congestion reduction and capacity expansion – a mobility orientation. Improvements to highways and roads account for 24% of funding and maintenance of roads 26%, adding up to 50% of the total of $3.8 billion in funding. For transit projects, the plan emphasizes concern for quality of life as well as air quality and discusses the need for improving both mobility and accessibility, although accessibility needs are not explicitly outlined. Projects focus on local, intercity, and interregional service, including commuter rail between the county and the San Francisco Bay Area to serve growing commuter demand. Although specific transit needs are not clearly defined, several planning studies are in progress. Complementary land use strategies are mentioned as important for the areas surrounding multi-modal transit stations. Transit gets a significant share of the funding pie, with 32% of total funding going toward local, intercity, and interregional bus service and another 11% going toward commuter rail. A bit of funding is directed toward bicycles, and transportation control measures are included in the plan, as required for air quality improvement efforts. In general, the accessibility-orientation of the transit projects included in the plan is not as clearly established as in the other RTPs reviewed.

Merced County

Merced County is located two counties south of San Joaquin County and is also experiencing "spillover" growth from the Bay Area, particularly from the Silicon Valley area. The 2000 population of 210,554 is expected to grow to 377,400 by 2025. The Merced County Association of Governments adopted its regional transportation plan in August 2004.

The plan outlines seven "vision themes" that provide the foundation for the plan. The vision themes stray beyond pure transportation concerns to emphasize economic development and environmental protection. The first goal listed is to provide "a good system of roads that are well maintained, safe, efficient and meet the transportation demands of people and freight." One objective under this goal is to "improve mobility and reduce congestion-related delays," while another is to "promote an efficient, linked system of interstate freeways, major streets, rail lines, public transit, bikeways and pedestrian paths that enhances accessibility and the movement of people and goods" (pg. 18). This goal thus mixes mobility and accessibility orientations. The second goal is to provide "a transit system that is a viable choice," with the objective of meeting "the individual needs of those who depend on public transit." This goal more clearly reflects an accessibility orientation. Another goal aims to "support orderly and planned growth that enhances the integration and connectivity of various modes of transportation" and aims to "provide a variety of transportation choices" (pg. 19). Thus, although traditional concerns with mobility are evident in the goals, they predominantly reflect an accessibility orientation.

The plan uses performance measures in 10 categories, including mobility, access, and connectivity (pg. 23). Mobility is measured as delay and peak hour level of service. Access is measured as time to destinations and time to transportation system. Although relatively simplistic, these measures are consistent with conventional measures of accessibility. Connectivity is measured as mode choice and land use integration. These measures also reflect an accessibility orientation, even if they are not labeled as such. In addition, access to employment centers is included as a measure of economic vitality. The performance measures thus mirror the orientation of the goals: some concern with mobility but a predominant if implicit concern with accessibility.

With respect to strategies, the plan emphasizes traditional mobility concerns but addresses accessibility concerns as well. Within the road strategies, the plan aims to maintain a level of service D on "all regionally significant roads," reflecting a mobility orientation. The funding allocation clearly favors roads, with $1.4 billion going to road and highway improvements, in contrast to $143 million for transit and $6 million for bicycles. However, policies in the plan address transit, passenger rail, non-motorized modes, and land use strategies, as well as roads. The plan outlines an "Unmet Transit Needs Process" to identify deficiencies in the transit system and recommends that customer services surveys be conducted every two to three years (pg. 52). One of the actions outlined within land use strategies is to assist local cities in developing and implementing design criteria "that make new commercial and residential developments friendly to pedestrian and bicyclists" (pg. 31), reflecting an accessibility orientation; another action within this category, however, is to help cities "identify necessary improvements that would improve traffic movement," reflecting a mobility orientation. The intent of the Regional Bikeway Plan, adopted in 2003, is to "connect major destinations throughout the County" (pg. 72), reflecting an accessibility orientation. Funding thus reflects a mobility orientation, while policies reflect a significant concern with accessibility.

CONCLUSIONS

With the exception of SACOG, these agencies do not seem to have articulated for themselves the distinction between planning for accessibility and planning for mobility. The plans do not define these terms, and they do not use them consistently throughout (Table 4). The MTC plan, for example, frequently uses the term "mobility," although its goals, performance measures, and strategies are largely accessibility oriented. The SJCOG plan doesn't say much directly about either mobility or accessibility, and while its performance measures are not accessibility oriented, some of its strategies are. The MCAG plan mixes accessibility objectives with mobility goals, but emphasizes accessibility in its performance measures and policies without really saying so. In contrast, SACOG defines these two terms, establishes specific goals for each, uses appropriate performance measures for each, and chooses appropriate strategies for each. The MTC, SJCOG, and MCAG plans may simply have a labeling problem, although the numerous inconsistencies in these plans suggest that planners have not made a clear distinction between the concepts of accessibility and mobility.

Together these plans offer a mixed assessment of the degree to which metropolitan planning organizations have adopted a planning-for-accessibility approach. A concern with accessibility is evident in all of these plans, although as an additional aim rather than as a replacement for a concern with mobility. All four plans reflect some concern with accessibility in their goals, their performance measures, and the kinds of strategies included in the plan. All four plans also reflect a concern with mobility in each of these elements as well. An emphasis on both accessibility and mobility is not necessarily contradictory: as noted earlier, good mobility can contribute to good accessibility, and a decline in mobility can reduce accessibility. What the plans do not quite commit to are the notions offered here that plans can enhance accessibility without increasing mobility and that providing for increased mobility has the potential to reduce accessibility.

It is perhaps not surprising that the size of the metropolitan region is negatively correlated with the share of the funding allocated to mobility-oriented projects (Table 4). The largest region, the San Francisco Bay Area, has the most significant congestion problems but also the most limited opportunities for expanding roadway capacity as a solution. Transit has long been a part of the arsenal of strategies for meeting transportation needs in the region. The smallest region, Merced County, faces a significant air quality problem but has only recently begun to feel the effects of congestion. Alternatives to driving are largely uncompetitive in this relatively low-density area. A larger survey of regional transportation plans would likely reveal a similar pattern: larger metropolitan areas are more likely to need – and more likely to see the benefits of – an accessibility-oriented approach.

A clearer understanding of these concepts and the distinctions between them will, I believe, lead to a further shift toward planning for accessibility. Besides the potential benefits outlined earlier in this paper, accessibility is a goal that most people can agree on and thus has important benefits for the planning process as well. A focus on accessibility helps move the policy discussion away from contentious issues associated with suburban growth, such as the right to choose a suburban lifestyle, and shifts the discussion towards a more productive focus on how transportation and land use policies can help make it easier for residents to go about their daily lives. The strategies that emerge from a planning-for-accessibility approach tend to have less environmental impact, may be significantly cheaper than those that aim to enhance mobility, and they offer a greater range of choice. In the long run, a further shift toward planning for accessibility rather than for mobility can produce better transportation systems for all these reasons.

REFERENCES

Capital Area Metropolitan Planning Organization (CAMPO) (2000). *CAMPO 2025 Transportation Plan*. Austin, TX. Adopted June 12. Available: http://www.campotexas.org/pdfs/2025adoptedplan.pdf. Accessed 11/3/04.

Chicago Area Transportation Study (CATS) (2003). *2030 Regional Transportation Plan for Northeastern Illinois*. Chicago, IL. Adopted October 9. Available: http://www.sp2030.com/2030rtp/index.html. Accessed 11/3/04.

Handy, S. (1993). A Cycle of Dependence: Automobiles, Accessibility and the Evolution of the Transportation and Retail Hierarchies. *The Berkeley Planning Journal*, **9**, 21-43.

Handy, S. (2002). *Accessibility vs. Mobility-Enhancing Strategies for Addressing Automobile Dependence in the U.S* Prepared for the European Conference of Ministers of Transport, Paris, France, May.

Handy, S. and D. Niemeier. (1997). Measuring Accessibility: An Exploration of Issues and Alternatives. *Environment and Planning A*, **29**, 1175-1194.

Hansen, W.G. (1959). How Accessibility Shapes Land Use. *Journal of the American Planning Institute*, **25**, 73-76.

Merced County Association of Governments (MCAG) (2004). *Regional Transportation Plan for Merced County*. Merced, CA. Adopted August 19. Available: http://www.mcag.cog.ca.us/publications/2004/RTP04final.pdf. Accessed 11/3/04.

Metropolitan Transportation Commission (MTC) (2001). *Performance Measures Report for the 2001 Regional Transportation Plan for the San Francisco Bay Area*. Oakland, CA. August. Available: http://www.mtc.ca.gov/projects/rtp/downloads/PM/Performance_Measures.pdf. Accessed 11/3/04.

Metropolitan Transportation Commission (MTC) (2002). *2001 Regional Transportation Plan for the San Francisco Bay Area*. Oakland, CA. Adopted December 2001/Amended November 2002. Available: http://www.mtc.ca.gov/projects/rtp/rtpindex.htm#rtp. Accessed 11/3/04.

Metropolitan Transportation Commission (MTC) (2003). Regional Transportation Plan (RTP). Available: http://www.mtc.ca.gov/projects/rtp/rtpindex.htm. Accessed 11/2/04.

Oxford English Dictionary (2002). *OED Online*. Oxford: Oxford University Press. Available: http://dictionary.oed.com/entrance.dtl. Accessed 5/9/02.

Pickett, Joseph, P. et al., editors (2000). *The American Heritage® Dictionary of the English Language: Fourth Edition*. Boston: Houghton Mifflin. Available: http://bartleby.com/61/. Accessed 5/9/02.

Sacramento Area Council of Governments (SACOG) (2002). Final Environmental Impact Report for the MTP for 2025. Available: http://www.sacog.org/mtp/eir/eir.pdf. Accessed 11/3/04.

Sacramento Area Council of Governments (SACOG) (2002). *The MTP for 2025: A Bold First Step for Mobility in the Sacramento Region*. Sacramento, CA. Adopted July. Available: http://www.sacog.org/mtp/pdf/mtpfor2025.pdf. Accessed 11/3/04.

San Joaquin Council of Governments (SJCOG) (2004). *Regional Transportation Plan: Vision 2030*. Stockton, CA. Available: http://www.sjcog.org/sections/trans-planning/rtp_pdf.php. Accessed 11/3/04.

Access to Destinations
D.M. Levinson and K.J. Krizek (editors)

CHAPTER 8

CURRENT DETERMINANTS OF RESIDENTIAL LOCATION CHOICES: AN EMPIRICAL STUDY IN THE GREATER COLUMBUS METROPOLITAN REGION

Moon Jeong Kim and Hazel Morrow-Jones, The Ohio State University

INTRODUCTION

In basic urban economics models, commuting cost has been considered the primary determinant of residential location relative to land cost. Giuliano and Small (1993) indicate that in the standard model residential locations are selected by households' tradeoffs between commuting cost and land cost. They note, however, that commuting is a limited factor in residential location, and other factors such as housing style and neighborhood characteristics are more important. The modern lifestyle is complex, dynamic, and diverse. Automobiles have become more affordable, and improved transportation systems have made commuting less of an issue when choosing a housing location. Many households choose a home based on two adults who commute to different work places. In addition, urban structure in most American metropolitan areas has changed from monocentric to polycentric (Cervero, 1998; Levine, 1998; Nelson, 1997; Filion, 1999) and the change has affected accessibility to jobs. Accordingly, residential location decisions might be influenced by a variety of factors, for example, school quality, individual housing preferences, and amenities of the neighborhood.

In this paper, we approach the determinants of residential location decisions from the point of view of individual preferences in order to develop an understanding of what criteria households are using to make their location decisions. We examine a sample of homeowners who moved recently to determine the importance of commuting and other accessibility features to their decisions (relative to other factors) and we examine which factors are affecting households' residential location choices. In addition, we explore the spatial differences in the importance of commuting and other factors to the movers' decisions.

LITERATURE REVIEW

In the standard model of urban economics, commuting cost has been considered an important determinant of residential location along with land cost (Alonso, 1964; Mills, 1972; Kain, 1961; Muth, 1969). Kain (1961, 1962, 2004) used a residential location model in which households determine their residential locations by a tradeoff between housing costs and commuting costs. The basic model was adopted and developed further by various people (James, 1974). Alonso (1964), Muth (1969), and Mills (1972) explain that as distance from the CBD increases, land rents decrease and transportation costs increase with income elasticity in a monocentric city, thereby locating higher income households in the suburbs (Mohan, 1979; Nelson, 1997). This model was simplistic even for the 1950s and 1960s, but when we consider the complex and diverse nature of modern life, affordable automobiles, and changing urban structure, commuting seems likely to be less important in today's residential location decisions. Muth (1985) in extending Kain's model noted that in real urban phenomena, non-CBD employment, nonresidential concentrations, and transportation costs (not just those caused by distance) modify the traditional monocentric model.

The original assumption of a monocentric city is also more unrealistic today. Most American metropolitan areas started from a monocentric city with a single central business district (CBD), containing most of the jobs. After World War II, American cities experienced aggressive suburbanization, and as purely residential suburbs grew, the commute to downtown became longer (Hayden, 2003). The suburbanization of jobs as well as housing led the monocentric city to a more polycentric city form, possibly changing commuting times as well. Levinson and Kumar. (1994) found that in a polycentric city, the Washington metropolitan region, commuting times were stabilized or reduced by "rational locators" (those who make rational location decisions in a city by balancing total costs and benefits in their decisions), despite increases in commuting distance and trip speed. In Levinson's later article (1998), he suggested that the suburbanization of jobs made commuting time stable by balancing accessibility and that urban structure "as measured by jobs and housing accessibility" would be an important factor in residential location.

The additional burdens placed on African Americans and other minorities by the combination of residential segregation and the movement of jobs to suburban areas are well known (Massey, 1993). People of color often face discrimination in both the housing and job markets and might be expected to have longer commutes. Ellis *et al.* argue that much more attention has been paid to the geography of residential location than of work location (Ellis, 2004) and the commuting relationship also deserves more attention.

Vandersmissen *et al.* (2003) argued that Levinson's results would be limited to large metropolitan areas. Their results for a medium sized city, Quebec City, indicated that commuting distances increased for both males and females, while commuting durations increased for males and decreased for females between 1977 and 1996. They concluded that the change from a monocentric city to a dispersed city is responsible for the increase of commuting time, in Quebec City, though they also noted spatial and social factors for the changes.

Several authors have explored the spatial decentralization of employment and found that it has not decreased commuting time and distance (Cervero, 1998; Levine, 1998). In their recent study, Levinson and Wu (2005) also found that commuting time increased in Twin Cities despite the stability in drive-alone commuting times in the Washington metropolitan region, and concluded that commuting time would depend on the spatial structure of the region. Even though the evidence on changes in commuting times and distances is mixed, it is clear that the change in urban form has affected the relations between residential location and commuting, and has made the determinants of residential location more complex.

While arguing the lessened impact of commuting, some researchers have tried to identify other determinants of residential location decisions. Giuliano and Small (1993) insisted that commuting would be a limited factor and other factors such as housing style and neighborhood characteristics would be more important. Filion *et al.* (1999) suggested that in dispersed cities residential location decision would be associated with a decline in the importance of space and an equivalent rise in that of place and proximity: "space (defined here as metropolitan-wide accessibility), place (home and neighborhood features) and proximity (the possibility to reach activities)." They showed that place features such as a nicer dwelling and better neighborhood, and proximity (convenient location) were the most important reasons for people's most recent moves in the Kitchener Census Metropolitan Area in Canada, while space features (e.g., change in employment) were relatively unimportant. The Ohio Housing Research Network (OHRN) (1994) argued that the major reasons for mobility in Ohio's seven major metropolitan areas were "schools, safety, property values, city services, and the preference for a larger, newer, or new home." Krizek and Waddell (2003) addressed the different lifestyles of households and clustered them according to travel patterns, activity participation, and residential location.

This brief discussion of some example from the literature indicates a wide range of opinions and ideas on the relationship between commuting and residential location choice. Most of the research results agree that commuting is less important than it used to be and that other factors may have more influence on residential location. In this paper we examine the extent to which commuting time changes when a household moves and whether the household considers commuting to have been important in its location decision. We agree with Giuliano and Small (1993) and Filion *et al.* (1999), among others, that other factors are more important. We test the role of proximity to family/friends and shopping, of housing characteristics and of neighborhood quality measures to try to discover what factors play an important role in residential location decisions.

METHODS

Data

In order to examine factors affecting residential location decisions we have selected a data set collected in the greater Columbus (Ohio) metropolitan region. Columbus is the 15th largest city in the US and the largest city in Ohio. The greater Columbus MSA is comprised of 6 counties —Delaware, Fairfield, Franklin (the central county), Licking, Madison, and Pickaway—and its population in 2000 was 1,540,157 people (U.S. Census Bureau). Our analysis also includes Union County (immediately adjacent to and northwest of Franklin County) because conceptually and geographically the county has been considered part of the region. As with most US metropolitan areas, the Columbus MSA has experienced strong suburbanization trends since World War II. For decades the metropolitan area remained essentially a monocentric city, even though several small cities in surrounding counties acted as local centers. However, the greater Columbus metropolitan area has been experiencing increasing urban sprawl, and has been transformed to a polycentric city with most of its growth to the North. The change of urban form in the region is making work-residence patterns increasingly complex.

The data used in this analysis are taken from a 1998 Homeowner Survey conducted by The Ohio State University and funded primarily by the Ohio Center for Real Estate Education and Research. The data are based on deed transfer records filed when a property is sold. Buyer and seller names from these deed transfers were matched and a data set created in which each case is a household that sold one home and bought another within the seven county metropolitan area during 1998. The data set included a great deal of information on the properties bought and sold, but nothing on the households. To remedy this problem a mail survey was sent to a sample of households in 1999. The present analysis is based on the responses to that survey (for more on the survey click on the housing tabs on the Center for Urban and Regional Development (CURA) web site: http://cura.osu.edu).

Bier and Howe (1998) found that this process of name matching works for 50 to 55 % of the repeat homebuyer cases in the Ohio metropolitan areas they studied (including Columbus). An examination of survey returns in Franklin County for this same survey in an earlier year (1996) indicted that the returns matched the geographic and price distribution of the homes sold quite well (no significant differences) but they slightly over represent higher priced and more suburban homes purchased (Morrow-Jones, 2000).

The survey data set provides information on a variety of factors that could explain homeowners' moving decisions. These factors include those related to housing, neighborhood, and public school characteristics as well as accessibility issues (e.g., commuting time). The survey excludes renters and first time buyers. The sample population is made up of people who were moving into at least their second owned home. Survey questions ask about the importance of characteristics of both the former (pre-move) and the current (post-move) home, neighborhood (including accessibility features) and schools (Appendix, Table A). The survey questions asked respondents to rate the importance of each factor on a 1 to 7 Likert scale, where 1 is very unimportant and 7 is very important. There were 4301 cases in the total buyer-seller matched data set (used in the maps and some of the descriptive analyses for this paper) and 2,080 households (48%) returned the survey (used in our analysis of reasons for movement and changes in commuting times).

The relatively high survey response rate is typical of this set of surveys (see the CURA web site for details). The surveys were mailed in Ohio State University (OSU) envelopes with cover letters on OSU stationery. There was an initial mailing and two follow up mailings to those who had not yet returned the survey. Those who returned the survey were entered into a drawing for a prize package that included a pair of OSU football tickets. These tickets are highly priced in the metropolitan area (even by those who have no interest in football) and this package almost certainly contributed to the high response rates.

Analytical Methods

To understand the region spatially, we followed Levinson and Kumar (1994) and divided the greater Columbus metropolitan region into three roughly concentric rings around the CBD: a central city ring, an older suburban ring, and a newer suburban ring (Figure 1). The older suburban ring was designated as those suburbs inside Interstate 270 (a perimeter freeway) that includes such communities as Upper Arlington, Worthington, Bexley, Grandview Heights, and Whitehall. For some of the following analyses, the communities in the three rings were further sub-divided to form seven groups based on their location and growth characteristics; Group 1 (CBD Area), Group 2 (Remainder of the City of Columbus), Group 3 (Older Suburbs Close to Job Opportunities), Group 4 (Typical Older Bedroom Suburbs), Group 5 (Newer Suburbs with More Residential Characteristics), Group 6 (Newer Suburbs with Employment Centers), and Group 7 (The Far Outlying Suburbs). Because the original survey data often identify each household by the name of its postal area, not necessarily of its community, zip codes approximating a community were designated as the community.

For the central city ring, zip code areas within or close to a 3-mile radius including the CBD were categorized as Group 1 (ZIP codes for each group are listed in the appendices). Because of the irregular shape of the City of Columbus (shown in the darkest color in Figure 1), all Columbus cases except the CBD were classified in Group 2. Bexley and Grandview Heights were classified as a separate group, Group 3. These two older suburbs are very close to downtown Columbus, and have not grown since WWII. Different factors might attract residents to these communities compared to the other three older suburbs. Upper Arlington, Whitehall, and Worthington were categorized as Group 4. These three communities are typical of older suburbs that grew dramatically after World War II but are now nearly landlocked.

Figure 1. Concentric Rings in Greater Columbus Metropolitan Region

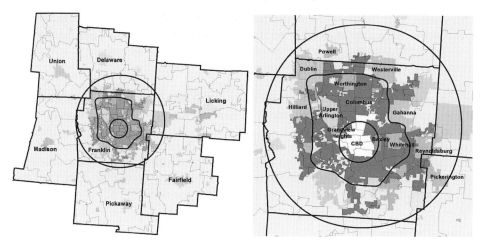

Note: Map was created using ESRI spatial data and Survey data set.
Coordinate System: NAD 1983 StatePlane Ohio South 3402 (Feet)

The newer suburban ring is within a 15 mile (24 km) radius primarily along and outside of Interstate 270 where growing employment centers are located. The area is experiencing vigorous growth, because of expanding employment centers (especially along the perimeter interstate) since the 1990s. Gahanna, Hilliard, and Powell have more residential characteristics and were aggregated into Group 5. Dublin and Westerville have seen very active growth with strong employment centers and were categorized as Group 6. Finally, the far outlying suburbs were categorized into Group 7. The Group 7 cases are sometimes omitted from the following analysis.

As the first step in our analysis, average commuting times before and after moving were calculated and compared for the whole region and each of the seven groups. Secondly, we created maps at the ZIP code level for the seven county region (these maps were created using Avenue Script, ArcView 3.3, and ArcGIS 8.3). Out of the total 4301 buyer-seller matched cases (not just those who returned the survey), 4072 cases (94.68%) were successfully mapped (the rest had errors during geocoding).

We also conducted statistical analysis based on the survey respondents' stated housing preferences, neighborhood preferences, views on the importance of accessibility issues, and school quality preferences using SAS v. 8.02 to discover what factors contribute to residential location decisions and whether determinants of residential choice vary across space. We calculated the 75th percentile for all variables in the whole region as well as each of the 7 groups. Finally, discriminant analysis was chosen to determine which factors discriminate among the three rings: the City of Columbus (Groups 1 and 2 combined), the Older Suburbs

(Groups 3 and 4 combined), and the Newer Suburbs (Groups 5 and 6 combined). Group 7 was omitted for the discriminant analysis. In this analysis, all 33 survey questions were used as independent variables and the ring location was the dependent variable. Although there are a variety of assumptions made when using discriminant analysis, it is a relatively robust technique and failing to meet all of the assumptions should not be fatal to the analysis (Klecka, 1980).[1]

ANALYSIS & RESULTS

Descriptive Statistics

Table 1 shows the proportion of all cases in each group and a description of commuting time as estimated by the survey respondents. (Detailed descriptive statistics of independent variables are shown in Table A of the Appendix). Almost 75% of households moved to homes outside the older suburban ring. Based on the valid survey responses, the average (post-move) commuting time of male respondents was 21.6 minutes and that of females was 17.8 minutes. In the region as a whole, commuting time for males was significantly longer than for females (t-test, $p<0.001$). As expected, post move commuting times of households living in the CBD area were the shortest: 10.8 minutes for males and 10.6 minutes for females. Except for the CBD area, all subgroups showed commuting times for males that were longer than for females (t-test, $p<0.05$), and as the distances of the communities from the CBD increased, average commuting times also generally increased as well.[2]

Estimated commuting times before moving and after moving were compared for those households who reported both (Table 2). Average commuting time in the seven counties remained stable for males before and after the move and that for females decreased by about 2 minutes after the move (t-test, $p<0.001$). In the CBD area, actual commuting time of male respondents decreased by half and that of females by over half. Also, within the older suburban ring, average commuting time for both males and females decreased after moves in each subgroup. The large decrease in commuting time for movers who chose the older suburban ring as well as the CBD may result from movers into the areas being especially interested in the benefits of being close to their jobs. On the other hand, commuting time for males of households who moved to newer suburbs, even with employment centers, showed no significant differences, while commuting time for females significantly decreased (t-test, $p<0.05$).

Table 1. Data Description & Commuting Time for the New Location

| | Category | Characteristics | No. of Cases | Percentage | Average Commuting Time (minutes) | | | |
					No. of Respondents	Male (SD*)	No. of Respondents	Female (SD)
Group 1	Central City Ring	CBD	59	1.37%	18	10.8 (11.4)	12	10.6 (7.3)
Group 2		Columbus	571	13.28%	184	17.9 (10.4)	183	16.1 (10.1)
Group 3	Older Suburb Ring	Close to CBD	124	2.88%	47	17.2 (10.9)	49	10.7 (8.2)
Group 4		Typical	395	9.18%	161	17.4 (13.3)	148	10.7 (9.6)
Group 5	Newer Suburb Ring	Residential	554	12.88%	202	20.5 (16.2)	210	17.1 (11.7)
Group 6		With Employment	632	14.69%	247	22.1 (13.2)	239	17.3 (12.9)
Group 7	The Rest	The Rest	1,966	45.71%	403	24.3 (14.2)	686	20.5 (14.4)
Total			4,301	100.00%	1,262	**21.6** (14.0)	1,527	**17.8** (13.0)

*SD: Standard Deviation

Table 2. Comparison between Previous and Current Commuting Times

| | Category | Characteristics | Male Commuting (minutes) | | | Female Commuting (minutes) | | |
			No. of Respondents	Previous (SD*)	Current (SD)	No. of Respondents	Previous (SD)	Current (SD)
Group 1	Central City Ring	CBD	10	14.3 (13.1)	7.2 (6.3)	6	22.5 (17.0)	9.2 (8.0)
Group 2	Older Suburb Ring	Columbus	96	21.3 (12.3)	18.0 (10.9)	112	20.4 (12.6)	17.2 (10.5)
Group 3		Close to CBD	19	22.4 (16.2)	16.6 (9.9)	33	16.8 (11.6)	9.8 (6.6)
Group 4		Typical	76	18.7 (11.9)	15.6 (11.0)	86	16.6 (12.0)	12.5 (9.1)
Group 5	Newer Suburb Ring	Residential	99	22.0 (17.3)	22.2 (19.8)	130	22.0 (12.9)	18.5 (11.7)
Group 6		With Employment	114	21.1 (11.3)	22.8 (14.1)	142	20.2 (13.8)	16.9 (11.4)
Group 7	The Rest	The Rest	335	22.1 (14.5)	23.7 (13.7)	403	21.3 (15.0)	21.6 (14.7)
Total			749	**21.4** (14.0)	**21.4** (14.4)	912	**20.5** (13.9)	**18.5** (13.0)

*SD: Standard Deviation

Spatial Analysis

Overall Moving Patterns

In addition to looking at numbers of movers to different areas, we wanted to be able to visualize the patterns created by the moves. To do this we created several maps of the whole seven county area. Our analysis focuses on the moves within a 15 mile (24 km) radius of the CBD. This covers most of the movement associated with Columbus and its suburban commuter-shed. The primary moves left out are focused on county seats in nearby counties (e.g. Newark in Licking County, to the northeast) that are still largely independent of the metropolitan area.

Figure 2 shows the overall patterns of mobility in the seven counties based on zip codes in 1998. Lines represent moving volume between the zip codes and grey colors represent moving volume within the same zip code area. Darker colors and/or thicker lines mean more movement. Because of the complexity of the patterns created by the lines, small volumes of moves (1 and 2 moves) were disregarded in showing the movement between the different zip code areas. Keeping in mind that small volume flows were disregarded, the downtown area and central part of the City of Columbus showed a small number of moves in 1998, while the surrounding areas of the City had a higher frequency of moves. Vigorous moving activity existed between the older suburb ring and the outer suburb ring. The boundary between Franklin County (in the center of the map) and Delaware County (directly north of Franklin) was especially active indicating the major direction of suburbanization in the region. Active suburbs include Westerville and Gahanna in the Northeast, Worthington and Powell in the North, and Upper Arlington and Hilliard in the Northwest. While there are relatively few moves in the South or Southeast, Pickerington and Reynoldsburg (Southeast) showed relatively larger amounts of movement.

Figure 3, net sales, delineates the moving trends in the region, showing that net sales were negative in most of the zip codes in Franklin County and positive in Delaware County, immediately adjacent to the north. (Darker colors mean that the number of repeat home buyers purchasing a home was larger than the number selling one. Lighter colors mean that more repeat home buyers sold a home than purchased one in the zip codes. In dotted areas there was no repeat homebuyer movement in the zip codes.) While housing sales were more active in the northern part of Franklin County, the net sales map shows that more homes were purchased than sold in Delaware County, particularly along the boundary of Franklin and Delaware Counties.

Losses of repeat homebuyers do not necessarily mean losses of population. After all, someone bought the homes these movers sold. Perhaps the purchasers were first time homebuyers or landlords adding to their rental portfolio. Or perhaps the homes were sold to investors who plan to use them for something other than housing. However the loss of repeat homebuyers does mean a loss of relatively high income, often family, households and can indicate problems for areas with net losses. Areas with large gains are usually seeing a lot of building and rapid growth, often with a different set of accompanying problems. For example, the zip code area which shows the largest gain in the region is the northern part of Westerville where the development has created the need for multiple high schools in the Westerville School District and has caused a great deal of traffic congestion.

Figure 2. Moving Patterns in Seven Counties in 1998

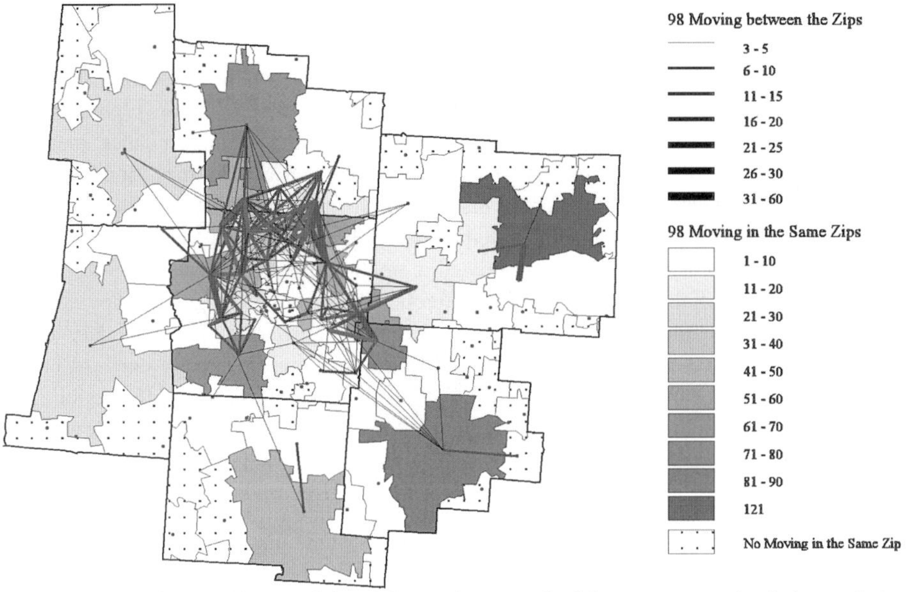

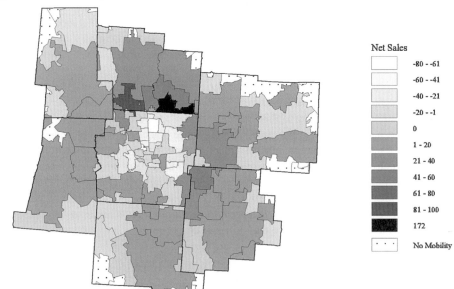

Note: Map was created using ESRI spatial data & Survey data set and polylines were created with Avenue Script.
Coordinate System: NAD 1983 StatePlane Ohio South 3402 (Feet)

Figure 3. Net Sales of Houses in Seven Counties in 1998

Note: Map was created using ESRI spatial data and Survey data set.
Coordinate System: NAD 1983 StatePlane Ohio South 3402 (Feet)

Statistical Analysis

Accessibility Considerations in Residential Location Choices

In order to examine the role of commuting in residential location decisions more directly, we analyzed the households' survey responses. Two questions in the survey asked respondents to rate the importance of distance to work. The first was in relation to the decision to sell their former home and the second in relation to the choice of their new home's location. Respondents were asked to rate the importance on a seven point scale from 1 (not at all important) to 7 (very important) with 4 specifically labeled as neutral. We assume that those who chose a number equal to or less than 4 (that is, "Neutral" or less important than neutral) can be regarded as people who do not consider commuting distance to be very important to their residential location choices. Those who answered "Not Applicable"[3] can also be thought of as moving for reasons other than commuting. They might be those who work at home, are retired from work, do not care about commuting, and so on.

Only 16.2% of respondents said that being closer to work was an important reason to sell their former home, while over twice that percentage of respondents (35.7%) said that it was not important and nearly 50% of people responded that their move was not related to distance to work (Table 3). On the other hand, when they chose their current home, 40.4% of respondents rated distance to work important (i.e. their response was greater than 4) and slightly more respondents (41.1%) considered distance to work not to be important. The survey also asked respondents if a change in job location was important in their move and 93.1% of respondents said that a change in job location was not associated with their move or

Table 3. Importance of Commuting in Moving Decision

Rating	Degree of Consideration	Wanted shorter distance to work (reason to sell previous home)	Percentage	Distance to work (reason to buy current home)	Percentage	Change in Job Location (reason to sell previous home)	Percentage
N.A.	Not Applicable	982	48.1%	374	18.5%	1,337	65.4%
1	Not Very Important	448	21.9%	206	10.2%	375	18.3%
2		56	2.7%	85	4.2%	43	2.1%
3		32	1.6%	77	3.8%	23	1.1%
4	Neutral	194	9.5%	461	22.8%	125	6.1%
5		121	5.9%	326	16.1%	34	1.7%
6		82	4.0%	225	11.1%	34	1.7%
7	Very Important	128	6.3%	265	13.1%	73	3.6%
Total		2,043	100.0%	2,019	100.0%	2,044	100.0%

it was not important. When people sell their homes and buy new ones in the same region, commuting distance or change in job location are not large factors in the decision to sell, but people tend to consider distance to work to be more important (though not overwhelmingly important) when they buy their new home.

Table 4 shows the ratings for accessibility features in choosing a new residential location. As mentioned above, the percentage of people who consider commuting distance to be important are roughly equal to those who do not at about 40% each. Distance to family and/or friends was important to about 40% of respondents as well. Over 60% of respondents said that being closer to more desirable shopping centers was not considered to be important in choosing their new home and only 18% rated the accessibility to shopping centers important. Thus, accessibility features do not appear to be very important in households' residential location choices.

Table 4. Accessibility Considerations

Accessibility Variables	Not Applicable	Not Important	Important	Total
Distance to work	374 (18.5%)	829 (41.1%)	816 (40.4%)	2,019 (100%)
Distance to family and/or friends	299 (14.8%)	924 (45.7%)	797 (39.5%)	2,020 (100%)
Wanted to be closer to more desirable shopping center	440 (21.8%)	1,218 (60.2%)	364 (18.0%)	2,022 (100%)

Factors Contributing to Residential Location Choices

The descriptive statistics in Table 5 (and Appendix A) show that people considered a variety of features when they chose their homes. Table 5 shows those features whose mean importance was at the 75th percentile or higher out of 33 items listed on the survey. The statistics are reported for the overall region and each of the seven geographic subgroups in the metropolitan area. For the region as a whole, housing characteristics such as quality of construction, floor plan, size of house, and other features are the most important factors in choosing a new home. Community and/or school features such as safety of the neighborhood and quality of schools' academic programs are ranked highly. None of the accessibility features appear among these most important factors for the region as a whole.

Table 5. Most Important Factors by Geographic Group

The Region		Group 1 CBD Area		Group 2 City of Columbus		Group 3 Older Suburbs Close to Work	
Features	Mean	Features	Mean	Features	Mean	Features	Mean
construction_house	6.19479	resale_value	6.38889	floor_plan	6.17176	sch_reputation	6.48214
floor_plan	6.12016	construction_house	6.05556	construction_house	6.15000	sch_academic	6.45455
size_house	6.10837	cost_house	6.05556	size_house	6.14231	neighbor_safety	6.34375
cost_house	6.04514	curb_appeal	6.00000	cost_house	6.00000	construction_house	6.23077
neighbor_safety	5.99062	size_house	5.55556	neighbor_safety	5.86640	resale_value	6.17460
resale_value	5.93831	neighbor_econ	5.41176	resale_value	5.73200	size_house	6.12500
sch_academic	5.74061	floor_plan	5.27778	curb_appeal	5.59375	cost_house	6.04615
curb_appeal	5.65863	neighbor_safety	5.18750	maintenance	5.36400	sch_safety	6.02041
sch_safety	5.55368	distance_work	5.12500	neighbor_econ	5.33884	floor_plan	5.98438

Group 4 Typical Older Suburbs		Group 5 Newer Suburbs with Residence		Group 6 Newer Suburbs with Jobs		Group 7 The Rest	
Features	Mean	Features	Mean	Features	Mean	Features	Mean
sch_academic	6.29730	floor_plan	6.30258	floor_plan	6.35714	construction_house	6.20307
sch_reputation	6.11446	size_house	6.30000	construction_house	6.30945	cost_house	6.04762
resale_value	6.05051	construction_house	6.22426	size_house	6.18447	size_house	6.03969
neighbor_safety	6.04737	neighbor_safety	6.12451	cost_house	6.11921	floor_plan	6.03842
size_house	6.04455	cost_house	6.02985	sch_academic	6.07522	neighbor_safety	5.93427
construction_house	6.00000	resale_value	5.98148	neighbor_safety	6.07192	resale_value	5.90315
cost_house	6.00000	curb_appeal	5.76981	resale_value	6.02280	curb_appeal	5.62980
floor_plan	5.93659	sch_reputation	5.76415	sch_reputation	5.94024	sch_academic	5.58454
sch_safety	5.79310	sch_academic	5.72589	curb_appeal	5.85479	sch_safety	5.49180

In order to compare importance among accessibility, housing characteristics, and community/school features, each of these nine most important factors were divided into three categories: "Not Applicable," not important, and important in Table 6 (using the same breakdown of the Likert scale used earlier). Over 90% of respondents rated the three housing characteristics (quality, floor plan and size) as important. Community features such as the safety of the neighborhood and whether the house is a good investment or will have good resale value also show high percentages (over 85%) of people regarding them as important.[4] Housing characteristics and community features are much more important in residential location choices than commuting and accessibility issues. School-related features appear somewhat less important because about 30% of respondents said that those issues are not applicable as a reason for their move. Those who answered "Not Applicable" probably do not have school-age children or their children attend private or parochial schools. Thus even though the percentage of respondents who consider school features to be important is lower than those who rate housing and community features highly, the quality of schools' academic programs and safety in schools should be considered to be important to specific markets (e.g., people who have school-age children). Housing characteristics and community/school features are more important than commuting and accessibility in the residential location choices of the movers in this sample.

Table 6. Possible Factors Contributing to Residential Location Choices

Housing Variables	Not Applicable	Not Important	Important	Total
Quality of construction of the house	19 (0.9%)	134 (6.5%)	1,899 (92.6%)	2,051 (100%)
Floor plan of house	15 (0.7%)	176 (8.6%)	1,863 (90.7%)	2,054 (100%)
Size of house	21 (1.0%)	173 (8.4%)	1,857 (90.5%)	2,051 (100%)
Cost of house	28 (1.4%)	208 (10.2%)	1,808 (88.5%)	2,044 (100%)
Curb appeal of house	43 (2.1%)	351 (17.2%)	1,641 (80.6%)	2,035 (100%)

Community/School Variables	Not Applicable	Not Important	Important	Total
Safety of the neighborhood	109 (5.4%)	183 (9.0%)	1,735 (85.6%)	2,027 (100%)
Good investment or resale value	57 (2.8%)	224 (10.9%)	1,770 (86.3%)	2,051 (100%)
Quality of schools' academic programs	607 (30.5%)	272 (13.7%)	1,112 (55.9%)	1,191 (100%)
Safety in schools	621 (31.3%)	312 (15.7%)	1,048 (52.9%)	1,981 (100%)

The 75th percentile for each of the seven geographic subgroups of Table 5 shows more interesting evidence related to the consideration of school features. Those respondents who chose the older suburbs (Groups 3 and 4) emphasized the importance of school reputation and quality of schools' academic programs along with neighborhood safety to their choices. In the Columbus metropolitan area, most of these communities are known for their good school districts and safe neighborhoods and 80% of these households contain school aged children (much higher than the CBD movers (20%) and somewhat higher than the other kinds of suburbs (between 60 and 70%)). Interestingly, good investment or resale value has the highest mean importance for the CBD group, and only those living in the CBD area (Group 1) included commuting distance as one of their most important factors. This goes with the fact that commuting time for both males and females who moved to the CBD decreased by over half (Table 2). Respondents who moved to newer suburbs (Groups 5 and 6) and to homes outside the 15 mile (24 km) ring (Group 7) showed the most interest in housing attributes such as floor plan, quality of construction, size, and cost of house.

In conclusion, commuting time and distance seem to be relatively unimportant and there are a variety of other factors affecting households' residential location decisions more strongly. Some previous studies have shown the importance of proximity to family/friends/relatives and other activities in residential location and/or neighborhood satisfaction (Brower, 2003; Lipsetz, 2000; Filion *et al.*, 1999), but we found that accessibility features were relatively unimportant to moving decisions in the Columbus region. In general housing features attract people when they choose their homes and community characteristics also affect households' decision. Presence of children in the household appears to play an important role for some households when they choose communities that have good school districts and high levels of safety. Households who move to the CBD are more likely to consider distance to work to be important and to shorten their commutes a great deal after the move.

Factors that Discriminate between Groups

We used discriminant analysis to determine how household ratings of important factors distinguish between the communities to which the households moved. Discriminant analysis is a technique related to linear regression that estimates a set of functions to show how well a group of independent variables (in our case the importance of a variety of reasons for locational choice) can distinguish between the categories of a categorical dependent variable (in our case, which of the three geographic rings the household chose for its home purchase). This analysis discriminates between the geographic subdivisions detailed above using a stepwise discriminant analysis (analogous to stepwise regression). Because of similarities in the factors and geography among some of the seven subgroups, we combined them to create three categories: those who moved to the City of Columbus (Groups 1 and 2 combined), those who moved to the Older Suburbs (Groups 3 and 4 combined), and those who moved to the Newer Suburbs (Groups 5 and 6 combined). We omitted Group 7 in order to focus on movers within and near Franklin County. This left the survey data set distributed across the three groups as: 61 in the City of Columbus (22.4%), 68 in the Older Suburbs (25%), and 143 in the Newer Suburbs (52.6%) (excluding cases with missing values on any variable). The final equations contain all variables that make a statistically significant contribution at the .05 level to discriminating between the three respondent groups. After eight steps seven independent variables remained.

As can be seen in Table 7, the importance of school reputation shows the highest statistical significance (F=12.61, p<0.0001) and the importance of curb appeal of house is shown as the second highest (F=7.69, p=0.0006). The remainder includes (in order) the importance of: mature trees, natural features, economic characteristics of the neighborhood, availability of community recreation and age of the housing unit. The importance of distance to work does not appear as one of the statistically significant variables distinguishing between location choices. This indicates that how important respondents feel commuting distance was to their choice of residential location does not differ by the home's location. These results imply that commuting issues do not distinguish between movers to different parts of this metropolitan area.

Table 7. Stepwise Selection Summary (City of Columbus, Older Suburbs, and Newer Suburbs)

Step	Number In	Entered	Removed	Partial R-Square	F Value	Pr > F	Wilks' Lambda	Pr < Lambda	Average Squared Canonical Correlation	Pr > ASCC
1	1	school_reputation		0.0857	12.61	<.0001	0.9143059	<.0001	0.0428471	<.0001
2	2	curb_appeal		0.0543	7.69	0.0006	0.8646799	<.0001	0.0687486	<.0001
3	3	mature_trees		0.0367	5.08	0.0068	0.8329779	<.0001	0.0858666	<.0001
4	4	natural_features		0.0526	7.38	0.0008	0.7891976	<.0001	0.1099251	<.0001
5	5	neighbor_econ		0.0307	4.2	0.016	0.7649305	<.0001	0.1242244	<.0001
6	6	com_rec		0.0270	3.67	0.0269	0.7442589	<.0001	0.1362176	<.0001
7	7	age_house		0.0246	3.31	0.0379	0.7259669	<.0001	0.1463028	<.0001

On the other hand, households who thought school reputation was very important to their move were more likely to move to older suburbs (school reputation has the largest discriminant function coefficient in the "Older Suburbs" column of Table 8), followed by newer suburbs with the central city last (this reflects the perception of low quality in the central city school district). Households who thought curb appeal was very important to their move were more likely to have moved to a newer suburb, but the central city was second. Households who wanted yards with mature trees were more likely to choose older suburbs, then the central city and the new suburbs were the least likely choice. Those who wanted to enjoy natural characteristics (ravine, waterfront, etc.) were more likely to move to newer suburbs as well, followed by the central city (where several highly regarded neighborhoods focus on ravines). Neighborhood economic characteristics were most likely to distinguish those who purchased in the central city, then the older suburbs. Households who valued community recreation resources were more likely to choose the older suburbs, followed by the newer ones. Finally those who found the age of the house important tended to buy in the newer suburbs, with the central city close behind. This may reflect an interest in new housing for the suburbanites and older housing for the city purchasers, since the central city has several very appealing gentrified older neighborhoods.

Table 8. Linear Discriminant Function Coefficients

Variable	City of Columbus	Older Suburbs	Newer Suburbs
Constant	-18.12552	-20.34063	-20.8013
school_reputation	0.55764	1.14293	1.02136
curb_appeal	2.60215	2.35534	2.67175
mature_trees	0.61741	0.85265	0.46717
natural_features	0.19931	0.02766	0.31444
neighbor_econ	2.19179	2.21175	2.02036
com_rec	0.03038	0.23563	0.17078
age_house	0.79039	0.58076	0.88064

The discriminant function provided adequate discriminating ability, though it was not exceptionally strong. The percent correctly classified showed over 50% error in the central city classifications (52.17%), 30.5% error in the older suburbs and 45.54% error in the newer suburbs.

Summary

Although some of our earlier analyses indicate that the Columbus metropolitan area is becoming more polycentric, it appears that to movers for whom commuting distance is important the metropolitan area is still focused on downtown. These households were more likely to locate in communities with easy access to the CBD (the CBD itself and the older suburbs) and they decreased their commuting duration significantly. However commuting itself was not considered to be important in their residential location decisions according to their responses in the survey results. If we asked where these people worked, we would probably find that most are employed downtown. The idea that the metropolitan area is spatially polycentric, but treated as a monocentric community by these movers brings up the interesting possibility that all households may treat their home areas as monocentric, but that each will have a different "center." Daily trip diaries would be a useful tool to use in studying this idea.

In the greater Columbus the region residential locations of repeat home buyers are determined more by households' preferences for housing features and community characteristics than by the trade-off between commuting cost and land cost. The results suggest that specific factors such as the presence of children can also affect households' decisions. Furthermore, although our results support the idea that the metropolitan area has become more polycentric other factors such as change of transportation mode and/or routes should also be examined in future research. Understanding these trends in the region will be critical for policy makers involved in land-use and transportation planning initiatives.

CONCLUSION AND POLICY IMPLICATIONS

As the analysis of the overall movement pattern analysis shows, the greater Columbus metropolitan area has been experiencing urban sprawl with continued outward movement of repeat homebuyers. The City has been transforming from a monocentric to a polycentric city and shows significant growth at the fringe, especially in the north. The change of urban form is contributing to making work-residence patterns diverse, and is associated with the general problems of sprawl. Comparing average commuting times before moving and after moving within the overall area, actual commuting time for males was the same (no statistical difference) and for females decreased by 2 minutes.

As we expected, when people chose their new residential locations, commuting and other accessibility features seemed to be less important than other aspects of the house and community for the majority of households. We found a variety of other determinants such as housing preferences (floor plan, quality of construction of the house, cost of house, and so on) and community characteristics (safety of the neighborhood and good investment or resale value). In addition, school features (school reputation and quality of schools' academic programs) are likely to affect households who have school-age children. Stepwise discriminant analysis provided evidence showing that school reputation, curb appeal of house, and several other factors distinguish between movers to different communities in the region.

It is important to keep in mind the limitations of the data. Our analysis is for one metropolitan area in one year and excludes all renters and first time home buyers. Arguably repeat home buyers may have the resources to be able to ignore commuting costs for the most part while renters and first time buyers may be more likely to take commuting costs into account in their residential location decisions. The name matching methodology loses some of the repeat home buyers, and in spite of the excellent survey response rate, the survey process loses more. The data set probably under-represents minority repeat home buyers and purchases in inner areas and of lower cost homes. These are all areas that future research should address. On the other hand, repeat buyers are a significant portion of any US housing market. They tend to have more resources and more knowledge and to be leaders in movement trends. Their patterns and attitudes have an impact on others in the housing market as well and their decisions have important things to tell us about access and residential choices.

The results of the study suggest that among repeat homebuyers the moving decision is complex and commuting is not necessarily one of the most important issues that households consider. The importance of housing features and community/school characteristics is not surprising but it has direct implications for land use-transportation planning and policy. By understanding growth patterns and the current determinants of these residential location decisions in the greater Columbus metropolitan area, policy-makers can better respond to the need for equity in development and the interests of households in the region. Research focused on minority, lower income or inner city/older suburban homebuyers is needed. Our data are not well suited to analyzing the first two, but they do allow us to say something about the last. Although we have not explicitly discussed the problems of older suburbs in this paper, Columbus' inner suburbs face some of the typical problems of these areas. Our results indicate that those suburbs also have strength on which they can capitalize (good schools, mature trees, safety and good locations for two earner households with access to both downtown and edge job centers). Most housing development is in newer suburbs and these areas suffer the growing pains of traffic congestion, rising taxes and heavy demands on infrastructure. Their attractions arise in part from newness, curb appeal and house characteristics.

Central cities have clear advantages for repeat homebuyers who are interested in short commuting times. This suggests an excellent opportunity for the CBD area to be marketed as a good home location for this group. Other research (Morrow-Jones, 2000) found that those interested in central city (not just CBD) purchases are more often childless and also appreciate the lower taxes of the central city over the suburbs. All of these features can be used as marketing tools for the CBD (for more on niche marketing of the city and related research see Morrow-Jones, Irwin and Roe, 2004).

This study finds that accessibility issues are less important than other features such as housing attributes and community/school characteristics, although we have not formulated a formal model of the role of all of these characteristics. However, the results of this study indicate a useful direction for research. Future research might focus on households' daily activity patterns and the extent to which those affect or are changed by moving decisions. The role of mode and route choice (e.g., freeways versus surface streets) should be considered as should the impact of gasoline prices and the implementation of major public policy initiatives such as light rail. These could have implications for outlying suburbs by changing commuting costs in significant ways. Inner communities, both older suburbs and central cities, should also consider capitalizing on their strengths to market themselves as good locations for repeat buyers since these are policies over which they have more control.

We started out with the hypothesis that commuting does not play a major role in housing location decisions. That hypothesis has been generally confirmed by our results with less than 20% of repeat home buyers having sold their home because of their commutes. Forty percent of buyers said that commuting distance was important in their subsequent purchase, however. Neither of these are among the most important reasons across the area as a whole. The most important reasons for selecting the new home appear to be house and neighborhood attributes. However it is also important to note the geographic variations we found and the likelihood of there being variations among subgroups of the population as well. We hope that this paper will contribute to the development of a more complete understanding of residential location decisions in US metropolitan areas.

REFERENCES

Alonso, W. (1964). *Location and Land Use.* Harvard University Press, Cambridge, MA.

Bier, T., and S. Howe (1998) Dynamics of Suburbanization in Ohio Metropolitan Areas. *Urban Geography,* **19,** 695-713.

Brower, S. (2003). *Designing for Community.* University of Maryland Press, College Park, Maryland.

Cervero, C., and K. Wu (1998). Sub-centering and Commuting: Evidence from the San Francisco Bay Area, 1980-1990. *Urban Studies,* **35,** 1059-11076.

Ellis, M., R. Wright, and V. Parks (2004). Work Together, Live Apart? Geographies of Racial and Ethnic Segregation at Home and at Work. *Annals of the Association of American Geographers,* **94,** 620-637.

Filion, P., T. Bunting, and K. Warriner (1999). The Entrenchment of Urban Dispersion: Residential Preferences and Location Patterns in the Dispersed City. *Urban Studies,* **36,** 1317-1347.

Giuliano, G., and K. A. Small (1993). Is the journey to work explained by urban structure? *Urban Studies,* **30,** 1485-1500.

James, F. J. (Ed.) (1974). *Models of Employment and Residence Location.* Center for Urban Policy Research. Rutgers University, New Brunswick.

Kain, J. F. (1961). *The Journey to Work as a Determinant of Residential Location.* Rand Corp, Santa Monica.

Kain, J. F. (1962). The journey to work as a determinant of residential location. *The Regional Science Association, Papers and Proceedings,* **9,** 137-160.

Kain, J. F. (2004). A Pioneer's Perspective on the Spatial Mismatch Literature. *Urban Studies,* **41,** 7-32.

Klecka, W. R. (1980). *Discriminant Analysis.* Sage Publication Inc., N.Y.

Krizek, K. J., and P. Waddell (2003). Analysis of Lifestyle Choices: Neighborhood Type, Travel Patterns, and Activity Participation. *Journal of the Transportation Research Board, Transportation Research Record,* **1807,** 119-128.

Levine, J. (1998). Rethinking Accessibility and Jobs-Housing Balance. *Journal of the American Planning Association,* **64.**

Levinson, D., and Y. Wu (2005). The rational locator reexamined: Are travel time still stable? *Transportation,* **32,** 187-202.

Levinson, D. M. (1998). Accessibility and the Journey to Work. *Journal of Transport Geography,* **6,** 11-21.

Levinson, D. M., and A. Kumar (1994). The rational locator: Why travel times have remained stable. *Journal of the American Planning Association,* **60,** 319-32.

Lipsetz, D. A. (2000). The Ohio State University, Columbus.

Massey, D. S., and N. A. Denton (1993). *American Apartheid.* Harvard University Press, Cambridge, MA.

Mills, E. S. (1972). *Studies in the Structure of the Urban Economy.* Johns Hopkins University Press, Baltimore.

Mohan, R. (1979). *Urban Economic and Planning Models.* Johns Hopkins University Press., Baltimore.

Morrow-Jones, H. A. (2000). In *The 2000 Thinning Metropolis Conference Cornell University*.

Muth, R. F. (1969). *Cities and Housing.* University of Chicago Press, Chicago, IL.

Muth, R. F. (1985). Models of Land-Use, Housing, and Rent: An Evaluation. *Journal of Regional Science,* **25,** 153-164.

Nelson, A. C., and T. W. Sanchez (1997). Exurban and Suburban Households: A Departure from Traditional Location Theory? *Journal of Housing Research,* **9,** 249-276.

OHRN. (1994) The Ohio Housing Research Network, The Ohio Urban University Program.

Vandersmissen, M., P. Villeneuve, and M. Theriault (2003). Analyzing Changes in Urban Form and Commuting Time. *Professional Geographer,* **55,** 446-463.

APPENDIX

ZIP codes and communities they are taken to represent:

Group 1 (CBD Area): 43201, 43211, 43215, 43203, 43205, 43206 and 43222

Group 2 (Remainder of the City of Columbus): ZIP codes not needed for this definition

Group 3 (Older Suburbs Close to Job Opportunities): Bexley (43209) and Grandview Heights (43212)

Group 4 (Typical Older Bedroom Suburbs): Upper Arlington (43220, 43221), Whitehall (43213) and Worthington (no zip code necessary)

Group 5 (Newer Suburbs with More Residential Characteristics): Gahanna (43230), Hilliard and Powell (no ZIP codes needed for the latter two)

Group 6 (Newer Suburbs with Employment Center): Dublin and Westerville (no ZIP codes necessary)

Group 7 (The Far Outlying Suburbs): No ZIP codes necessary

Table A. 1998 Home Owner Survey

Please circle the number that best represents how important the following items were when **choosing your current house and neighborhood**. For example, if the item was not very important when choosing this home or neighborhood, circle 1. If on the other hands, the item was very important when choosing this home or neighborhood, circle 7. If the item does not apply to you, circle 0. If you do not know, circle DK.

Independent Variables	No. of valid cases	No. of missing cases	Mean	Standard Deviation	Detailed Question
age_house	1,944	2,357	4.94	1.84081	Age of house
size_house	2,030	2,271	6.11	1.21112	Size of house
floor_plan	2,039	2,262	6.12	1.15645	Floor plan of house
construction_house	2,033	2,268	6.19	1.01680	Quality of construction of the house
curb_appeal	1,992	2,309	5.66	1.34390	Curb appeal of house
cost_house	2,016	2,285	6.05	1.17224	Cost of house
lg_yard	1,690	2,611	4.72	2.11907	House has larger yard
sm_yard	1,197	3,104	3.58	2.25301	House has smaller yard
landscaping	1,828	2,473	4.32	1.75450	Quality of landscaping
natural	1,455	2,846	4.37	2.04204	Other special natural characteristics (ravine, waterfront, etc.)
mature trees	1,625	2,676	4.50	1.98489	Yard has mature trees
with_deck	1,294	3,007	3.90	2.11388	House has a deck
accessibility	1,262	3,039	3.79	2.36031	House has accessibility features (for example, no stairs, lower counters, etc.)
resale_value	1,994	2,307	5.94	1.24866	Good investment or resale value
maintenance	1,968	2,333	5.36	1.60973	Ease of maintenance of house
sch_reputation	1,566	2,735	5.55	1.78495	Reputation of schools
traffic	1,870	2,431	5.39	1.43564	Traffic in the neighborhood

[5]neighbor_econ	1,872	2,429	5.39	1.36698	Economic characteristics of the neighborhood
neighbor_racial	1,573	2,728	3.48	1.77895	Racial composition of the neighborhood
neighbor_safety	1,918	2,383	5.99	1.15782	Safety of the neighborhood
property_tax	1,865	2,436	4.70	1.51676	Property taxes too high
local_services	1,865	2,436	4.99	1.57584	Local services (for example, garbage collection, fire or police protection, water, etc.)
distance_work	1,645	2,656	4.43	1.84216	Distance to work
distance_ff	1,721	2,580	4.40	1.84040	Distance to family and/or friends
closer_shopping	1,582	2,719	3.29	1.89724	Wanted to be closer to more desirable shopping areas
com_rec	1,607	2,694	3.87	1.88890	Community recreational opportunities
sch_academic	1,384	2,917	5.74	1.67450	Quality of schools' academic programs
sch_athletic	1,332	2,969	4.90	1.84950	Quality of schools' athletic or extracurricular programs
sch_racial_com	1,206	3,095	3.46	1.68960	Racial composition of school student body
sch_econ_stat	1,259	3,042	3.72	1.63770	Economic status of student body
sch_programs	1,279	3,022	4.62	1.88531	Special programs in schools (for example, gifted programs, arts, sciences)
sch_safety	1,360	2,941	5.55	1.72973	Safety in schools
sch_building	1,353	2,948	5.21	1.70434	Quality of school district's buildings and facilities

NOTES

[1] The data are not normally distributed (according to the Shapiro-Wilk W statistic). Since chi-square is significant at the 0.05 level, a within covariance matrix is used.

[2] Asking respondents to estimate their commuting time runs the risk of creating unreliable data if the respondents estimate inaccurately. However it can be argued that the perception of commute distance is more important than actual distance in making a residential location decision. Since we do not know work locations for the respondents we are unable to check the accuracy of the estimates, so the reader is cautioned to keep in mind the source of the data when interpreting the results.

[3] Descriptive statistics on this "not applicable" group support the possibility that it is largely made up of retirees. About half are over 60, 15% are widowed and 71% have no children at home. All of these figures are much higher than the data set's averages.

[4] We grouped good investment or resale value as community characteristics because they are so heavily influenced by neighborhood.

Access to Destinations
D.M. Levinson and K.J. Krizek (editors)

CHAPTER 9

EVALUATING MEASURES OF JOB-HOUSING PROXIMITY: BOSTON AND ATLANTA, 1980–2000[1]

Jiawen Yang and Joseph Ferreira Jr, Massachusetts Institute of Technology

INTRODUCTION

Balanced growth has been a major policy component of the 'smart growth' initiative. Increased congestion, particularly in suburban areas, has been linked to numerical imbalances and qualitative mismatches between jobs and housing. Balanced growth that improves job-housing proximity (that is, the spatial proximity between workplace and residence) is believed to have the potential to reduce commuting time and distance (Cervero, 1989).

Different studies, however, contain contrasting arguments on the commuting impacts of job-housing proximity, using different measures for the spatial relationship between workplace and residence. For instance, Guiliano and Small (1993), in their studies of Los Angeles, find that the impacts of job-housing proximity on commuting are weak, if job-housing proximity is measured with minimum required commuting (MRC). Shen (2000), using demand-adjusted job accessibility to represent residence's proximity to workplaces, finds that in Boston, average residence commuting time at the level of transportation analysis zones (TAZ) is strongly affected by job-housing proximity.

In this research, we hypothesize that different selection of job-housing proximity measures can bring about a different quantitative relationship between job-housing proximity and commuting. Before we ask questions about whether job-housing proximity can explain commuting or to what extent commuting length relies on job-housing proximity, we should ask a more fundamental one: How can we characterize current urban development pattern in terms of job-housing proximity? This question is important because a weak representation of job-housing proximity would result in not only a weak quantitative relationship between commuting and job-housing proximity, but also a poor performance of urban growth strategies guided by the inferior measure.

Existing studies of commuting length in American metropolitan areas have mainly used three categories of measures of job-housing proximity. They are the ratio of jobs to employed residents, JER (Cervero, 1996; Peng, 1997); job or labor accessibility (Levinson, 1998; Shen 2000; Wang 2001); and minimum required commuting, MRC (Guiliano and Small, 1993; Horner, 2002; Rodriquez, 2004). One can refer to Horner (2004) for a conceptual background of these measures and a summary of the existing studies using the three types of measures. Although existing studies have justifications of why a particular category of measure or a particular format is preferred, none of them has presented comparative empirical evaluation of different measures.

In order to address this issue, we offer a qualitative assessment and empirical examination of the three categories of measures, revealing their possible weakness regarding their ability to relate job-housing distribution to commuting. We first use the 2000 journey-to-work data from Census Transportation Planning Packages (CTPP) for the Boston metropolitan area. We compare the spatial patterns of job-housing proximity, represented by different measures, as well as their relationship to commuting length. Since different observations on the commuting impacts of job-housing proximity might stem from the selection of different regions or the selection of different years for the same region, we supplement the Boston 2000 data with consistent journey-to-work data for 1990 and 1980. Furthermore, we use the same CTPP datasets for the Atlanta metropolitan area, which is of almost the same size as Boston but with contrasting land development in terms of urban forms.

MEASURES OF JOB-HOUSING PROXIMITY

In broad terms, we define job-housing proximity as the spatial relationship between workplace and residence. So in this paper, job-housing proximity, low or high, reflects the geographical conditions of the job and labor markets. The distribution of jobs represents the demand for labor in a two dimensional space. The distribution of housing units approximates the supply of the labor force in terms of the workers' residences. This broad concept covers the terms that have been used to study the commuting impacts of urban development patterns, including urban spatial structure, job-housing balance, accessibility and spatial mismatch. The associated measures include different formats of JER, Accessibility and MRC. These three categories of measures vary in terms of the approach to measure the geographical conditions of job supply and labor supply. We will first define these measures, and then compute them for Boston and Atlanta.

Boston and Atlanta are two sizable but contrasting regions. Boston covers an area of 7,340 km^2 and Atlanta covers 11,470 km^2. In 2000, there are 2.3 million jobs and 2.1 million employed residents within the Boston metro area, and 1.9 million jobs and 2.0 million employed residents within the Atlanta metro area. During the past two decades, both Boston and Atlanta have experienced increased commuting duration while the economy and population have grown and their urbanized areas have expanded (Ferreira and Yang, 2004). The Atlanta region is ranked first and the Boston region is ranked seventh in terms of commuting time increase from 1990 to 2000 (McGuckin and Srinivasan 2003). Boston has kept a prosperous central city while new development, at an annual rate of 30 thousand jobs and workers, has dispersed to the inner and outer ring roads: Route 128 and I-495. Atlanta has grown much faster than Boston, with an annual growth rate of 64,000 jobs and workers. Atlanta has doubled its population—primarily through low-density suburban development. Atlanta is now almost the same size as Boston, but its growth has been less balanced with more striking differences between the central city and suburbs and between north and south parts of the region (Ferreira and Yang, 2004; The Brookings Institution, 2000). The boundary and major roads in Boston and Atlanta are illustrated in Fig. 1. Downtown Boston is around the intersection between I-93 and I-90. Downtown Atlanta is along the merged segment of I-75 and I-85, slightly north of I-20.

Figure 1. Boundary and major roads for the two metropolitan areas

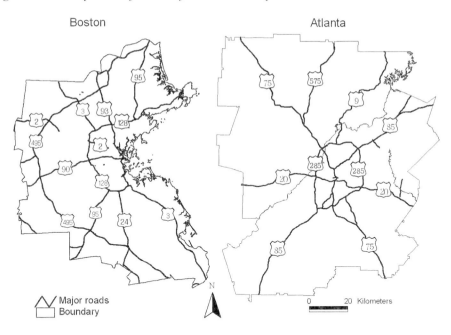

We use spatially disaggregated CTPP data for three decades (1980, 1990 and 2000) to compute the three categories of measures. We expect the same job-housing distribution, when represented by different measures, to appear different. To save space, we do not present maps and numbers for the two regions for each year. Instead, we use data for Boston 2000 as the major source to illustrate our points.

Ratios of jobs to employed residents (JER)

Among the three categories of job-housing proximity measures, JER is the easiest one to compute. It represents the workplace-home relationship with a simple ratio of jobs to employed residents. Information on number of jobs and employed residents are available in census data. The existing criticism of JER focuses on its selection of the geographical level of analysis units and the geographical scope of job and labor markets. For example, the selection of administrative units as the analysis units and the geographical scope of the job and labor markets is valued as a convenient way to offer land use planning information to policy-makers of local jurisdictions (Cervero, 1989). The weakness associated with this choice has been commented from three perspectives. First, coincidence of the boundary of the analysis unit and that of the labor and job market is never the reality (Peng, 1997). In addition, the analysis units are too large. Large analysis units tend to be self-contained by nature. For example, an entire metropolitan region is balanced by definition regardless of how its internal structure impacts its commuting pattern. At the county level, research by Giuliano (1991) reveals the sequential growth of population and employment moves toward balance over time, no matter how commuting length changes. Finally, variation in commuting length is significant at the neighborhood level. Measures grouped by local jurisdictions are too aggregated to reflect this neighborhood level variation (Shen, 2000).

In order to address the above criticism, improved JER measures reduce the analysis unit to a neighborhood level. Using GIS methods, a floating catchment area is constructed and attached to each analysis unit to represent the geographical scope of the job and labor markets. Each catchment area is defined as a buffer zone around the neighborhood, with a radius close to the average commuting distance (Peng, 1997). This improvement makes the measure more sensitive to the change of job-housing distribution, though the definition of the size of catchment areas is subject to arbitrariness and each catchment area is still viewed as self-contained.

For this paper, we compute JER for each census tract in two different ways: floating catchment areas composed of the 10 closest tracts, and catchment areas composed of tracts whose centroids are within 10 km buffer of the target census tract. Fig. 2 shows the results for Boston 2000.

As seen in Fig 2, the two formats of JER have similar spatial patterns[2]. Jobs are better supplied (higher JER) in the downtown as well as the areas around the first ring road, route 128, particularly the areas close to route 2 and route 3. Several places along route I-495 also have relatively high JER. This pattern represents the current development situation in Boston.

Figure 2. Job-housing proximity by JER

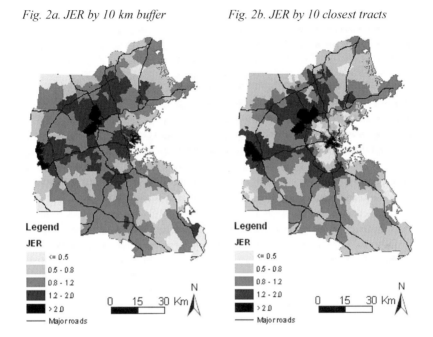

Fig. 2a. JER by 10 km buffer   Fig. 2b. JER by 10 closest tracts

Gravity type accessibility

As an alternative to measuring job-housing proximity, gravity type accessibility measures avoid the clear-cut artificial catchment areas of JER by weighting opportunities with a spatial decay function. Accessibility scores are usually computed at the neighborhood level with the region viewed as an integrated market of jobs and labor. These measures count the number of activities available at a given distance from the origin, and discount that number by the intervening travel cost. Exponential functions are commonly used to discount travel distance.

Accessibility measures raise issues with respect to the arbitrariness of what to include in the opportunity set, and which travel impedance function to use (Morris *et al.*, 1979). Nevertheless accessibility is an important measure in defining and explaining regional form and function (Wachs and Kumagai, 1973). Researchers, including Hanson and Schwab (1987), Shen (1998), and Wachs and Kumagai (1973), represent spatial structure by measuring the level of job accessibility using various gravity formulations. Recent advance in accessibility presentation, methodology and applications are well summarized by Kwan and her colleagues (2003). A typical way to compute gravity-type accessibility is presented with the following formulas.

$$DAA_i = \sum_j \frac{O_j f(C_{ij})}{UA_j}$$

$$UA_j = \sum_k P_k f(C_{jk})$$

$$f(C_{ij}) = \exp(-\beta * C_{ij})$$

Where DAA_i is the demand-adjusted job accessibility for zone i, opportunity O_j is the number of jobs supplied in zone j, UA_j is unadjusted labor accessibility (or competitive labor force) for zone j, P_k is the number of workers residing in zone k, β is the spatial decay parameter, and C_{ij} is the travel impedance (that is, travel cost measure) between zones i and j. (To compute demand adjusted **labor** accessibility, swap the job and labor terms in the formulae.)

We compute unadjusted job accessibility (job UA) and unadjusted labor accessibility (labor UA), demand-adjusted job accessibility (job DAA), and demand-adjusted labor accessibility (labor DAA) for each census tract. For travel impedance, C_{ij}, we compute the shortest route distance between each pair of census tracts, using the major road layer from ERSI. The spatial decay parameter is set to 0.1, which means that a worker's likelihood of working in a particular workplace decreases by 10% as the distance between the workplace and the worker's residence increases by one kilometer. Fig. 3 shows job-housing proximity as represented by job DAA and labor DAA. To save space, we do not present maps of UA because it is easy to envision that UA has the highest value in the urban core. In addition, UA measures accumulated opportunities rather than job-housing balance. Therefore, our evaluation of accessibility measure focuses on DAA. But we still present some UA result because UA is still a valuable measure to help understand urban spatial structure.

The spatial patterns by DAA are different from those by JERs. The core of the region has the peak value for both job DAA and labor DAA. DAA scores generally decrease for locations farther from downtown, with minor local variations. Both labor DAA and job DAA have scores greater than one at the urban core, and the job DAA score is higher than labor DAA since jobs tend to be less dispersed than housing.

Figure 3. Job-housing proximity by DAA

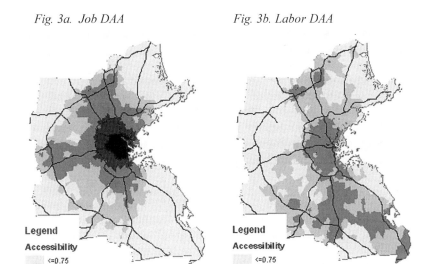

Fig. 3a. Job DAA   *Fig. 3b. Labor DAA*

Minimum required commuting (MRC)

MRC is the minimum commuting distance required by the underlying job-labor distribution. It was first introduced by White (1988). MRC relies on a linear optimization model that minimizes total commuting cost by matching jobs and employed residents. This model was widely used to study transportation in a polycentric urban space, and to understand why people commute more than required by the underlying land use patterns (Horner, 2002; Rodriguez, 2003). To compute MRC we need to determine the assignment of workers to jobs that minimizes the total travel cost across all assignments:

$$\text{Minimize } Z = \sum_i \sum_j c_{ij} x_{ij}$$

$$\text{Subject to: } \sum_j x_{ij} = N_i$$

$$\sum_i x_{ij} = E_j$$

$$x_{ij} \geq 0$$

Where Z is the total travel cost, N_i and E_i represents the total number of workers and number of jobs in zone i, X_{ij} is the number of workers living in zone i and working in zone j, and C_{ij} is the travel cost between zones i and j.

After solving this assignment model, the minimum value of Z, divided by the number of workers, is MRC at the metropolitan level. MRC_i for zone i can be obtained by averaging the travel costs for the minimum travel assignment, weighted by the commuting flow, from zone i to all other zones (when zone i is viewed as a home site), or from all other zones to zone i (when zone i is viewed as a job site).

Timothy and Wheaton (2001) points out that the resulting commuting pattern would have no cross commuting. Otherwise, one can always switch those crossed matched commuters to reduce total commuting cost. It is easy to envision that the MRC measure is sensitive to various features of urban sprawl, such as land use utilization rates, development discontinuities, and land use homogeneity. Lower land use densities, higher development discontinuity, and higher land use homogeneity will tend to generate larger values of MRC.

The above model can be expanded to account for additional spatial mismatch if we further break workers and jobs into subgroups. Horner (2002), Giuliano and Small (1993) and Kim (1995) all disaggregate the problem by worker characteristics, although the specifics employed for disaggregation can be different.

We compute two MRC measures: a general MRC that does not account for a job skill requirement, and an MRC that does address job type by disaggregating jobs and workers for each census tract into two groups: low skilled and high skilled. Again, estimated distance is used to represent travel cost. Fig. 4 shows the tract level MRC measures after stratifying jobs and workers by job skill. We exclude maps of the general MRC because its spatial pattern is similar.

On average, areas within route 128 have clearly lower values than other parts of the metropolitan areas. This reflects the density affect: areas within route 128 are more densely developed than areas outside it. The only exception is that some areas downtown have very high workplace MRC. This is primarily because the extremely high concentration of jobs downtown necessitates pulling labor in from a large area, even if jobs are matched with the closest available labor supply.

Figure 4. Job-housing proximity by MRC

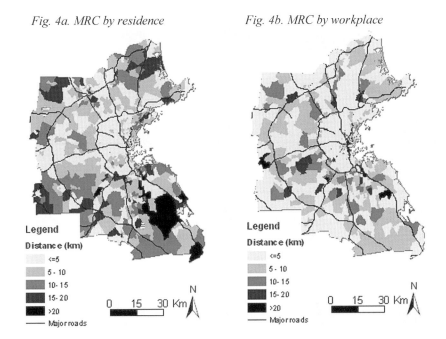

Fig. 4a. MRC by residence

Fig. 4b. MRC by workplace

Correlation between different measures

The above maps demonstrate that significant differences exist in the spatial patterns of job-housing proximity when it is described by different measure categories. In order to see whether these measures are essentially different, we further analyze the correlation among the three measure categories. Using Boston 2000 as the example, Table 1a shows the correlation coefficients of the measures when tracts are viewed as residence. Table 1b shows the correlation coefficients with tracts viewed as workplace.

Table 1a. Correlation coefficients of residence measures for Boston 2000

		MRC		Job accessibility		JER	
		General	By job skills	DAA	UA	10 tracts	10 km buffer
MRC	General	1					
	By job skills	0.89	1				
Job Accessibility	DAA	-0.51	-0.48	1			
	UA	-0.46	-0.44	0.97	1		
JER	10 closest tracts	-0.22	-0.25	0.34	0.31	1	
	10 km buffer	-0.43	-0.43	0.69	0.65	0.28	1

Table 1b. Correlation coefficients of workplace measures for Boston 2000

		MRC		Labor accessibility		JER	
		General	By job skills	DAA	UA	10 tracts	10 km buffer
MRC	General	1					
	By job skills	0.71	1				
Labor accessibility	DAA	-0.05	0.12*	1			
	UA	-0.10	0.12*	0.76	1		
JER	10 closest tracts	0.10	0.16	0.20*	0.24*	1	
	10 km buffer	0.05	0.24	0.42*	0.62*	0.28	1

Note: Numbers marked with * have unexpected signs.

If all measures are consistent in describing job-housing proximity, the expected signs of the correlation coefficients should have the following pattern: MRC by residence should be negatively associated with job accessibility since a higher MRC means a worse job supply and a higher job accessibility score stands for a better job supply. MRC by workplace should be negatively associated with labor accessibility since a higher MRC means a worse labor supply and a higher labor accessibility score stands for a better labor supply. For the same reason, MRC by residence should be negatively associated with JER, and MRC by workplace should be positively associated with JER. Job accessibility should be positively correlated with JER and labor accessibility should be negatively correlated with JER.

Checking the correlation coefficients by place of residence in Table 1a, we find that all coefficients have correct signs. However, the coefficients vary considerably across the different measures and the majority of coefficients are below 0.5.

Checking the correlation coefficients by workplace in table 1b, we find that labor accessibility is positively correlated with JER, and MRC (accounting for job skills) is positively associated with labor accessibility. These are not the signs that we expected. The other coefficients have expected signs, but they typically have a low value.

The correlation coefficients, therefore, tell that these measures are significantly different from, or even inconsistent with each other in describing the same scenario of job-housing distribution. Similar problems can be observed when we use Boston data for 1980 and 1990, or Atlanta data for all three decades. This lack of correlation partially explains why studies using different measures tend to have different conclusions about the commuting impacts of job-housing proximity. The correlation analysis, however, cannot tell which measure best captures the relationship between job-housing relationships. For that, we need to compare these measures with data about actual commuting (AC).

JOB-HOUSING PROXIMITY AND COMMUTING

The CTPP data provides information on actual commuting time between pairs of origin and destinations at detailed geographical levels. This information is estimated from commuting time in minutes self-reported on the Census long-form for each worker in a household. Census aggregates this information by residence and workplace.

Actual commuting

We use data for Boston 2000 to illustrate the spatial variation of actual commuting (AC) at the tract level. Fig. 5 shows that there is no dominant regional trend in the spatial variation of actual commuting time.

It is obvious that the spatial pattern of actual commuting does not match that of the DAA measures or of JER. Unlike DAA (Fig. 3), actual commuting time is not orderly sorted with the shortest commuting time in the central city. Neither is residence commuting time short where JER is high (Fig. 2). The MRC measures (Fig. 4) do seem to have a spatial pattern similar to that of commuting. However, with this visual examination alone, we cannot tell how well the local variation of MRC matches that of actual commuting times.

A subregional comparison between job-housing proximity indicators and actual commuting times implies that MRC might be the best measure. In the downtown area, for example, residence commuting time is the lowest, and workplace commuting is the highest. If the relationship between job-housing proximity and commuting holds, downtown measures should represent a better supply of jobs and a worse supply of labor compared to other areas. Checking the values of the measures, we find that MRC is the only one with the expected numeric result. Labor accessibility contradicts this expectation because it has the highest values downtown, rather than the lowest values as we would expect. In addition, downtown JER values are not among the highest in the region, suggesting that JER may not do well in explaining actual commuting.

Figure 5. Tract level actual commuting times for Boston 2000

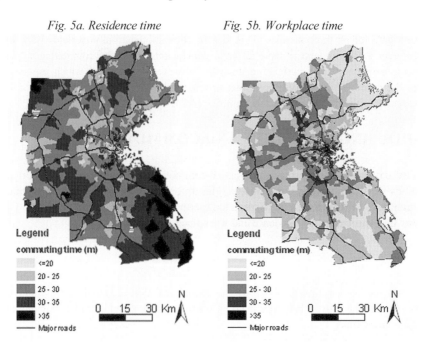

Fig. 5a. Residence time　　　　　*Fig. 5b. Workplace time*

Let us continue to analyze the correlation between actual commuting time and job-housing proximity measures. We would also expect the following: a better supply of jobs, as represented by job-housing proximity measures, should be associated with shorter residence commuting times, and a better supply of labor should be associated with shorter workplace commuting. Measures violating this criterion would be regarded as inferior to those adhering to it. Table 2 shows the correlation coefficients for Boston 1980, 1990, 2000 and Atlanta 2000. Correlation coefficients with unexpected signs are marked with an asterisk, *.

The correlation between actual commuting and job-housing proximity varies according to different measures. The measures that have a consistent relationship to actual commuting are MRC and JER. Higher MRC is associated with higher commuting time both by residence and by workplace. Higher JER is associated with a shorter residence commuting time and a longer workplace commuting time. The relationship between labor accessibility and workplace commuting time is not what we expected. In all three years for Boston, and in year 2000 for Atlanta, the empirical results show a positive relationship between labor accessibility and workplace commuting time.

Table 2. Correlation between proximity measures and commuting

Commuting time		Boston 1980		Boston 1990		Boston 2000		Atlanta 2000	
		Res.	Work	Res.	Work	Res.	Work	Res.	Work
MRC	General	0.41	0.08	0.33	0.12	0.34	0.10	0.36	0.11
	By job skills	0.45	0.19	0.35	0.09	0.37	0.21	0.42	0.17
Accessibility	DAA	-0.15	0.65*	-0.17	0.45*	-0.12	0.55*	-0.32	0.52*
	UA	-0.27	0.33*	-0.07	0.60*	-0.21	0.35*	-0.35	0.22*
JER	10 closest tracts	-0.33	0.38	-0.13	0.48	-0.33	0.35	-0.21	0.30
	10 km buffer	-0.04	0.52	-0.30	0.35	-0.14	0.45	-0.32	0.43

Notes:
1. Numbers marked with * have unexpected signs.
2. For MRC, residence MRC is used to correlate residence commuting time and workplace MRC is used to correlate workplace commuting.
3. For accessibility, job DAA and job UA is used to correlate residence commuting time, and labor DAA and labor UA is used to correlate workplace commuting time.

Accessibility vs. commuting

How can we explain the unexpected relationship between labor accessibility and actual commuting time? One possibility is that some coincidence of other factors overshadows the relationship between accessibility and commuting length. For example, congestion in the downtown is the highest and workplace commuting is also the longest. To further analyze this possibility, we fit a regression model that controls the mobility factors as well as other socio-economic variables. The model uses actual workplace commuting time as the dependent variable. Independent variables include labor DAA, percentages of mode share, percentage of female workers, percentage of black workers and percentage of Hispanic workers. In addition, we include driving speed, which is obtained by dividing tract level average commuting distances by average commuting times for those who drive alone to the workplace. The data is for Boston 2000. The basic analysis units are census tracts and the regression results are shown in Table 3.

Table 3. Regression model for workplace commuting times

Dependent variable: workplace commuting time	Coefficients	Standardized Coefficients	Sig.
Constant	30.945		0.000
Percentage of driving along	2.420	0.058	0.246
Percentage of carpool	3.478	0.023	0.402
Percentage of transit	37.980	0.574	0.000
Percentage of non-motorized transport	-24.251	-0.228	0.000
Percentage of female workers	-12.565	-0.222	0.000
Percentage of black workers	3.629	0.053	0.073
Percentage of Hispanic workers	-7.722	-0.083	0.002
Average drive speed	-0.130	-0.218	0.000
Demand-adjusted labor accessibility (labor DAA)	2.080	0.062	0.084
R square	0.46		

As seen in table 3, labor DAA still has the unusual positive relationship with commuting time and the estimate for labor DAA is marginally significant at the 8.4% level. Other census tract characteristics are more strongly correlated with actual commuting times. The four most significant factors (average driving speed and the percentages of transit, non-motorized transport, and female workers) account for almost half the variation in actual commuting time. This problem persists in models using data for Boston 1980 and 1990, as well as models using data for Atlanta. In addition, we have developed models using spatial lag variables to control for the spatial autocorrelation problem. We define spatial proximity matrix with buffers of 2 km, 4 km and 8 km³. None of these models generate a significant and negative estimate for labor DAA.

Alternatively, one might doubt the quality of the data. However, we have no reason to believe that the sampling and estimation of CTPP data has such a systematic problem that it creates the unexpected relationship between labor accessibility and workplace commuting in each year and in different regions.

We propose another hypothesis—the presence of centrality advantage—to explain this unusual quantitative relationship. To reveal the problem, let's start with a simulation in a hypothesized region. The region is circular with a radius of 25 km and is composed of cells sized at 1*1. Each cell has 100 jobs and 100 workers, i.e., a uniform distribution.

If we calculate job DAA, it is easy to envision that the central places have a higher accessibility than the peripheral location. The numeric result in Fig. 6 is computed with a spatial decay parameter of 0.1. Even though jobs and labor are uniformly distributed, the accessibility score is one-third lower at the edge of the region, 25 km from the center. Job accessibility is greater than one in the central place and lower than one at the periphery. Given the symmetric nature of the mathematics, it is easy to envision that spatial patterns of labor DAA should be the same as job DAA.

Figure 6. DAA in a hypothetical circular region

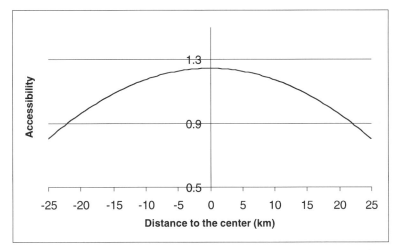

Note that a job DAA score greater than one means that, at that locality, jobs are over supplied relative to labor competition, and a labor DAA score greater than one means, at that locality, labor is over supplied relative to employment competition. It is illogical to have both labor DAA and job DAA greater than one at one locality. (One cannot have a>b and b>a simultaneously.) Our interpretation of the inconsistency is that DAA represents not only workplace-residence proximity, but also centrality advantage. Central location is naturally advantaged compared with other places in terms of job access and labor access. Centrality advantage is meaningful when the accessibility measures are used to interpret job and hiring opportunities. However, centrality *advantage* in the DAA acts as measurement *noise* if we are trying to interpret commuting.

One may soon notice that this centrality noise comes from the mathematics that favors central location and the periphery is assigned a lower value because of the boundary effect. Although jobs and workers are evenly distributed across the metro region, they drop to zero past the metro boundary. So locations nearer the boundary are accessible to fewer and fewer jobs (and workers). In a given region, the extent of the centrality effect depends on the spatial decay function. Only two extreme spatial decay functions (corresponding to $\beta=0$ and $\beta\rightarrow\infty$) can entirely eliminate the centrality noise. For realistic betas, the centrality noise will be noticeable unless the boundary impacts a negligible fraction of the commuters. For metropolitan areas such as Boston and Atlanta, typical commuting distances cover a sizable portion of the metro area. Hence, the boundary effect remains and the centrality effect persists.

One may wonder why this problem is not noticeable in the correlation between job DAA and residence commuting. The answer is that although gravity type job DAA has the same centrality advantage as labor DAA, the centrality noise in job DAA is not so strong as that in labor DAA. In a metropolitan area, jobs are typically more spatially concentrated than residents. This is true for both Boston and Atlanta. Consequently from central city to periphery, job DAA declines faster than labor DAA (Fig. 3). The extent of the centrality advantage, however, is the same for both job DAA and labor DAA because it is determined by the spatial decay function in computation. Thus, the presence of the centrality noise is relatively stronger in labor DAA than in job DAA, resulting in an unexpected relationship between workplace commuting and labor DAA. It is easy to see the centrality noise also exists in the computation of UA.

The existence of the centrality noise, therefore, is the likely factor bringing about unexpected correlation between labor accessibility and workplace commuting. Our observation, however, can be challenged because there are many variations of accessibility computation and many alternatives to specify the regression model. For example, Shen (2000) incorporates mode split into accessibility computation and uses job DAA to interpret the spatial variation of residence commuting, although that paper does not compute labor DAA for interpreting workplace commuting. Levinson (1998) uses both job UA and labor UA in his regression models to interpret both workplace and residence commuting time. These studies help in understanding metropolitan spatial structure. However, it is the purpose of this paper to seek job-housing proximity measures that are consistent in interpreting commuting patterns. The unusual result and the centrality advantage (noise) revealed here surely challenge planners who wish to use accessibility measures to interpret commuting patterns and monitor urban growth.

JER vs. commuting

Based on the correlation coefficients, one may argue JER is a good measure. However, JER falls short in terms of providing guidance for urban growth strategies. As pointed out by Peng (1997), a higher JER means a better job supply, which means shorter residence commuting. A higher JER, however, also means a poorer labor supply, which in turn means a longer workplace commuting. Therefore, strategies that improve labor supply for job rich areas will not only decrease workplace commuting time, but also increase residence commuting time. To examine the net effect, we might consider averaging commuting times across both residents and workers to see if, on balance, the net commuting impacts are better or worse. Toward this end, we took the actual commuting times for workers and residents in each tract - that is, the underlying data used to compute the two by-residence and by-workplace correlations in Table 2 - and computed an overall average commuting time for each tract (across both residence and workplace). The resulting correlation between the JER measures and this single average commuting time is disappointingly low. For example, for Boston 2000, the coefficients are 0.11 and 0.19, respectively, for the JER measure using the 10 closest tracts and the JER measure using the 10 km buffer. The corresponding numbers for Atlanta are 0.10 and -0.05.

These results suggest that JER measures tell us little about the net effect on commuting of local changes in the ratio of jobs and workers.

The point can be further illustrated with examples of land development. For instance, a proportional densification of jobs and workers tends to result in shorter commuting for both workplace and residence because of the associated improvement in job-housing proximity. However, the way JER measures job-housing proximity results in the same pre- and post-change indicators, which means no change in job-housing proximity. In addition, in a region-to-region comparison, the commuting differences, when caused by land use differences such as densification and infill development, would be hardly captured by JER measures.

MRC vs. commuting

The numeric result tends to suggest that MRC is the best of the three classes of job-housing proximity measures. In Both Boston and Atlanta, the spatial correlation of MRC and AC has the expected signs. Replacing labor accessibility with MRC in the regression model presented in Table 3, we find that commuting time decreases by 0.25 m when MRC is reduced by 1 km. Models using data from Boston 1980, 1990 and Atlanta have similar significant estimates.

In addition, unlike JER and accessibility, which measure workplace-home relation with ratios or spatially discounted opportunities or both, MRC measures job-housing proximity with explicit commuting cost. This feature makes it convenient to link urban development patterns to commuting. The neighborhood level variation of commuting can be compared with the neighborhood level MRC. Neighborhood level MRC can be easily aggregated to the municipality or county level to provide indicators for monitoring land development trends.

Using MRC, we can explain not only the spatial variation of commuting from one neighborhood to the other, but also the temporal change of commuting from one decade to the other. In Boston, for example, MRC (measured with accounting for job skills) is 5.9 km in 1980. It increases to 6.2 km in 1990 and further to 6.8 km in 2000. The associated commuting time increases from 23.1 minutes to 23.8 minutes and further to 27.6 minutes. In a region-to-region comparison, in 2000, Atlanta's MRC is 10.4 km, compared to 6.8 km in Boston. The associated commuting time is 30.5 minutes, compared to 27.6 minutes in Boston. The region with the higher MRC values has longer commuting times.

In contrast, JER and accessibility measures aren't helpful in making region-to-region or year-to-year comparisons because one can hardly explain the meaning of aggregating JER and accessibility measures from census tract levels up to the regional level. In addition, in many circumstances, inconsistent or unreliable quantitative relationships can be easily envisioned because of the previously mentioned centrality problem. For example, comparing two metropolitan areas of different sizes, the center of the larger one generally has a longer workplace commuting. However, that center also tends to have a higher labor UA, pointing to an inconsistent relationship between UA and commuting. The corresponding value of labor

DAA at the center of the larger region can be either lower or higher than that in the smaller region, depending on the magnitude of job decentralization relative to labor decentralization. This points to a possibly unreliable relationship between DAA and commuting.

Furthermore, in a single growing region, UA scores increase as the result of job and labor growth. This means that a typical place in the region becomes better supplied with jobs and labor, represented by the increase in accessibility scores. While this increased accessibility is desirable, it does not imply that commuting time will decline over time. Rather, in most cases, commuting duration increases. From 1990 to 2000, every American metropolitan area with over one million population saw an increase in commuting time (McGuckin and Srinivasan, 2003).

METHODOLOGICAL IMPROVEMENT BEYOND USING MRC

Despite all the above arguments about the virtues of MRC relative to JER and Accessibility, MRC has its own weaknesses in representing job-housing proximity in decentralizing urban spaces. To get a handle on this limitation, first let us observe that MRC and AC can be viewed as two among many possible commuting scenarios that are possible for any given job-housing distribution, the difference between them stems from the relative importance of travel cost in location decisions. The assumption underlying MRC is that commuting cost plays a dominant role. In reality, actual commuting (AC) choices are not determined solely by travel cost. Indeed, we might expect a decreasing role for commuting cost in location decisions. As income rises relative to transportation costs, commuting cost takes a decreasing share of real income, and the effects of commuting cost may be overshadowed by other factors involving household characteristics, preferences and location amenities (Giuliano, 1995).

The decreasing role of commuting cost would result in an increasing portion of AC that is classified as excessive commuting, and possibly a weaker relationship between MRC and AC (Guiliano and Small, 1993). For a certain workplace, for example, the geographical scope of labor pools for actual commuting will tend to be much larger than that for MRC. This is particularly true in the high mobility suburban areas where significant overlap of commuting sheds is common. Therefore, compared to the sizes of job and labor pools for AC, job-housing proximity characterized by MRC tends to focus on local rather than the regional aspects of urban development patterns. The increasing gap between MRC and AC, resulting from the decreasing role of travel cost in the decentralizing region, leaves a large portion of commuting unexplained by MRC. The excessive part accounts for 20%-80% of AC (Giuliano, 1995). A possible strategy to address this problem is to supplement MRC with another measure that characterizes the regional aspects of job-housing distribution. One possible improvement is the maximum commuting concept proposed by Horner (2002), although the maximum commuting is an untested measure at the intraurban scale.

In addition, while the quantitative relationship between MRC and commuting length is consistent, the results should be interpreted carefully. The strength of the linkage between

MRC and AC depends on whether job-housing imbalance, represented by MRC, actually imposes constraints on location decisions of job and home sites.

First, the role of MRC in residential location decisions is a function of mobility. AC has a higher reliance on MRC when mobility is lower and relies less on MRC when mobility is high. The empirical results seem to confirm this. Fig. 7 plots AC against MRC in two mobility sectors: one where driving speeds are < 25 km/hour and the other where driving speeds are > 45 km /hour. The tract level indicators are computed using data for Boston 2000.

Fig. 7 shows that the relationship between MRC and commuting distance is stronger in the low mobility sector than in the higher mobility sector. In the low mobility sector, decreasing MRC by one km decreases commuting distance by 0.65 km, and 43% of the variation of AC is explained by MRC. In the higher mobility sector, however, decreasing MRC by 1 km reduces AC by 0.31 km and only 23% of the variation of AC is explained. Therefore, quantitative studies of the effectiveness of urban growth strategies must consider the intervening forces of mobility conditions.

Figure 7. AC vs. MRC in different mobility sectors

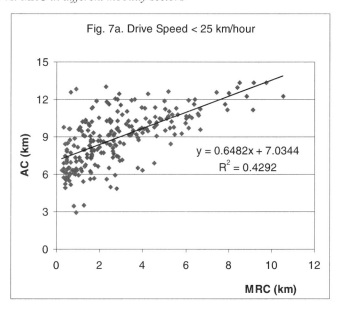

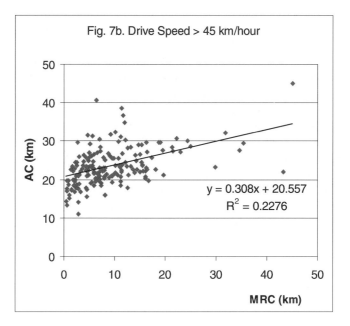

Fig. 7b. Drive Speed > 45 km/hour

$y = 0.308x + 20.557$
$R^2 = 0.2276$

Second, the role of MRC in location decisions depends on the magnitude of MRC itself. It is easy to envision that when MRC is low, that is to say when jobs and housing are already well balanced, the change of AC would not mirror the change of MRC. An increase of MRC from 3 km to 5 km may have no impacts on residence location decision because job-housing proximity is so good that it is not an important constraint in location decisions. When MRC is higher, for example, 30 km, a reduction from 30 km to 20 km is more likely to bring down AC because 30 km MRC is much higher than average commuting, and likely to impact location decisions more. Therefore, discussion about the effectiveness of job-housing balance strategies for congestion relief should not neglect the existing conditions of job-housing proximity.

CONCLUSIONS

The imperfect but important relation between commuting and job-housing proximity challenges today's researchers and planners in quantifying the commuting impacts of current urban growth trends. We focused on three categories of measures that have been widely employed to relate commuting and job-housing proximity. They are JER, accessibility and MRC. The qualitative and quantitative assessment has identified MRC as the most consistent measure to characterize job-housing proximity in order to reveal the commuting impacts of urban development patterns.

We are aware that various accessibility and JER measures can be valuable in characterizing settlement patterns and job opportunities. Our focus has been on understanding spatial and temporal changes in **commuting** patterns. Rather than criticize all use of these common measures, our intention is to help identify better methods for interpreting **commuting** patterns in relation to job-housing proximity. Toward this end, we recognize that MRC should be supplemented with other measures that capture the regional aspects of job-housing distributions. One possible addition is the 'maximum commuting concept' proposed by Horner (2002). Another possibility that we are considering is to assign workers to jobs in proportion to worker and job densities. We call this measure 'proportionally matched commuting' (PMC) and preliminary results suggest that, taken together, MRC and PMC can explain much of the observed variations in actual commuting. As part of our University Transportation Center study of job-housing proximity in Atlanta and Boston, we are continuing to develop methods that can better characterize the spatial relationship between workplace and residence, and assist in explaining commuting patterns consistently across space, over time and between different regions.

REFERENCES

Cervero, R. (1989). Jobs-housing balancing and regional mobility. *Journal of the American Planning Association*, **1**, 136-150.

Cervero, R. (1996). Job-housing balance revisited: trends and impacts in the San Francisco Bay Area. *Journal of the American Planning Association*, **62**, 492-511.

Ferreira, J. and J. W. Yang (2004). Basic trends of urban growth and commuting in Boston and Atlanta. Working paper in Region One University Transportation Center, MITR16-2a.

Guiliano, G. (1991). Is jobs housing a transportation issue? In: *Achieving a Jobs-Housing Balance: Land Use Planning for Regional Growth*. The Lincoln Institute.

Giuliano, G. (1995). The weakening transportation-land use connection. *Access*, **6**, 3-11.

Giuliano, G. and K. A. Small (1993). Is the journey to work explained by urban structure? *Urban Studies*, **30**, 1485-1500.

Hamilton, B.W. (1989). Wasteful commutingagain. *Journal of Political Economy*, **97**, 1498-1504.

Hansen, S. and M. Schwab (1987). Accessibility and intraurban travel. *Environment and Planning A*, **19**, 735-748.

Kim, S. (1995). Excess commuting for two-worker households in the Los Angeles metropolitan area. *Journal of Urban Economics*, **38**, 166-182.

Horner, M.W. (2002). Extensions to the concept of excess commuting. *Environment and Planning A*, **34**, 543-566.

Horner, M.W. (2004). Spatial dimensions of urban commuting: a review of major issues and their implications for future geographic research. *The Professional Geographer*, **56**, 160-173.

Kwan, M. P., A. T. Murray, M. O'Kelly and M. Tiefelsdorf (2003). Recent advances in accessibility research: representation, methodology and applications. *Journal of Geographical Systems*, **5**, 129-138.

Levinson, D. (1998). Accessibility and the journey to work. *Journal of Transport Geography*, **6(1)**, 11-21.

McGuckin, N., and N. Srinivasan (2003). Journey-to-work trends in the United States and its major metropolitan areas, 1960–2000. Report prepared for US Department of Transportation.

Morris, J. M., P. L. Dumble and M. R. Wigan (1979). Accessibility indicators for transportation planning. *Transportation Research A*, **13**, 91-109.

Peng, Z. R. (1997). Jobs-housing balance and urban commuting. *Urban Studies*, **34**, 1215-1235.

Rodríguez, D. (2004). Spatial choices and excess commuting: A case study of bank tellers in Bogotá, Colombia, *Journal of Transport Geography*, **12**, 49-61.

Shen, Q. (1998). Location characteristics of inner-city neighborhoods and employment accessibility of low-wage workers. *Environment and Planning B*, **25**, 345-365.

Shen, Q. (2000). Spatial and social dimensions of commuting. *Journal of the American Planning Association*, **66**, 68-82.

The Brookings Institution Center on Urban and Metropolitan Policy (2000). *Moving beyond sprawl: The Challenge for metropolitan Atlanta.* Brookings Institution, Washington, DC.

Timothy, D. and W. Wheaton (2001). Intraurban wage variation, employment location and commuting times. *Journal of Urban Economics*, **50**, 338-366.

Wachs, M. and T. G. Kumagai (1973). Physical accessibility as a social indicator. *Social Economic Planning Science*, **7**, 437-456.

Wang, F. (2001). Explaining intra-urban variations of commuting by job accessibility and worker characteristics. *Environment and Planning B*, **28**, 169-182.

White, M. J. (1988). Urban commuting journeys are not wasteful. *Journal of Political Economy*, **96**, 1097-1110.

NOTES

[1] The authors acknowledge partial support for this work from the Region One University Transportation Center through project MITR-16-2. We appreciate the helpful comments from Professors Moshe Ben-Akiva and Ralph Gakenheimer at MIT. We also thank Professors Harvey Miller and Mark Horner for their informative comments.

[2] To make these two maps comparable, the same classification has to be applied to each map. Therefore, all maps in this paper are shaded with manual classification, which is a slight adjustment from the standard quintile classification.

[3] We used Geoda, a software developed by the Spatial Analysis Laboratory (SAL) in the Department of Geography at the University of Illinois.

Access to Destinations
D.M. Levinson and K.J. Krizek (editors)
© 2005 Elsevier Ltd. All rights reserved.

CHAPTER 10

EXAMINING THE SPATIAL AND SOCIAL VARIATION IN EMPLOYMENT ACCESSIBILITY: A CASE STUDY OF BUS TRANSIT IN AUSTIN, TEXAS

Mark W. Horner, Florida State University
Jessica N. Mefford, The Ohio State University

INTRODUCTION

As efficiency, reliability, and cost effectiveness persist as goals of transportation systems, ensuring equitable access to job locations, shopping centers, and other important destinations must be given more attention in transport policy agendas (Kwan *et al.*, 2003). This presents real challenges for public transit systems facing increasingly constrained financial resources and pressure to increase and extend service (Murray, 2001). Moreover, the growing complexity of modern day urban form and its resultant travel demand have made achieving equitable accessibility to urban opportunities a tenuous proposition (Miller and Meyer, 2001).

Arguably, providing accessibility to job locations is the most important function of public transit systems. People must be gainfully employed to sustain their life activities, and for many, transit is the means by which this is achieved (Hanson, 1995; Horner, 2004). Contemporary land use patterns do not make this easy because even though the number of workers and jobs are generally balanced for a given metropolitan area (Giuliano and Small, 1993), the spatial patterns of firms and residences *within* the area are far from commensurate with one another (Scott *et al.*, 1997). Recognizing the possible difficulties this portends for the poor, mobility-challenged, and other vulnerable groups without sufficient transport resources, some have advocated the need for achieving greater localized *jobs-housing balance*, or the conscientious mixing of residential and commercial land uses (e.g. Cervero, 1989; Sultana, 2002). Strategic long-term land use planning approaches targeted at improving jobs-housing balance have been proposed (e.g. Horner and Murray, 2003), but due to the political and financial difficulties of implementing such plans, delivering workers to disparate job locations has fallen on the shoulders of transport infrastructure providers.

Like others, we take the view that public transit is a promising mode of transportation because it is potentially a means by which cities can become more sustainable (Wu and Murray, 2005). It is uniquely positioned to help ameliorate such jobs-housing divides, but care must be taken to account for the divergent needs of transit patrons, their variable socioeconomic statuses and activity schedules, as well as other factors. As notions of 'spatial mismatch' figure prominently into the planning vernacular (see Kain, 1992; Preston and McLafferty, 1998; Sanchez *et al.*, 2004a), so too must planners be cognizant of the racial and ethnic makeup of patrons and seek to ensure minorities are not at a disadvantage in terms of how service is provided. Despite the importance of these issues to planners as well as researchers interested in accessibility, there has been relatively little exploration into the nature of the accessibility provided by public transit. Our work attempts to fulfill that need by examining the spatial and social variation in people's access to employment.

We examine how an urban bus transit system provides accessibility to residents through a case study of the bus service in Austin, Texas. Austin is a medium sized urban area with a diverse economy and well-developed bus transit system. The research focuses exclusively on the transit mode, work commute trips, and those workers and jobs accessed by transit. Data are the 2000 Census Transportation Planning Package (CTPP, 2000) which include the locations and detailed characteristics of workers and their jobs. They are resolved to the traffic analysis zone (TAZ) level and available publicly from the Bureau of Transportation Statistics (BTS). A detailed transit network is constructed in a geographic information system (GIS) to facilitate accurate estimates of the bus travel times between residences and workplaces. Given these data, we develop several GIS-based measures of accessibility and use them to explore the landscape of transit service provision. Specifically, the social dimensions of accessibility are investigated by computing indices controlling for patrons' racial, ethnic and employment characteristics. Spatially, we visualize accessibility across the Austin area, and produce statistical summaries of how accessibility varies at multiple scales (i.e. a systems level vs. a neighborhood level).

First, initial background is provided on land use, transit and accessibility issues. Then basic models of accessibility are reviewed and extended to model the transit case. As previously mentioned, accessibility indices are developed at two spatial scales; disaggregate measures are designed to summarize the opportunities available from specific locations (i.e. a neighborhood level), while aggregate measures are developed to capture the nature of opportunities from all locations (i.e. a systems level). The latter of these measures may be used to compare the overall level of transit service provided to a specific worker group, while the former of these measures lends itself to visualization in a GIS. We then close by discussing the implications of our findings and suggesting directions for future research.

BACKGROUND AND LITERATURE REVIEW

Journey to work questions, particularly those involving how to link workers with jobs are important considerations for planners (Blumenberg, 2002). Certain workers face structural barriers in reaching and obtaining employment (Immergluck, 1998; Preston and McLafferty, 1999). On the land use side, the configurations of jobs and housing in a given area are often such that there are no suitable jobs within a reasonable proximity of workers (Cervero, 1989; Immergluck, 1998; Levine, 1998). Along these lines, Ong and Blumenberg (1998) point out that it is people in poor neighborhoods who often face the most difficulties in finding employment because of jobs-housing imbalances.

At the same time, social considerations driven by the historical remnants of forced segregation and persistent housing and employment discrimination compel researchers to investigate the ways in which home-work issues differ between minorities and their counterparts (see Kain, 1968; Ihlanfeldt and Sjoquist, 1998; Sanchez *et al.*, 2004a). The *spatial mismatch hypothesis* is at the heart of these concerns, as it posits that due to residential and/or employment discrimination, minorities are at a disadvantage in terms of finding employment (see Kain, 1992 for a discussion and review).

Public transportation has an important role to play in mitigating home-work imbalances that make it difficult for minorities to navigate job opportunities (Sanchez, 1999). It is well understood that those without access to a car are at a disadvantage both in terms of finding work and earnings potential (Raphael and Rice, 2002). For this 'transit dependent' market, as well as other potential riders, the placement of transit infrastructure, such as the locations of stops, is a key consideration for a sustainable system (e.g. Ong and Houston, 2002). Transit service quality is also a factor, especially for those who might use the system to access employment (Cervero, 2002).

One way to systematically examine the convergence of employment opportunities, urban structural constraints, and broader social phenomena is to model the accessibility provided by transit systems. In a sense, the success of transit systems is determined largely by the *accessibility* they provide. Accessibility is a term used to describe the ease with which certain activities can be reached from a particular location using the transportation system (Handy and Niemeier, 1997; Kwan, 1998; O'Sullivan *et al.*, 2000). An accessible transportation system achieves the objective of transporting people to their desired destination in an efficient manner. To capture efficiency, transportation accessibility may be viewed from a spatial perspective, incorporating the spatial distribution of destinations, their desirability, and the ease with which individuals can reach those destinations from their respective origins (Hanson, 1995; Handy and Niemeier, 1997). Our interest then is to learn more about how accessibility differs among prospective worker groups.

Measures of accessibility typically account for the travel costs of reaching destinations, as well as their attributes (Hansen, 1959; Miller, 1999), which are characterized by their attractiveness. For example, if one wanted to measure accessibility to retail locations it might be appropriate to use locations' total retail floor space or retail employment as a proxy for attractiveness (O'Kelly and Horner, 2003). Perhaps the most conventional opportunity type considered in accessibility modeling is that of employment. Some have explored the relationship between accessibility and journeys to work, or commutes (Levinson, 1998), and others have considered accessibility measures while modeling employment outcomes (e.g. Immergluck, 1998; Sanchez *et al.*, 2004b). Though their specific applications are diverse, in general accessibility measures characterize destinations by attributes designed to meet the study needs (Handy and Niemeier, 1997).

The pioneering work of Hansen (1959) is largely credited as the impetus for accessibility research over the last decades. Several new approaches to conceptualizing accessibility have evolved over time (e.g. Handy and Niemeier, 1997; Miller, 1999; Kwan and Weber, 2003; O'Kelly and Horner, 2003); each has been defined and operationalized in different ways, depending on the context of the application (see Kwan *et al.*, 2003 for a review). Regardless of the modeling need, each approach includes both transportation and activity elements, but they differ in the units of analysis and the degree to which they reflect travel behavior. One fundamental distinction can be made between types of accessibility measures; that is whether they are based on individuals, or places/locations (Kwan, 1998; Horner 2004). Place-based accessibility measures follow from the Hansen framework and characterize locations; specifically how easily locations (and their activities) can be reached using a given transportation system from other places (e.g. Shen, 1998, 2000; O'Kelly and Horner, 2003). Such measures involve aggregate data sets and are most suited for analyzing accessibility across an entire transportation system. Another important development is that of individual-level accessibility, which is derived directly from the space-time approach to accessibility developed by Hägerstrand (1970). Individual measures take into account both the physical accessibility of locations, as well as the space-time budgets of individuals (O'Sullivan *et al.*, 2000; Kwan and Weber, 2003). Given our objectives and need for general indices, place-based accessibility measures derived from the Hansen framework are most appropriate for the planned transit system evaluation. Among the factors that govern the transit planning process, the spatial distribution of population, which cannot be separated from place, is perhaps most significant (Pas, 1995). For this reason, we will focus on location-based indices of accessibility via the transit system.

Recently, the incorporation of multiple axes of dissimilarity (e.g. income, worker class, race/ethnicity) was identified as an important direction for accessibility research that requires further investigation (Kwan *et al.*, 2003). This work attempts to address that need by producing a better understanding of whether disparate levels of accessibility are provided to worker groups. Because it is integral to broader equity, land use, transportation, and sustainability concerns, we focus solely on the transit mode.

ACCESSIBLITY MODELS

The approaches utilized to estimate accessibility are derivatives of the Hansen (1959) model. This group of models shares linkages to the gravity-based spatial interaction model (see Fotheringham and O'Kelly, 1989) as well as the notion of attraction found in traditional four-step transportation planning (see Domencich and McFadden, 1975). The general Hansen formulation is as follows:

$$A_i = \sum_{j=1, i \neq j}^{m} O_j f(C_{ij})$$
(1)

where
A_i is the accessibility at point i,
O_j is the opportunities at point j,
$f(C_{ij})$ is a function of the generalized travel cost from i to j.

Because this study seeks to measure accessibility to employment, the *total number of jobs* in each zone represents the element of attractiveness in the urban landscape. Thus, we let $O_j = E_j$ and note that C_{ij} represents the travel cost between zones i and j via the transit network

The Hansen framework is flexible and can accommodate many forms of $f()$. To satisfy the objectives of this analysis, the general model has been modified to include an exponential function which can account for experimental β values. The function of β is to determine the effect of travel costs on accessibility values; relatively small β values lessen the effect of costs on the overall accessibility score, while relatively large β values achieve the opposite (Fotheringham and O'Kelly, 1989). This formulation is:

$$A_i = \sum_{j=1}^{m} E_j \, \exp(-\beta C_{ij})$$
(2)

A power-based model is also utilized. This model is similar to Equation 2, though the power function supplants the exponential function. Mathematically speaking, both accessibility models perform a spatial smoothing operation on a set of raw input data. Letting p represent the parameter governing the deterrent effect of costs, the model is formulated as follows:

$$A_i = \sum_{j=1, i \neq j}^{m} E_j \, C_{ij}^{-p} \tag{3}$$

In addition to exponential and power-based accessibility models, a cumulative opportunities measure is implemented. A cumulative opportunity measure counts the opportunities, in this case jobs, available within a predetermined travel time from a specific origin (O'Kelly and Horner, 2003). The model is formulated as follows:

$$A_i^S = \sum_{j \in N_i} E_j \tag{4}$$

where
$$N_i = \{ j \mid C_{ij} \leq S \}$$
S is some pre-specified travel time.

Summing over the destinations j within each origin set N_i that meet the condition $C_{ij} \leq S$ ensures that any destination j with travel cost less than S from origin i is counted.

This measure provides two interesting contrasts to the gravity-type accessibility formulations of (2) and (3). First, (4) can be considered a discontinuous measure because a given destination j is not counted into the ith origin's accessibility score if the cost of traversing C_{ij} exceeds S. No such cost/distance limitations are placed on the accessibility indices in (2) and (3). Secondly, the units of A_i in (2) and (3) are a mix of travel cost and employment value, so they will be somewhat less interpretable than the value returned by (4), which simply returns the total number of accessible jobs from a given location. However, similar across all measures is that larger values of A_i imply greater accessibility at a location.

An interesting and useful extension of the basic cumulative opportunities measure (4) is to control for the number of resident workers in each zone. This allows a direct assessment of the workers/jobs match and is achieved by dividing the cumulative employment opportunities available to a given zone by its total resident workers. Essentially, the inclusion of resident workers produces a measure that more accurately captures *accessibility* to employment, controlling for the effects of the origin zone worker population. It can be viewed as a per-capita measure of access by zone. Formally, it is

$$A_i^{S*} = \sum_{j \in N_i} E_j \, / \, R_i \tag{5}$$

where
R_i are the resident workers at zone i

The presented accessibility models are readily implemented in GIS. In this case TransCAD version 4.5 is used to compute all indices via matrix multiplication. Travel cost, C_{ij}, is input into the appropriate accessibility model, along with the employment vector of interest. In the next section, we describe approaches for analytically summarizing these scores at a systems level. On the visualization side, accessibility across the study area is mapped by joining the indices to the attribute table of the traffic analysis zones. The resultant maps spatially depict the accessibility of each zone, which are then subject to visual interpretation.

DATA AND ANALYSIS

Detailed data on employment and residential locations are incorporated from the Census Transportation Planning Package (CTPP, 2000). The CTPP is a large transportation dataset available for all major metropolitan areas of the US. The dataset is produced by the Bureau of Transportation Statistics (BTS) and is readily available to the public (Horner, 2002). The CTPP contains detailed data on journeys to work, and is organized into three parts at multiple spatial resolutions. This study uses data captured at the traffic analysis zone (TAZ), which is the most disaggregate unit of geography available. Several variables, representing potential axes of dissimilarity across worker groups, stratify the accessibility models. These variables focus on race/ethnicity, industry of employment, and income. These were chosen because they address the core equity issues we seek to explore. One limitation of our study is that the CTPP data were preliminary at the time of the research, as they were newly released. Thus, some attribute values, particularly on the residential side, were masked by BTS until the counts could be finalized. This was not especially problematic, however, as there was sufficient data to produce a reasonable picture of accessibility in the study area.

The study area is the service area of Capital Metropolitan Transportation Authority (CapMetro) in Austin, Texas. The Austin metropolitan area is representative of rapidly growing cities inhabited by an economically and culturally diverse population. Population growth presents significant challenges to transit service in Austin; its population of approximately 1.25 million people is expected to double within the next twenty-five years (Capital Area Metropolitan Planning Organization, 2000). CapMetro provides bus transit service to the city and surrounding suburbs within its jurisdiction. Currently there are over 100 routes, and roughly 3,000 stops in the service area, providing service to approximately 130,000 one-way bus trips per day (Capital Metropolitan Transportation Authority, 2005). For purposes of this analysis, only TAZs that contained bus stops in 2003 are included because zones not containing stops have no direct access to the transit system. This selection produces a study area where some TAZs are not contiguous with others. The reason for these "island" zones is that the bus routes serving the distant TAZs (i.e. express routes) do not have intermittent stops along each segment of the route, thus the route cannot be accessed from some zones, even though the routes pass through it. Data on 2003 route and stop locations were obtained from CapMetro. U.S. streets data (2000) enhanced by Caliper Corporation serve as the line layer underlying the transit network. The transit network and study area appear in Figure 1.

Figure 1. Study area and route configuration

Travel time by bus (C_{ij}), was determined based on a GIS transit network. The network was generated from spatial data containing CapMetro's route and stop locations. This was performed in TransCAD and served as the input to all accessibility models. For the purposes of this study, TAZs are assumed analogous to neighborhoods. Trips ending and beginning within each neighborhood, or TAZ, are modeled such that they end at a TAZ centroid. The aggregation of travel data to zone centroid is a process common to zonal-based GIS analysis (Horner and Murray 2004). Data on trip destinations are connected to the transit network by means of connecting TAZ centroids to the nearest bus stop. The access links from TAZ centroids to bus stops are Euclidian distances between the two points, and are assigned a walking speed, as physical access to bus stops is typically gained by walking (Murray and Wu, 2003). Once data on origins and destinations are connected to the transit network, a shortest path algorithm imbedded in TransCAD is used to create a cost matrix between zones via the transit network, using travel time as impedance. The network module in TransCAD allows the modeler to account for wait times, transfer penalties, etc. in the cost matrix. The final output is a single matrix containing travel time between all zones served by transit.

RESULTS

We now present the results of our accessibility modeling in GIS. This analysis produced zonal-based indices that depict accessibility from residential locations to employment locations via the transit network. Generally, the indices measure the accessibility provided to resident workers of a given zone to job opportunities in the urban area. To maintain consistency, we controlled for worker/job type throughout the analysis. Taking this step facilitates a more thorough investigation of accessibility differentials as it allows opportunities to be contrasted across worker groups.

Each type of accessibility measure introduced in the previous section produces results facilitating varying levels of interpretation. Figure 2 shows accessibility for all worker groups to all jobs using examples of the three basic measures; one power-based gravity measure ($p = 2$), one exponential-based gravity measure ($\beta = 2$), and one cumulative opportunities measure ($s = 30$ minutes). Parameters used in the gravity formulations, equations (2) and (3), are based on suggestions presented in past research (Fotheringham and O'Kelly, 1989; Sanchez, 1999). Unfortunately, due to space limitations, other parameter choices are not presented. With respect to S used in the cumulative opportunities models, the maximum travel time to employment opportunities was selected based on empirically observed average transit travel times within the study area. From the 1990 CTPP (Part III), mean transit travel time for this city is approximately thirty-two minutes (we note that CTPP 2000 Part III was not available at the time of this research). Because transit service has substantially increased since that time, a slightly lower travel time, thirty minutes, was chosen.

The two gravity model formulations visualized in Figure 2 reveal different assessments of accessibility to jobs. The power-based gravity measure does not exhibit the same magnitude of accessibility near the urban core as compared to the exponential-based gravity model. This is partly a function of how the data and model are structured, as in this research, we have made $C_{ij}=0$. Obviously, in the power-based model, since 0^{-p} is undefined, there can be no inclusion of employment opportunities within workers' home zone in A_i when $i=j$. However, the exponential-based accessibility model does handle the case when $i=j$ because $exp(0) = 1$. Of course, one way around the difficulty with the power model is to use an arbitrary or observed nonzero value for C_{ij} (see O'Kelly and Horner, 2003 for a fuller discussion of intrazonal travel costs). Given our data structure and need to account for local jobs, the exponential model is the better assessment of accessibility.

Figure 2. Three accessibility measures computed for all job/worker types

Power-Based Gravity Model (p=2)

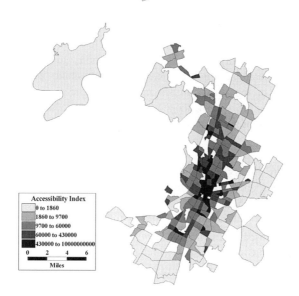

Exponential-Based Gravity Model (B=2)

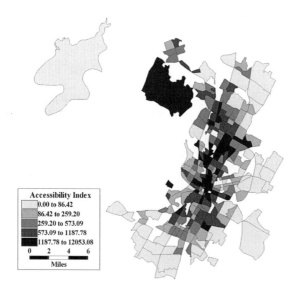

Figure 2 (continued)

Total Jobs Accessible Within 30 Minutes

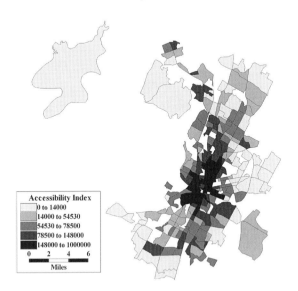

The two gravity-based accessibility models equations (2-3) and the basic cumulative opportunities model equation (4) (see Figure 2) produced spatial patterns of accessibility that reflect zones of high employment; this is evidenced by a pattern of high accessibility near the urban core, and progressively lower accessibility moving outward from center. This smoothing effect potentially limits the interpretability of maps using such measures, as it is difficult to distinguish between employment and distance decay effects. In addition, because the numeric values produced by the gravity models do not represent any specific unit of measurement, many types of scaling or adjustments would be meaningless. Because of these limitations, cumulative opportunities measures, scaled by the number of residences of a given type in each zone, are subjected to further analysis (see Figure 3). These measures capture the total number of jobs accessible from each zone; this number can then be scaled by the number of residences of a given worker type in each zone by simply dividing the cumulative opportunities by the number of resident workers (see equation 5). The resultant measures are perhaps more appropriate for modeling accessibility to employment because they control for zonal population and address the notion of worker/job mismatch.

Figure 3. Per capita accessibility for all jobs/workers

Total Jobs Accessible Per Resident Worker Within 30 Minutes

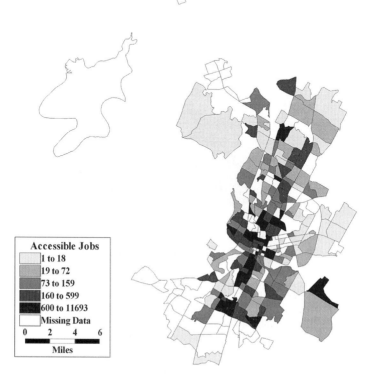

The worker groups examined for disparities in access include the three most populated race/ethnicity categories in the study area; they are non-Hispanic white, non-Hispanic black, and Hispanic white (or all Hispanic workers where Hispanic white is unavailable). Each of these groups is further stratified by industry for the purpose of comparing accessibility to different types of employment. Industry arguably plays a determining role in individuals' socioeconomic status, and is thus an appropriate attribute for this analysis. The selected industries are professional, retail, and entertainment. Differences in accessibility between professional and the latter two industry groups might indicate some disparity in accessibility provided to individuals of higher vs. lower socioeconomic status, respectively. A third attribute, poverty status, is included to capture the accessibility provided to workers who are likely to be transit dependent, by virtue of their economic status. Accessibility analysis by income levels is not possible due to this attribute not being directly reported in the CTPP.

To compare aggregate accessibility for each worker group across the entire study area, (i.e. for the system itself), a weighted mean accessibility is calculated (see Table 1). Given a travel time threshold, S, the *weighted* mean accessibility, ω, represents the average employment opportunities available to a given worker group within the *entire* study area, controlling for the number of resident workers of that group. The weighted mean can be interpreted as the total number of potential jobs accessible per resident worker. Interpreting the weighted mean identifies any large disparities in accessibility produced by the transit system overall. In notation this is

$$\omega = \frac{\sum_i A_i^S R_i}{\sum_i R_i} \qquad (6)$$

Considering accessibility of the three primary race/ethnicity groups to all types of jobs available to them, there is little difference in the accessibility provided to the two most populous groups, the non-Hispanic, white and Hispanic, white workers (97.22 and 93.42 accessible jobs per worker, respectively). The similarity in weighted mean accessibility provided to these two groups indicates that, on average across the entire system, there are not large differences in accessibility. In other words, the transit service provides comparable access to job sites. However, there is a large difference in accessibility between the two former groups and non-Hispanic, black workers (73.36 accessible jobs per worker). This suggests that, system-wide, non-Hispanic black workers are provided far less accessibility than their counterparts. It is important to consider that in this study area, non-Hispanic, black workers are the lesser populated minority group (15,185 workers), while there are more Hispanic, white workers (21,009 workers).

With regard to commuters' industry type, sharp differences are observed within the same race/ethnicity group. This is particularly interesting because it indicates that, holding race constant, workers of different industries and thus different socioeconomic status are not served in the same fashion by transit. Considering the three industry classes used in this analysis, it seems that the transit network would be configured to provide the greatest accessibility to workers in the retail or entertainment industry, as wages are typically lower than in the professional sectors, and thus, these workers would be more in need of transit service. However, the weighted mean accessibility for Hispanic, white professional workers, for example, is 116.05 employment opportunities per resident worker, and only 84.15 and 75.89 employment opportunities per retail and entertainment resident worker, respectively. This finding is counterintuitive because conventional thinking states that lower-income people would be more likely to use transit. Because of their potential propensity to be dependent on transit, it is reasonable to surmise that workers in retail and entertainment industries would be provided greater accessibility than workers in professional industries, who are more apt to choose other modes of transportation. However, this is not the case. Professional workers have greater accessibility to employment, as evidenced by their higher number of accessible employment opportunities.

Table 1. Weighted Mean and Mean Accessibility to Employment

Worker Group	Employment Type	Cumulative Opportunities	Total Resident Workers	Weighted Mean (ω)	Mean ($\bar{x}$)
Non-Hispanic, White	All Available jobs	9,446,149	97,166	97.22	1124.04
Non-Hispanic, Black	All Available jobs	1,113,944	15,185	73.36	183.02
Hispanic, White	All Available jobs	1,962,777	21,009	93.42	250.21
Non-Hispanic, White	Professional Jobs	1,528,165	15,113	101.11	363.37
Non-Hispanic, Black	Professional Jobs	90,066	1,172	76.85	32.18
Hispanic, White	Professional Jobs	201,007	1,732	116.05	79.68
Non-Hispanic, White	Retail Jobs	1,392,567	10,974	126.90	294.52
Non-Hispanic, Black	Retail Jobs	126,576	1,508	83.94	59.76
Hispanic, White	Retail Jobs	180,847	2,149	84.15	72.11
Non-Hispanic, White	Entertainment Jobs	163,862	2,159	75.89	182.37
Non-Hispanic, Black	Entertainment Jobs	72,107	592	121.80	37.08
Hispanic, White	Entertainment Jobs	163,862	2,159	75.89	61.43
Non-Hispanic, White Below Poverty Level	All Available Jobs	526,973	7,083	74.39	160.61
Non-Hispanic, Black Below Poverty Level	All Available Jobs	100,373	1,247	80.49	36.60
All Hispanic Below Poverty Level	All Available Jobs	415,896	6,704	62.03	105.73

Accessibility as experienced by workers living below the poverty level is also low relative to total accessibility. This suggests that workers living below the poverty level are not provided adequate accessibility to employment. Workers living below poverty are likely to be transit dependent because of their economic status. To further probe these findings, we can consider mean zonal accessibility and spatial patterns of accessibility (see Table 1). Mean zonal accessibility is calculated as the mean of all zonal accessibility indices (cumulative opportunities) for a particular worker group. In contrast with weighted mean accessibility, mean zonal accessibility simply averages the cumulative opportunities measure estimated for each zone, irrespective of its number of workers. In notation, this is simply

$$\overline{x} = \frac{\sum_i A_i^S}{n}$$

$$(7)$$

where n = the total number of zones.

The inclusion of weighted mean accessibility and the interpretation of spatial patterns of accessibility allow an additional scale of analysis. When we consider mean accessibility, non-Hispanic, black workers experiences the lowest accessibility in each employment category. This finding strengthens the preliminary conclusion that this worker group is not provided the same level of accessibility as other race/ethnicity worker groups. The following maps allow us to consider the spatial distribution of the accessibility indices that produce the mean accessibility values.

Non-Hispanic, white mean accessibility to professional jobs is 363.37 cumulative opportunities per resident worker, while that for non-Hispanic, black professional workers is 32.18. The spatial patterns of accessibility of these groups (Figure 4) are helpful in explaining this relatively large disparity. High areas of accessibility to professional jobs for non-Hispanic, white workers (see Figure 4) are clustered about the urban core, along several linear paths which are routes serving these workers' neighborhoods. In contrast, high accessibility for non-Hispanic, black workers is dispersed throughout the service area, with few striking clusters of high accessibility neighborhoods. This dispersed pattern reflects the lack of routes traveling from non-Hispanic, black neighborhoods to high concentrations of professional jobs.
The primary difference between the weighted mean (equation 6) and mean accessibility (equation 7) is that mean accessibility reflects how the transit system serves *neighborhoods*, rather than the entire service area. Weighted mean accessibility will produce sharp contrasts only if an entire group is provided substantially more or less accessibility than another; the weighted mean thus produces a smoothing effect. The mean accessibility index explicitly considers each neighborhood's accessibility score. As a result, neighborhoods experiencing extremely high or low accessibility for a given worker group will have measurable influence on the mean accessibility.

Figure 4. Comparison of accessibility to professional jobs

Total Jobs Accessible Per Non-Hispanic, White Resident Worker Within 30 Minutes

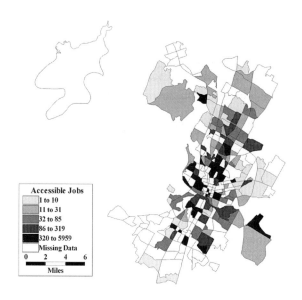

Total Jobs Accessible Per Non-Hispanic, Black Resident Worker Within 30 Minutes

Given the properties these summaries measure, it is not surprising that the there is greater disparity among mean accessibility values, as opposed to weighted means. An interesting point to consider is that most groups have higher mean accessibility values than weighted mean accessibility, and only a few groups exhibit the opposite relationship. In each employment category, non-Hispanic, black workers have lower mean accessibility than weighted mean accessibility, and Hispanic workers also have lower mean accessibility in three of five employment categories. Conversely, non-Hispanic, white workers experienced higher mean than weighted mean accessibility in each category. The relatively low mean accessibility values for the minority groups indicate that there are a significant number of neighborhoods where these worker groups have low accessibility; these values cause the mean accessibility value to drop far below the system-wide weighted mean. The opposite effect occurs for non-Hispanic, white neighborhoods; higher mean accessibility values are driven by zones with relatively high accessibility. While non-Hispanic, white workers as a whole do not achieve far greater accessibility than their minority counterparts, it is evident that there are some neighborhoods where they receive relatively high levels of service, relative to system-wide levels.

This pattern of low minority mean accessibility and high non-minority mean accessibility is repeated throughout each group; non-Hispanic, white workers, emerged as having the greatest mean accessibility in each employment category, while Hispanic-white workers were second each time, and non-Hispanic, black workers had the lowest mean accessibility. The uniformity in this pattern is strong evidence that the transit network favors non-Hispanic, white neighborhoods relative to minority neighborhoods.

Between minority groups, the Hispanic, white worker group consistently emerges as having greater accessibility than the non-Hispanic, black worker group. To analyze accessibility between these groups, it is important to consider whether they share the same neighborhoods. To do so, we can examine the spatial patterns of these race/ethnicity groups within the same industry. Figure 5 depicts the spatial accessibility of Hispanic, white and non-Hispanic, black entertainment workers. It is evident that these two minority groups do not share the same residential and employment neighborhoods, as their concentrations of high and low accessibility, even within the same industry, is not alike. Because the spatial accessibility patterns of these two groups are dissimilar, we can conclude that underserved non-Hispanic, black and Hispanic neighborhoods are not duplicate low-service areas; rather both minority groups are separately underserved.

In general, these findings identify observable disparities in accessibility relative to the socioeconomic status of worker groups. In most cases, worker groups who are most likely to be dependent on transit achieved lesser accessibility than non-minority or professional workers.

Figure 5. Comparison of accessibility to entertainment jobs

Total Jobs Accessible Per Non-Hispanic, Black Resident Worker Within 30 Minutes

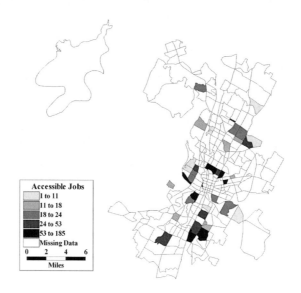

Total Jobs Accessible Per Non-Hispanic, White Resident Worker Within 30 Minutes

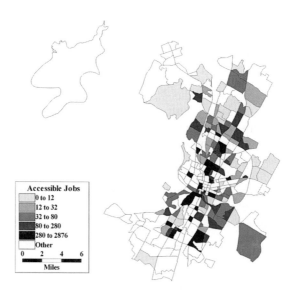

DISCUSSION AND CONCLUSIONS

As we move toward developing more comprehensive understandings of accessibility, questions of equity should be at the forefront of research efforts. There is complex interplay between contemporary land use patterns, limits on service provision, and the persistent disadvantages faced by people in housing and labor markets. It is in this milieu that public transit systems operate, and they must remain cognizant of these challenges so as to deliver accessibility in an equitable way.

This analysis of bus transportation in Austin, Texas demonstrated divergence in the levels of accessibility people experienced. Variations were revealed based on socioeconomic status and location relative to urban employment opportunities, as well as the nature of the transit service itself. The relatively low number of employment opportunities accessible to Hispanic white, and especially non-Hispanic, black workers in the lower-paying service, retail, and entertainment industries suggest that minority residents in this study area may be underserved by the transit system. Clearly our empirical findings can only speak to conditions in Austin at the time of our study, though our work has implications for broader accessibility issues.

First, researchers and planners must be aware of the multi-scale nature of accessibility. Returning to our research, the picture of accessibility produced at the neighborhood level was very different than the system level assessment. If scalar variation were not explored, then these deficiencies would go unnoticed. As we have shown, our multi-scale approach is one means of addressing this problem, but even still, issues involving the flexible definitions of neighborhoods (e.g. Horner and Murray, 2004), and the impacts of location-based accessibility on specific individuals (Kwan and Weber, 2003), among others, are indicative of the prominent role scale, and more broadly geography, plays in accessibility research.

Secondly, by stratifying workers in our analysis by race/ethnicity and industry we were able to detect disparities in the accessibility provided to various groups. We stated at the outset that the issue of race and ethnicity had previously not been explored in the context of accessibility provision. And more generally, others have indicated the need for transportation research that is attuned to issues of race (e.g. Giuliano, 2003). Researchers have argued that minority workers face barriers to economic success in the form of discrimination in housing and labor markets (Kain, 1992; McLafferty and Preston, 1999); and poor accessibility to employment greatly decreases their likelihood of experiencing economic and social success (Kain, 1968). Thus, transportation systems must be carefully planned to foster the social inclusion of diverse urban populations.

In conclusion, the results of this study provide empirical evidence that conditions of accessibility, especially in minority and low-income neighborhoods, require greater attention and resources. Equity should remain an important consideration in accessibility research; it is of particular importance for mass transit, where a lack of accessibility has the potential to marginalize people.

REFERENCES

Blumenberg, E. (2002). On the way to work: Welfare participants and barriers to employment. *Economic Development Quarterly,* **16**, 314-325.

Capital Area Metropolitan Planning Organization (2000). *CAMPO 2025 Transportation Plan.* Available on www.campotexas.org

Capital Metropolitan Transportation Authority (2005). Available on www.capmetro.org.

Census Transportation Planning Package (CTPP) 2000. Bureau of Transportation Statistics, available on www.bts.gov. Accessed May 14, 2004.

Cervero, R. (1989). Jobs-housing balancing and regional mobility. *Journal Of The American Planning Association,* **55**, 136-150.

Cervero R., O. Sandoval and J. Landis (2002). Transportation as a stimulus of welfare-to-work: Private versus public mobility. *Journal Of Planning Education And Research,* **22**, 50-63.

Domencich, T. and D. McFadden (1975). *Urban Travel Demand: A Behavioral Analysis.* North Holland Publishing Company, New York.

Fotheringham, A. S. and M. E. O'Kelly (1989). *Spatial Interaction Models: Formulations and Applications.* Kluwer Academic, Dordrecht.

Giuliano, G. (2003) Travel, location, and race/ethnicity. Transportation Research A, 37, 351-372.

Giuliano, G. and K. Small (1993). Is the journey to work explained by urban structure? *Urban Studies,* **30**, 1485-1500.

Hägerstrand, T. (1970). What about people in regional science? *Papers of the Regional Science Association,* **24**, 7-21.

Handy, S. L. and D. A. Niemeier (1997). Measuring accessibility: an exploration of issues and alternatives. *Environment and Planning A,* **29**, 1175-1194.

Hansen, W. (1959). How Accessibility Shapes Land Use. *Journal of the American Institute of Planners,* **25**, 73-76.

Hanson, S., ed. (1995) *The Geography of Urban Transportation.* The Guilford Press, New York.

Horner, M. W. (2002). Extensions to the concept of excess commuting. *Environment and Planning A,* **34**, 543-566.

Horner, M. W. (2004). Spatial dimensions of urban commuting: A review of major issues and their implications for future geographic research. *The Professional Geographer,* **56**, 160-173.

Horner, M. W. and A. T. Murray (2004). Spatial representation and scale impacts in transit service assessment. *Environment and Planning B,* **31**, 785-797.

Ihlanfeldt, K. and D. Sjoquist (1998). The spatial mismatch hypothesis: A review of recent studies and their implications for welfare reform. *Housing Policy Debate,* **9**, 849-892

Immergluck, D. (1998). Job proximity and the urban employment problem: Do suitable nearby jobs improve neighbourhood employment rates? *Urban Studies,* **35**, 7-23.

Kain, J. (1968). Housing segregation, Negro employment, and metropolitan decentralization. *Quarterly Journal of Economics,* **82**, 175–97.

Kain, J. (1992). The spatial mismatch hypothesis: Three decades later. *Housing Policy Debate*, **3**, 371–460.

Kwan, M. P. (1998). Space-time and integral measures of individual accessibility: A comparative analysis using a point-based framework. *Geographical Analysis*, **30**, 191-216.

Kwan, M.P. and J. Weber (2003). Individual accessibility revisited: Implications for Geographical analysis in the twenty-first century. *Geographical Analysis*, **35**, 341-53.

Kwan, M. P., A. T. Murray, M. E. O'Kelly, and M. Tiefelsdorf (2003). Recent advances in accessibility research: representation, methodology and applications. *Journal of Geographical Systems*, **5**, 129-38.

Levine, J. (1998). Rethinking accessibility and jobs-housing balance. *Journal of the American Planning Association*, **64**, 133-149.

Levinson, D. M. (1998). Accessibility and the journey to work. *Journal of Transport Geography*, **6**, 11-21.

Meyer M. and E. Miller (2001). *Urban Transportation Planning: A Decision Oriented Approach*, McGraw-Hill, New York.

Miller, H. J. (1999). Measuring space-time accessibility benefits within transportation networks: basic theory and computational procedures. *Geographical Analysis*, **31**, 1999, 187-212.

Murray, A. T. (2001). Strategic analysis of transport coverage. *Socio-economic Planning Sciences*, **35**, 175-88.

Murray, A. T. and X. Wu (2003). Accessibility tradeoffs in public transit planning. *Journal of Geographical Systems*, **5**, 93-107.

O'Kelly, M. E. and M. W. Horner (2003). Aggregate accessibility to population at the county level, U.S. 1940-2000. *Journal of Geographical Systems*, **5**, 5-23.

Ong, P. and E. Blumenberg (1998). Job access, commute and travel burden among welfare recipients. *Urban Studies*, **35**, 77-93.

Ong P. and D. Houston (2002). Transit, employment and women on welfare. *Urban Geography*, **23**, 344-364.

O'Sullivan, D., A. Morrison and J. Shearer (2000) Using desktop GIS for the investigation of accessibility by public transport: an isochrone approach. *International Journal of Geographical Information Science*, **14**, 85-104.

Pas, E. I. (1995). "The Urban Transportation Planning Process," in *The Geography of Urban Transportation*, S. Hanson, ed. Guilford Press, New York.

Preston, V. and S. McLafferty (1999). Spatial mismatch research in the 1990s: progress and potential. *Papers in Regional Science*, **78**, 387-402.

Raphael, S and L. Rice (2002). Car ownership, employment, and earnings. *Journal Of Urban Economics*, **52**, 109-130.

Sanchez, T. (1999). The connection between public transit and employment - The cases of Portland and Atlanta. *Journal Of The American Planning Association*, **65**, 284-296.

Sanchez, T., R. Stolz, and J. Ma (2004a) Inequitable effects of transportation policies on minorities. *Transportation Research Record*, **1885**, 104-110.

Sanchez, T., Q. Shen, and Z. Peng (2004b) Transit mobility, jobs access and low-income labour participation in US metropolitan areas. *Urban Studies*, **41**, 1313-1331.

Scott D., P. Kanaroglou and W. Anderson (1997). Impacts of commuting efficiency on congestion and emissions: Case of the Hamilton CMA, Canada. *Transportation Research Part D-Transport and Environment*, **2**, 245-257

Shen, Q. (1998). Location characteristics of inner-city neighborhoods and employment accessibility of low-wage workers. *Environment and Planning B-Planning & Design*, **25**, 345-365.

Shen, Q. (2000). Spatial and social dimensions of commuting. *Journal of the American Planning Association,* **66**, 68-82.

Sultana, S. (2002). Job/housing imbalance and commuting time in the Atlanta metropolitan area: Exploration of causes of longer commute time. *Urban Geography*, **23**, 728-749.

Wu, C. and A. Murray (2005). Optimizing public transit quality and system access: the multiple-route, maximal covering/shortest-path problem. *Environment and Planning B: Planning and Design*, **32**, 163–178.

CHAPTER 11

PARCEL-LEVEL MEASURE OF PUBLIC TRANSIT ACCESSIBILITY TO DESTINATIONS

Brian Ho-Yin Lee, University of Washington

INTRODUCTION

Researchers and practitioners have long recognised accessibility as an important concept for consideration in transportation and land-use planning (Hansen, 1959; Willier, 1947). There has been much research conducted to operationalise this concept as practical and meaningful measurements (Kwan *et al.*, 2003). There remains, however, a need for measurement techniques that can capture the multiple dimensions of accessibility as well as produce easily interpretable results. Although measures of the different components of accessibility—transportation, spatial, and temporal (Burns, 1979)—have been developed to various degrees in isolation, attempts to relate these parts and assess the whole have been rare until the recent past (Handy and Niemeier, 1997; Kwan *et al.*, 2003). This is particularly true for the public transit mode.

The translation of accessibility from a concept into a measurement is especially complicated for public transit because of its inherent spatial and temporal constraints. With the exception of demand-responsive bus service, public transit is provided in a specific network of routes on a repetitive, fixed-schedule basis with vehicles stopping to pick up and deliver passengers at designated locations. Each route, however, may have deviations in its travel pattern (e.g., complete and partial routes) and different types of service (e.g., local and limited stop service) on a discretionary basis. The frequencies of service usually differ during different periods of the day and week and can often change with the seasons. Furthermore, there are significant out-of-vehicle cost components, such as waiting time and transfers, to be considered.

The multi-modal nature of public transit further compounds the spatial and temporal constraints. Transit stops and stations usually do not coincide precisely with the trip origins and destinations. For most users, access to and from the public transit network requires travel by at least one other mode, of which the most common is walking. For some transit stations, bicycles and private vehicles may contribute significantly to the transit ridership but most of these users would still have to walk at the other end of their trips. A measure of accessibility by public transit, therefore, necessitates the consideration of different complementary and supporting transportation sub-networks, particularly for pedestrians.

The existing methods to measure public transit accessibility are typically coarse and do not systematically address all of the issues described above. Most of the research has concentrated on the spatial dimension while the temporal dimension is often not explicitly recognised or incorporated into the measurement techniques (Polzin *et al.*, 2002). The development of the spatial measures is complete as most studies found in the transit planning literature focus on access to and from the public transit network with no regard towards actual destinations (Hsiao *et al.*, 1997; Javid and Seneviratne, 1994; Kerrigan and Bull, 1992; KFH Group, 1999; Murray, 2001; O'Neill *et al.*, 1992; Pike and Vougioukas, 1982; Polzin *et al.*, 2002; Pratt and Park, 2000; Pulugurtha *et al.*, 1999; Rastogi and Rao, 2003; Ruppert, 1979; Zhao *et al.*, 2003). Aggregate spatial units are often used to estimate transit service coverage. These estimates, however, are known to be scale sensitive and best represented by disaggregate spatial information (Horner and Murray, 2004). Furthermore, public transit accessibility may vary significantly within large areas when pedestrian access is considered. In theory, smaller zones or disaggregate units of analysis at the parcel or household levels should produce more accurate estimates of accessibility (Handy and Niemeier, 1997).

The goals of this research are twofold. The primary objective is to develop a procedure to measure public transit accessibility at the parcel-level that accounts for time considerations as well as access to destinations and activities. This involves the use of Geographic Information Systems (GIS), known for their capacity to integrate different spatial data and represent spatial relations (Peng and Dueker, 1995), in combination with other database management and analysis tools that can handle temporal data. The secondary objective is to represent the measurement results for the purposes of communicating with planners, policy makers, and stakeholders as well as making connections with other spatial, demographic, and socio-economic variables. In these respects, GIS software is useful for producing effective maps and making informative links between land parcels and a transit system.

This paper reports on a prototype of this parcel-level public transit accessibility measurement method and presents one test case to validate its functions. Of course, a single test of this kind cannot cover the whole range of public transit planning problems for which this technique is applicable. Nevertheless, there is clear evidence for many new prospects that could be developed from this technique.

The next section provides a review of the measurement techniques that have been developed to date and points out some of their limitations. Section three presents the conceptual framework of the proposed accessibility measure and a step-by-step description of the analysis process. Section four describes the data needs and structure for the proposed technique; data from the King County Metro transit system was used as an example for the prototype. Section five illustrates a test case application from the City of Seattle and discusses the results. The final section concludes with a summary of the capabilities and limitations of the technique and makes recommendations for further developments.

EXISTING MEASUREMENT TECHNIQUES

This section reviews existing public transit accessibility measurement techniques and their development. First, there is an examination of the selection of trip ends. Then, two broad categories of techniques are surveyed: those that measure access to a public transit network and those that measure public transit accessibility to destinations. Following this is a summary of the practical and research implications and the need for measurement advancements.

Selecting trip ends

By far the most commonly used trip end in existing measurement techniques is the home (Handy and Niemeier, 1997). Public transit accessibility is generally measured between a residential location and a destination or an access point in the transit network. Home-based accessibility measures can be related to the demographic and socio-economic characteristics of the residents (Peng and Dueker, 1995). This enables the identification of concentrations of vulnerable social groups who are relatively more dependent on public transit service. Inferences can also be made in regards to how well the transportation needs of these people are being met. The importance of home-based trips, however, has been in decline as non-home-based trips as well as trip chaining increase in proportions (Polzin and Chu, 2003). This may signify a growing need for non-home-based accessibility measurement techniques.

Some measures have been developed for evaluating the accessibility to employment opportunities in terms of the number of jobs or workers within a transit catchment area (Kittelson and Associates and URS, 2001) or the transit travel costs between employment and residential zones (Thompson, 2001). Work-based accessibility measures can serve as indicators to monitor the progress in bringing the labour force closer to jobs (Cervero *et al.*, 1999) and help explain the influence of location on commuting behaviour (Levinson, 1998; Sultana, 2002). The public transit accessibility to and from employment locations is especially important from a social welfare point of view for providing good quality service to transit dependent households. Transit service can act as a lifeline to sources of income and help to correct spatial mismatch between job and home locations.

Equity in transportation, however, does not only apply to travel for employment. It is also important to consider non-work related transit trips, as they may be particularly relevant for one- or no-vehicle households (Thompson, 2001). It should be a social goal to help transit dependent people travel to food retail stores, medical and child-care services, schools, community centres, and civic activities. With the decentralisation of jobs in the central business districts as well as the increasing dispersion of urban activities away from core cities, there is also a need to customize accessibility measures to individual land use and travel patterns (Thakuriah et al, 2003).

Measuring access to a public transit network

For the purposes of ridership forecasting and service performance evaluation, the concept of accessibility in transit planning often only refers to the assessment of transit stops and stations in terms of their catchment potentials (Murray, 2001). There are usually no regards to the destinations and activities that the users can actually reach. The underlying assumption is that access to the public transit network acts as a proxy for access to a range of specific destinations (Kerrigan and Bull, 1992). This, however, may be reasonably realistic only if the network is relatively dense and well-integrated.

One of the first measurement techniques to evaluate the extent to which a public transit network meets the needs of its users is the area buffer method. This method is represented by catchment area maps where circles are drawn around transit stops and stations of the network. The commonly accepted radius is 400 metres, or approximately one quarter mile, which has been found to be the maximum distance that a person is likely to walk to use transit service (Murray, 2001; Pratt and Park, 2000). From these maps, the proportion of an area having access to the transit system can be determined. Subsequently, the proportion of the population, workers, and activities within walking distance to transit service can be inferred. This simple technique is still widely used by transit agencies, particularly those in small urban areas with limited resources (KFH Group, 1999), despite the fact that it suffers from some serious and well-recognised flaws (Zhao *et al.*, 2003).

Early on in the development of transit assessment tools, Ruppert (1979) acknowledged that catchment area maps give the false impression of "a high degree of accessibility" when the majority of an area falls within the catchment. The buffers only establish a spatial link between an area and the transit network. They do not account for circuitous routes or provide any information on the temporal dimension of the service. The catchment area, therefore, shows geographically that "the lowest limit of accessibility" (i.e., a spatial link) has been established and nothing more (Ruppert, 1979). The area buffer method is also cumbersome when route density is high and the service area is large.

Other critics of the area buffer method pointed out the dangers of using a Euclidean distance (also referred to as air- or straight-line distance) to define the catchment area (Horner and Murray, 2004; O'Neill *et al.*, 1992) and area ratios to estimate the number of people or jobs being served (Javid and Seneviratne, 1994; Zhao *et al.*, 2003). The Euclidean distance represents the farthest reach of actual walking distance from a stop or station. The difference between that and the actual reach may vary significantly depending on the connectivity and configurations of the pedestrian network as well as the density of those connections. Circular catchment areas ignore physical barriers to walking and tend to overestimate the spatial service coverage of a transit system (Horner and Murray, 2004); Figure 1 compares catchment areas based on a Euclidean distance and a ground travel distance through the street network. Likewise, the use of area ratios to derive population and employment estimates is subject to representation problems. Catchment areas are usually overlaid with census tracts, Transportation Analysis Zones (TAZ), zip code areas, or some other areal aggregation system with socio-demographic information (Peng and Dueker, 1995). Residents and workers are assumed to be evenly distributed within these units even though they are often too large to capture variations in the local access and land use patterns (Hess *et al.*, 2001; Krizek, 2003). Aggregate spatial estimates such as these are sensitive to scale and area definition issues collectively known as the Modifiable Areal Unit Problem (MAUP), in which the estimates and their relations to other variables may change significantly with changes in zone size and boundary locations (Armhein, 1995; Hess *et al.*, 2001, Horner and Murray, 2004).

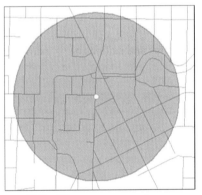
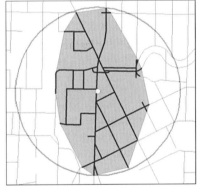

A) Euclidean measurement   B) Network measurement

Figure 1. Catchment areas based on Euclidean and network measurements of the same distance

To improve upon the area buffer method and area ratio calculations, O'Neill *et al.* (1992) proposed a procedure that uses network ratios, which partly accounts for the layout of the underlying street network. Analogous to using area ratios, population and jobs can be allocated using the proportions of street length within a Euclidean buffer to the total street length in an area. The fundamental assumption is that the number of residents and workers on a street is proportional to its length. It further assumes that they are uniformly distributed across all streets in an area, which the authors admitted to be a weak assumption but one no weaker than that of the area buffer method. The major contribution of this work is that

O'Neill *et al.* (1992) were able to show that network ratios is better suited for measuring access to a transit system than area ratios. Applications of network ratios, compared to area ratios, in evaluating transit market potentials and walk-transit access by other researchers lend support to this claim (Hsiao *et al.*, 1997; Pulugurtha *et al.*, 1999).

Recognising the shortcomings that still existed in the network ratio method, Zhao *et al.* (2003) employed disaggregate land use data to depict population distribution and explicitly considered barriers, such as bodies of water and community walls, in the pedestrian network. Parcel-level GIS data, detailing information on the exact locations and the spatial distribution of residential units, were integrated with street-level (i.e., network) GIS data, depicting barriers, to estimate the number of people living on properties within network walking distance of transit stops and stations. The total number of bedrooms in these properties, extracted from local tax assessment databases, multiplied by a room occupancy rate (or different rates for different types of residential units) was found to provide more realistic estimates of residents who can access transit. When comparing these results with those derived from area and network ratios, Zhao *et al.* (2003) found that the latter methods consistently overestimated the transit catchment potentials. In a publication of the same period, Horner and Murray (2004) similarly concluded that a disaggregate approach is best for service coverage analysis in transit studies.

Same as the area and network ratio methods, the parcel-level measurement technique neglects the time dimension of transit access. Acknowledging the need to incorporate temporal characteristics, the Transit Level of Service (TLOS) software, first developed in 1999, claims to be the first measurement tool that accounts for both space and time components of public transit (Kittelson and Associates and URS, 2001). The TLOS performance measure is percent person-minutes served, defined as the percentage of time that an average person has transit service available. The premise is that within a given area, only certain locations are within walking distance of a transit stop. Of the locations served, transit service is only provided at certain times of the day and week. The TLOS software rewards greater service frequencies and service coverage areas with higher values for performance. Using area- or network-based buffer ratios, TLOS calculates the total number of people and jobs served by transit each minute during a specified time interval. Working along similar lines, Polzin *et al.* (2002) developed a "time-of-day-based transit accessibility analysis tool" that recognises the temporal dimension of both the supply and demand sides of public transit service. On the supply side, the frequency of service for a route and the average wait time are used to determine the actual time duration in a period when transit service is available. This supply availability is then weighted by a time-of-day distribution that places greater significance to service in periods of high demands. As the TLOS software and this time-of-day tool use area or line buffers, they both sacrifice spatial clarity with area or network ratios to relate time and space. Their calculations based on zonal coverage have the same spatial distribution and area definition drawbacks as mentioned above.

Measuring accessibility to destinations by public transit

The techniques cited for measuring access to a public transit network do not specify actual destinations and activities. Of the ones in the literature that do, most are found in applications where gravity-based accessibility measures are used to analyse transit travel times between origin and destination TAZs. Recent examples of these include work by Grengs (2001), Thompson (2001), Pendyala *et al.* (2002), and Beimborn *et al.* (2003). This type of measure weights the quantity of activity opportunities at the destination zones by impedance as a function of travel time from the origin zones. The main principle of gravity-based measures is that the more there is of an activity and the shorter amount of time it is from an origin, the more it will contribute to the accessibility of that zone. This definition of accessibility was first formulated by Hansen (1959) as

$$A_{ij} = S_j \cdot f\left(T_{ij}\right),$$ (1)

where
A_{ij} is the accessibility at zone i to an activity within zone j,
S_j equals the size of the activity at zone j,
T_{ij} equals the travel time/distance between the two zones, and
f is a function that describes the effect of travel between zones.

In his analyses, Hansen (1959) defined

$$f = \frac{1}{T_{ij}^x},$$ (2)

where
x is an empirically determined exponent that describes the importance of an activity.

The accessibility at zone i, A_i, to an activity within all other zones is simply

$$A_i = \sum_j A_{ij}.$$ (3)

The use of gravity-based measurement techniques is common in regional transportation travel demand and mode choice models for calculating the accessibility by different travel modes. For public transit, there are different ways to account for transit accessible opportunities and the proportion of the population in each TAZ that can access them, as shown by the previous section. Most researchers and practitioners, however, use the crude approach of creating Euclidean buffers around transit stops and stations and using area ratios (Grengs, 2001; Pendyala *et al.*, 2002; Beimborn *et al.*, 2003) or simply assign the same transit travel times for everyone in each zone (Thompson, 2001). Most measurement tools also only consider a few broad categories of activities, such as service, industrial, and commercial jobs for accessibility to employment.

The TLOS software does have another component that can estimate the number of residents and jobs located within a certain transit travel time of a specific location (Kittelson and Associates and URS, 2001). Similarly, the isochrone approach devised by O'Sullivan *et al.* (2000) can describe the accessibility of one analyst-defined origin. Both of these cumulative measures identify the total catchment area of a location within a certain amount of time. As mentioned before, however, the TLOS performs area calculations at an aggregate level by using either Euclidean or network buffers. Likewise, O'Sullivan *et al.* (2000) employed route-level transit data, proximity polygons to represent walk access, and line buffers in their calculations. These measures, therefore, suffer from aggregation problems that also face the gravity-based TAZ techniques.

Practical and research implications

The accessibility measurement issues outlined present some limitations for practical transit planning, policy evaluation, and transportation research work. Simplistic measures of a transit system such as stop or station service coverage tend to exaggerate the potentials for existing and planned service to benefit the population (Horner and Murray, 2004; Kwan *et al.*, 2003). Aerial buffers around transit access points or along linear representations of transit routes overestimate ridership forecasts while the disregard for access to actual destinations and activities distorts the utilities of specific routes and services. Consequently policy makers and the public may receive unrealistic information regarding transit service evaluations and proposals. The trend for planners of new rail transit systems across North America to overestimate ridership may be partly attributable to ill-defined accessibility measures and coarse measurement techniques (Pickrell, 1992).

As for transportation research, zonal approaches and emphasis on home- and work-based transit trips provide an incomplete picture of transit user behaviours and travel choice considerations. It is recognised that the accessibility of a place to and from other places by public transit can affect how people make transportation- and location-related decisions. For example, Thompson (2001) found that transit accessibility could explain automobile ownership rates. Yet, the application of aggregate measures of accessibility to decision making processes that are best described at the disaggregate individual or household level leaves much room for error. The increasingly scattered distribution of dwelling units, work places, and other land use activities also call for measures at finer levels.

An accurate measurement of accessibility to specific destinations by public transit at the parcel-level would help the analysis of policies targeting those who are transit-dependent or willing to make a mode switch. It may also contribute significantly to transportation and land use planning and research by increasing the sophistication of evaluation and forecast models, such as those for travel mode choice, automobile ownership, and residential location.

Advancement needs for measurement techniques

Walk access is critical for most transit users and of great significance to the measurement of public transit accessibility. When walking is involved, measurement techniques should be sensitive to local conditions and a small zoning system or a disaggregate approach is needed to reveal any variations. The parcel-level method proposed by Zhao *et al.* (2003) is promising, but there remains a need to link such disaggregate spatial techniques with actual destinations or activities as well as the temporal dimension to achieve a more complete translation of the accessibility concept. This is true for transit planning and policy evaluation purposes and for travel model applications. Disaggregated accessibility variables are still not adopted in traditional aggregated travel demand forecasting and modal split models (Zhao *et al.*, 2002). Their effects on model outcomes are not yet verified. The first hurdle to overcome is the problem of defining and measuring the effects of land use and urban design on transit at the disaggregate level. Coarse measurements may present impediments in model developments: at least one study has found that more precise specifications of transit service lead to models with higher predictive powers when compared with conventional transit representations (Beimborn *et al.*, 2003).

Improvements in the availability of spatial data and computational power should also allow for the advancement of disaggregate measurement approaches. Parcel-level GIS data on land use and detailed databases of street networks, transit stops and stations, and transit schedules have become increasingly available in the recent decade. In addition, temporal considerations should no longer be neglected as public transit service is greatly time dependent.

CONCEPTUAL FRAMEWORK

The proposed technique uses a classification system of land uses at the parcel-level and measures the accessibility, in terms of the total travel time, to each location of a destination type from other land parcels by public transit. The accessibility of a single type of destination may serve as an indicator by itself for a particular purpose. Conversely, it may be combined with the accessibility measurements of other types of destinations to produce a composite measure. The "other" land parcels (i.e., the origins) may also be categorised by the same land use classification system to specify certain types of trips (e.g., home- or work-based trips). Ultimately, this technique can be used to measure the accessibility between any two types of land uses. It is up to the analysts or stakeholders to decide which pairs are of importance.

For example, the Food Policy Planning Council of a city may wish to evaluate the accessibility of food retail stores by socio-economically marginalised people. This population may include households with low income, low automobile ownership, single parents, and racial/ethnic minorities; people with significant physical challenges that limit their ability to drive; people who are too young or too old to drive; plus other non-drivers. In this case, all land parcels classified as being food retail stores could be identified as one type of destination. Alternatively they could be sub-categorised as different types of destinations according to the variety or types of foods sold or the store size. The origins of interest would be homes. The residences could be further broken down by types of dwelling units (e.g., single- or multi-family) and tenure (e.g., own or rental). Residential land parcels that are not accessible to any food retail stores by public transit, or ones that are only accessible to stores with a limited variety of foods can be identified. Then, the size of the vulnerable population not having access to an adequate supply of foods can be estimated using the technique proposed by Zhao *et al.* (2003) to link the dwelling units in these parcels with demographic and socio-economic information from other sources.

The proposed technique will take into account of in- and out-of-vehicle transit travel times to compute the accessibility measure. This includes travel times in transit vehicles; walk access from the origins to the transit network, as well as from the transit network to the destinations; transfer walk times if the transfers do not occur at the same stops; and average waiting times for transit vehicles to arrive at the stops or stations. One or more boundary conditions concerning the maximum travel time and the maximum number of transfers can be imposed to limit the extent of travel in the transit network. In addition, a condition may be set to identify land parcels deemed close enough to the destinations where it would not make sense to take transit to reach them. For the food retail store example, the Food Policy Planning Council could set a threshold as a policy judgement for the total travel time, say 45 minutes, above which it would consider unacceptable for residents to travel to reach a major grocery store. In addition, the Council could decide to differentiate between residential areas that are accessible to food retail stores by public transit with no transfer and those that require a maximum of one transfer because of the added burden of making transfers while carrying potentially heavy loads. Furthermore, it may consider any locations within one kilometre of a food retail store to be accessible on foot.

Figure 2 illustrates the analytical flow process for measuring the accessibility by public transit to one type of food retail store destination (e.g., grocery stores) with no transfer. The process by trips with transfers would be similar but more complicated. There would be additional steps in the calculation of the transit component to account for additional segments of in-vehicle travel time. A passenger alighting from one transit vehicle to transfer to another at the same or a different stop location would be considered a random arrival and a separate average waiting time would be included for each transfer. For transfers that include a walk component, additional walking times would be considered. The last step in this process is an example application for food planning and policy evaluation purposes. This process can be easily adapted for other types of destinations and can be used for calculating the accessibility by public transit to a set of destinations. Different commercial GIS packages and extensions such as TransCAD or Network Analyst in ArcGIS can be used to help employ this process. Most programs, however, may need the support of other tools such as Access, Oracle, MySQL, or Python for database management purposes.

Figure 2. Steps for measuring accessibility by public transit to one type of destination

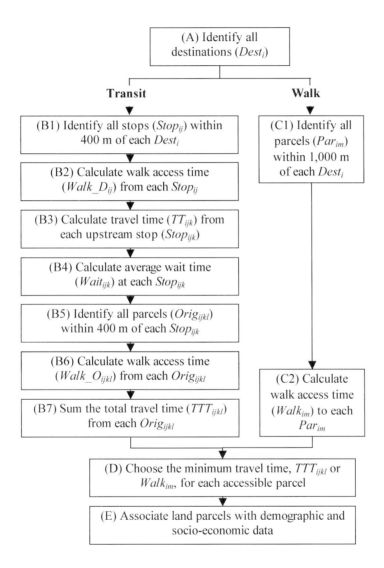

The following list provides a brief description for each step of the pseudo-code algorithm.

A) Identify all locations of the type of destination of interest ($Dest_i$), which in this case are all parcels having "grocery stores" as their recorded land use

To identify the origins accessible by public transit:

B1) For each $Dest_i$, identify all stops ($Stop_{ij}$) within a 400-m network distance. For each set of $Stop_{ij}$, keep only the closest one per transit route per direction:
   i) Calculate the distance (d_{ij}) to each $Stop_{ij}$ from each $Dest_i$
   ii) Sort each set of $Stop_{ij}$ by d_{ij} in ascending order
   iii) Examine each set of $Stop_{ij}$ and only keep the closest one per transit route per direction

B2) Calculate the walk access time ($Walk\_D_{ij}$) to each $Dest_i$ from each $Stop_{ij}$ by multiplying d_{ij} by an average walking speed, which depends on the segment of the population of interest (e.g., 1.2 m/s for a physically-able adult)

B3) For each $Stop_{ij}$, identify all stops that precede it ($Stop_{ijk}$) and calculate the scheduled travel time for all $Stop_{ijk}$ (TT_{ijk})

B4) Calculate the average wait time ($Wait_{ijk}$) for a transit vehicle to arrive at each $Stop_{ijk}$ by considering the day of the week and the time of the day of concern and dividing the average scheduled headway in half (e.g., the average wait time is assumed to be 5 minutes if the schedule headway between transit vehicles is 10 minutes)

B5) For each $Stop_{ijk}$, identify all land parcels ($Orig_{ijkl}$) within a 400-m network distance

B6) Similar to step B2, calculate the walk access time ($Walk\_O_{ij}$) from each $Orig_{ijkl}$ to each $Stop_{ijk}$

B7) Sum the total travel time (TTT_{ijkl}) for each $Orig_{ijkl}$ to access $Dest_i$

To identify the origins accessible by walking:

C1) For each $Dest_i$, identify all land parcels (Par_{im}) within a 1,000-m network distance

C2) Similar to step B2, calculate the walk access time ($Walk_{im}$) from each Par_{im} to each $Stop_{ijk}$

D) For each land parcel that has been selected as accessible to $Dest_i$ by public transit ($Orig_{ijkl}$) and/or walking (Par_{im}), choose the minimum travel time, TTT_{ijkl} or $Walk_{im}$. As a policy judgement, also exclude any parcel that has a travel time greater than 45 minutes.

E) Food planning and policy evaluation application: identify at risk parcels where it is not practical to access any grocery stores by public transit or walking. If the people living in these parcels exhibit certain demographic and socio-economic characteristics, such as low household income or low automobile ownership, then there may be a need for interventions.

The public transit accessibility, A_{ix}, to any location of the destination type i from any origin x is, therefore,

$$A_{ix} = \min\left(TTT_{ijkl}, Walk_{im}\right), (4)$$

where

A_{ix} is the public transit accessibility to i from x in terms of the total travel time,
TTT_{ijkl} is the total travel time by public transit to i from l as previously defined,
$WALK_{im}$ is the walking time to i from m as previously defined, and
l and m are both subsets of x (i.e., $l \in x$ and $m \in x$).

To calculate the scheduled in-vehicle travel times and the average waiting times, it is necessary to consider the day of the week and the time of the day. It is possible to measure the accessibility for every hour of the day and for each day of the week. But since most transit services have the same or similar schedule patterns for all weekdays and certain periods of the day, it is reasonable to only consider selected periods. One example is
    Weekday 06:00-08:59 (AM Peak);
    Weekday 09:00-14:59 (Midday);
    Weekday 15:00-17:59 (PM Peak);
    Weekday 18:00-21:59 (Evening);
    Saturday 08:00-21:59 (Saturday); and
    Sunday 08:00-21:59 (Sunday).
In this case, the weekday AM and PM Peak periods were determined by the morning and afternoon rush hour schedules set by the King County Metro for the Seattle region. For transit operations with different peak periods, the hours can be adjusted to match the schedules. In other cases, the public transit accessibility can be measured for any periods of interest to match observed travel demand patterns or activity schedules.

This measurement process identifies all land parcels within 45 minutes of total travel time to one or more grocery stores. As part of their attributes, each of these parcels will have a field that denotes the minimum total travel time, by transit or walking, to the closest store. Their spatial distribution and that of the grocery stores can both be mapped and they are linked temporally by the travel time. Those land parcels excluded from this selection can be identified as not having practical access to any grocery stores by public transit or walking— further analysis could identify any socially vulnerable residents that may be affected. Furthermore, the total travel time attribute, or a dummy variable denoting whether a parcel is accessible or not to a grocery store, may be used for modeling purposes. This would be especially useful for discrete choice and activity-based types of travel models.

DATA

The prototype for this accessibility measurement technique was developed and tested in the context of Seattle, Washington. Since buses are the only form of public transit service of any significance in this region, they are the focus of the test case. In theory, the transferability of this research to other forms of public transit is high and the methodology should be readily adaptable to other types of fixed-route service and to other cities.

Table 1 summarises the data that were used in this test case, their sources, and any significant preparatory processing or transformations. These types of data should be available in most urban centres with scheduled transit service, though the data structures will vary. Instead of detailing the data processing and transformation procedure, the general tasks are described to identify the actions and their purposes. Idiosyncratic details are not included here as there may be considerable variations for other transit systems.

Table 1. Data Used

Data	Source	Processing/Transformation
King County street network	King County (via WAGDA[1])	Deleted non-pedestrian links to create walking network
Bus stop table (shows stop point locations)	King County Metro[2]	Related to each other to identify all stops for each pattern in sequence and by direction.
Bus time-point (TP) table (shows TP point locations)		
Bus time-point interval (TPI) table (a TPI is a directional path between 2 TP)		
Bus pattern table (a pattern is an ordered set of TPI and is a path that a route follows)		
Bus route network	King County Metro[2]	Related to pattern table to identify routes
Bus trip table (a trip is a scheduled run on a pattern)	King County Metro[2]	Related to each other and to pattern table to calculate frequencies and travel times
Bus event table (an event is a scheduled pass at a TP)		
King County tax assessment land parcels	King County Department of Assessments (via WAGDA[1])	Land uses were categorised according to a classification system
Census population	US Census	n/a

[1] Washington State Geospatial Data Archive, University of Washington Libraries

[2] Management Information and Transit Technology, King County Metro Transit Division

The most important part of the data structure is the correct association between the stops and the transit patterns and routes that they serve. All stops must be ordered according to the direction of travel. This will ensure that the origins and destinations are properly linked by the transit network and enable the scheduled travel times to be interpolated between the time-points. King County Metro, for example, serves approximately 9,500 bus stops in the Seattle region with close to 250 bus routes traveling on more than 1,200 different patterns. The association and ordering of the bus stops and the patterns and routes represents a laborious task for a large transit system if these relationships have not already been established. Any accessibility measurement technique, however, is only as accurate and precise as the representation of the transit network itself.

TEST CASE APPLICATION

Continuing with the grocery store example, the following application illustrates the results of the parcel-level measure of public transit accessibility to destinations. For simplicity, the first part of the test case used one location of a grocery store in the City of Seattle as the destination and residential parcels as the origins to capture the accessibility from homes to this particular food store. The four weekday and two weekend periods mentioned previously were considered. The number of people that can be served by this location using public transit or walking was estimated with population data from the 2000 US Census (US Census Bureau, 2004). The results from the proposed method are compared to two other methods. The advantage of reviewing the accessibility to one location is that the precision of each method can be distinguished and compared with relative ease. The second part of the test case illustrates the public transit accessibility to all grocery stores in Seattle. The proposed measurement process was repeated for each location and maps were created to show how the results for all locations of a destination type may be communicated.

Public transit accessibility to one grocery store location and population estimates

The two other transit accessibility measurement and population estimation techniques that are being compared are the area ratio method using Euclidean buffers around the bus stops ("area ratio") and a simplified version of the proposed parcel-level method ("simplified parcel"). Both of these techniques only account for spatial accessibility in terms of access between the residential origins and the grocery store destination by public transit. The simplified version differs from the proposed parcel method by not accounting for travel times and the period when travel is being considered. Similar to the proposed version, however, it does measure accessibility and estimate population at a disaggregate level using network distances.

Figures 3 and 4 show the graphical results of the area ratio and simplified parcel methods, respectively. The area ratio buffers in Figure 3 radiate 400 m from each stop and they convey an extensive sense of coverage without revealing any information regarding the land uses within these buffers and the people and activities they contain. In contrast, the simplified parcel method in Figure 4 identifies the locations of residential uses and accounts for walking distances within the pedestrian network. It is easy to see, from the individual parcels highlighted, that many stops far west of the destination do not serve large concentrations of residential land uses. On the other hand, the areas immediately north and south of the grocery store, along the transit routes that can access the store, are mainly residential neighbourhoods. This type of information can be useful in determining how public transit is meeting the needs of certain trips by linking specific land uses at the origins and destinations. The parcels can also be categorised by dwelling type or the number of units to highlight the intensities of these uses. Nevertheless, Figure 4 does not reveal anything regarding the temporal component of accessibility.

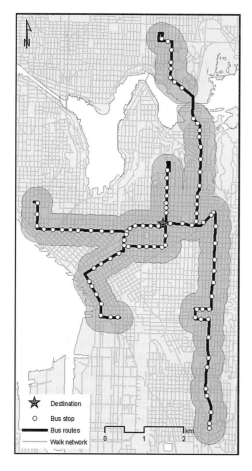

Figure 3. Public transit accessibility by the area ratio method

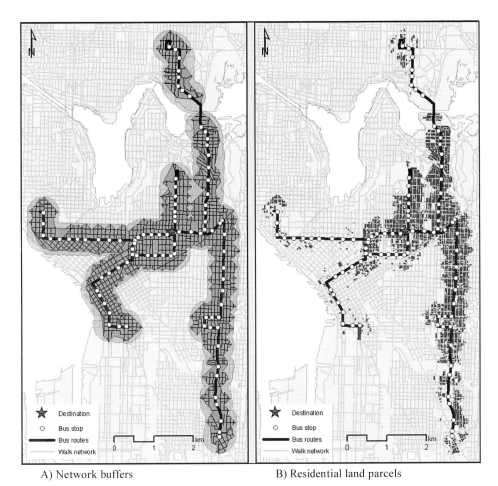

A) Network buffers   B) Residential land parcels

Figure 4. Public transit accessibility by the simplified parcel method

Figure 5 shows the results of the proposed parcel method for the weekday midday period. The residential parcels are categorised by the total travel times in minutes, using public transit or on foot, to the destination. An important distinction between this and the previous figures is that seven stops on a route southwest of the destination were not included because service from those stops was not available during the weekday midday period. Consequently, a number of parcels that were shown to be served by public transit in Figure 4 are not included in Figure 5. Although the difference was small because the southwest route did not serve a large concentration of residential land uses, this case illustrates how the proposed parcel method may make a significant difference in other neighbourhoods or for trips from other land uses. Another distinction is that numerous parcels in Figure 5 are shown to be more than 45 minutes of travel time away from the selected grocery store. According to a boundary condition set earlier, this would exclude them from being considered accessible. It is also easy to discern that the routes from north of the destination provide more frequent and/or faster service than the route from the south. Overall, the proposed parcel method allows for more precise specifications and measurements of accessibility.

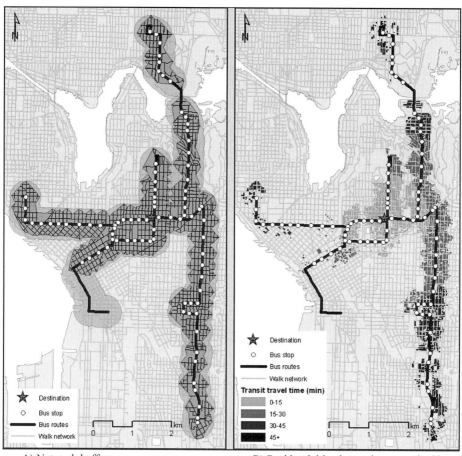

| A) Network buffers | B) Residential land parcels categorised by total travel time by transit to destination |

Figure 5. Public transit accessibility by the proposed parcel method for the weekday midday period

The populations that could be served by public transit to access this grocery store were estimated at the census block group level in order to examine the relative precision of each of these methods. For the area ratio method, the proportion of each block group covered by the Euclidean buffer was multiplied by the total population of that group. For the simplified and proposed parcel methods, the ratio of the number of residential units in the selected parcels of each block group to the total number of units in that group was used instead. In total, there

Table 2. Estimated Census Tract Populations Accessible to the Destination by Various Methods

Census Tract	Total Population	Estimated Populations Served			Ratio	
		Area Ratio [A]	Simplified Parcel [B]	Proposed Parcel [C]	[A]/[C]	[B]/[C]
4400	1,597	142	10	10	14.28	1
4500	923	3	0	0	n/a	n/a
5200	3,260	874	396	391	2.23	1.01
5301	8,509	5,558	4,949	4,949	1.12	1
5302	2,528	960	348	348	2.76	1
5802	2,016	3	0	0	n/a	n/a
6200	3,800	2,276	3,037	2,974	0.77	1.02
6400	3,201	3,114	2,681	2,631	1.18	1.02
6500	1,511	915	831	788	1.16	1.05
7000	4,744	2,355	2,183	2,131	1.10	1.02
7100	1,809	1,158	1,183	1,183	0.98	1
7200	3,084	2,439	1,483	1,483	1.64	1
7300	2,218	1,653	1,283	1,271	1.30	1.01
7400	8,932	6,838	6,313	6,307	1.08	1.00
7500	5,554	5,242	5,334	5,323	0.98	1.00
7600	3,261	3,261	3,261	3,176	1.03	1.03
7700	3,990	3,760	3,616	3,573	1.05	1.01
7800	4,577	759	443	393	1.93	1.13
7900	4,232	2,626	2,106	1,932	1.36	1.09
8001	3,410	2,039	1,543	1,543	1.32	1.00
8002	1,618	393	212	212	1.85	1
8100	3,477	3,168	3,329	1,840	1.72	1.81
8200	2,875	1,810	1,411	1,406	1.29	1.00
8300	1,882	1,742	1,793	1,793	0.97	1
8400	3,838	3,160	3,133	3,133	1.01	1
8500	4,838	565	0	0	n/a	n/a
8600	2,993	259	240	229	1.13	1.05
8700	2,256	628	563	549	1.14	1.03
8800	3,506	3,219	3,075	3,046	1.06	1.01
8900	4,596	3,228	2,816	2,741	1.18	1.03
9000	2,134	576	514	514	1.12	1
9100	2,083	933	1,012	0	n/a	n/a
9200	1,836	1,765	1,836	0	n/a	n/a
9300	1,627	86	667	0	n/a	n/a
9400	2,045	143	211	211	0.68	1
9500	5,142	3,054	2,384	2,331	1.31	1.02
10000	1,954	469	94	86	5.47	1.09
10100	2,296	111	12	9	12.69	1.39
Total	124,152	71,283	64,301	58,505	1.22	1.10
Percent	100%	57.4%	51.8%	47.1%		

were 117 block groups in 38 census tracts that intersected one or more of the area ratio buffers. The calculations for each block group are shown in the Appendix. Table 2 shows a summary of these calculations at the census tract level and it confirms the proposed parcel method's advantage in precision. In most individual block groups and also in the aggregate, the area ratio method overestimated the number of people that are served when compared to the other two methods. There are also block groups where the area ratio method made significant underestimations. Compared to the proposed parcel method, the simplified version also overestimated the population served. In total, the estimate by the simplified method was 10% larger.

Although it is not possible to say with certainty which set of results is the closest to the actual number of people that can be served by public transit to the destination, it is reasonable to argue that the proposed parcel method produced the most realistic estimates. Accessibility measures are indicators of access to opportunities. It is recognised that the aggregate measures can exaggerate the quantity of that access (Horner and Murray, 2004). In this application, a disaggregate approach at the parcel-level and the incorporation of temporal components resulted in more conservative measures of accessibility. These conservative estimates reflect the fact that the frequencies, travel times, and the availabilities of transit service vary throughout the day and week. Also, by attaching a specific type of destination to the accessibility measure, a better understanding of the utilities from which a traveler may derive can be achieved. This specification provides greater meanings to the measure and it can better translate the concept of accessibility.

Public transit accessibility to all grocery store locations

This section shows the results of the proposed parcel method for the weekday midday period when all grocery store locations in the City of Seattle are considered. The measurement process was repeated for each location and all residential land parcels deemed accessible to a grocery store by the previously defined criteria are identified. If a parcel is within reach of more than one grocery store, then only the shortest total travel time is retained. The proposed parcel method, therefore, considers accessibility to be access to the closest location of a destination type. Figure 6 maps the residential land parcels that are accessible to grocery stores by public transit according to the total travel times. It also identifies all other residential parcels that are deemed not transit accessible to grocery stores. This accessibility map would be useful for planners, policy makers, and social services providers to help identify areas that may be at risk with food access and hunger issues. It can be overlaid with demographic and socio-economic information to reveal populations that do not have practical access to any grocery stores and exhibit characteristics such as low household income or low automobile ownership. Figure 7 highlights one such area in the southwest part of the City where there are known high proportions of socially vulnerable people. Such maps can help justify funding for interventions and service provision.

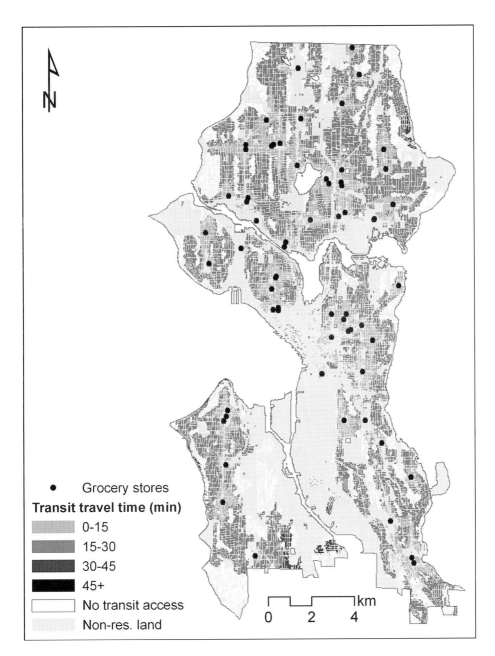

Figure 6. Public transit accessibility to all grocery store locations from residential land parcels in the City of Seattle

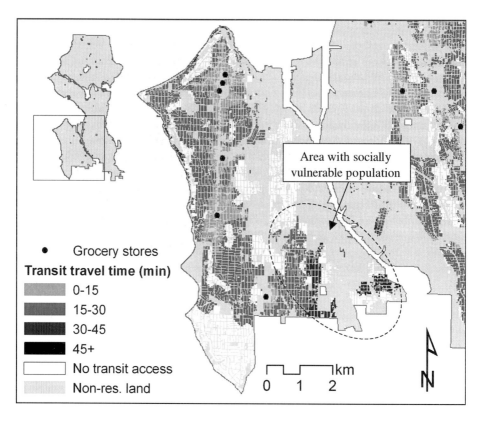

Figure 7. Public transit accessibility to all grocery store locations from residential land parcels in the southwest part of the City of Seattle

CONCLUSIONS, RECOMMENDATIONS, AND LIMITATIONS

The prototype of the parcel-level measure proposed here recognises the complexity of the transit accessibility measurement problem and builds upon the work of others. It accounts for the transportation, spatial, and temporal components together. It also has the flexibility to specify the locations or the type of land uses for the origins and destinations. It makes use of available GIS and data processing technologies as well as transit and land use datasets that are becoming more widely available. In the application shown, it produced graphic results that were more informative and numerical estimates that were conservative, but more realistic, compared to other methods. This parcel-level measure of public transit accessibility to destinations shows great promise for planning, evaluation, and modeling applications and its development should be continued with further testing in each of these areas. It is recommended that the development process include streamlining of the data management and processing procedures. The prototype shows that this measure can be useful and applicable, but its efficiency as a program system remains to be proven.

There are some limitations to this proposed parcel-level measure and they include the land use classification system, the estimation of the average waiting times, and the disregard of monetary costs and other transit service characteristics. The specification of the origins and destinations is only as meaningful as the classification system. If this measure is applied for comparison purposes between multiple places and multiple transit services, then it is imperative that the same land use classification system is used for each measurement. In addition, it is necessary to balance between using a fine or a coarse classification system. The more destination types there are, the more calculations and processing are necessary for composite measures of accessibility, but the more descriptive the activity or the purpose of the trips can be. As for the estimation of the average waiting times, the current proposed procedure does not account for any information that a traveler may receive in terms of published schedules or real time transit vehicle arrival announcements. In fact, the use of half of the headway as the average waiting time is based on an even distribution assumption of passenger arrivals at the stops, which may be applicable for heavily used and frequent services but not so for others. Finally, like other accessibility measurement techniques that use travel time as the measure, this technique does not account for any monetary travel costs or other characteristics such as comfort and security. When making decisions regarding travel mode choice or residential location, travel times to various destinations may be an important consideration but it is most likely not the only factor of importance. It is, therefore, essential to recognise that this measure does not account for any other characteristics of travel by public transit.

REFERENCES

Armhein, C. (1995). Searching for the elusive aggregation effect. Evidence from statistical simulations. *Environment and Planning A*, **27**(1), 105-119.

Beimborn, E. A., M. J. Greenwald and X. Jin (2003). Accessibility, connectivity, and captivity: Impacts on transit choice. *Transportation Research Record*, **1835**, 1-9.

Burns, L. (1979). *Transportation, Temporal, and Spatial Components of Accessibility.* Lexington, MA: Heath and Company.

Cervero, R., T. Rood and B. Appleyard (1999). Tracking accessibility: employment and housing opportunities in the San Francisco Bay Area. *Environment and Planning A*, **31**(7), 1259-1278.

Grengs, J. (2001). Does public transit counteract the segregation of carless households. *Transportation Research Record*, **1753**, 3-10.

Handy, S. L. and D. A. Niemeier (1997). Measuring accessibility: an exploration of issues and alternatives. *Environment and Planning A*, **29**(7), 1175-1194.

Hansen, W. G. (1959). How accessibility shapes land use. *Journal of American Institute of Planners*, **25**(2), 73–76.

Hess, P., A. V. Moudon, and M. G. Logsdon (2001). Measuring land use patterns for transportation research. *Transportation Research Record*, **1780**, 17-24.

Horner, M. W. and A. T. Murray (2004). Spatial representation and scale impacts in transit service assessment. *Environment and Planning B*, **31(5)**, 785-797.

Hsiao, S., J. Lu, J. Streling and M. Weatherford (1997). Use of geographic information system for analysis of transit pedestrian access. *Transportation Research Record*, **1604**, 50-59.

Javid, M. and P. N. Seneviratne (1994). *Application of GIS in estimating transit accessibility to people with disabilities.* Paper presented at the Compendium of Technical Papers, 64th ITE Annual Meeting, Dallas, Texas.

Kwan, M.-P., A. T. Murray, M. E. O'Kelly, and M. Tiefelsdorf (2003). Recent advances in accessibility research: Representation, methodology, and applications. *Journal of Geographical Systems*, **5(1)**, 129-138.

Kerrigan, M. and D. Bull (1992, 14-18 September). *Measuring accessibility—a public transport accessibility index.* Paper presented at Seminar B held at the PTRC Transport, Highways and Planning Summer Annual Meeting, University of Manchester Institute of Science and Technology, England.

KFH Group, Inc. (1999). *Transit cooperative research program report 54: Management toolkit for rural and small urban transportation systems.* Federal Transit Administration, Washington, DC .

Kittelson and Associates, Inc. (2003). *Transit Capacity and Quality of Serivce Manual.* Retrieved 13 September 2004, from http://trb.org/news/blurb_detail.asp?id=2326

Kittelson and Associates, Inc. and URS, Inc. (2001). *Transit level of service (TLOS) software user's guide.* Retrieved 13 September, 2004, from http://www.dot.state.fl.us/transit/Pages/TLOS%20Software%20Users%20Guide.pdf

Krizek, K. (2003). Operationalizing neighborhood accessibility for land use–travel behavior research and regional modeling. *Journal of Planning Education and Research*, **22(3)**, 270-287.

Levinson, D. M. (1998). Accessibility and the journey to work. *Journal of Transport Geography*, **6(1)**, 11-21.

Murray, A.T. (2001). Strategic analysis of public transport coverage. *Socio-Economic Planning Sciences*, **35(3)**, 175-188.

O'Neill, W., R. D. Ramsey and J. Chou (1992). Analysis of transit service areas using geographic information systems. *Transportation Research Record*, **1364**, 131-138.

O'Sullivan, D., A. Morrison, and J. Shearer (2000). Using desktop GIS for the investigation of accessibility by public transport: An isochrone approach. *International Journal of Geographical Information Science* **14(1),** 85-104.

Pendyala, R. M. and I. Ubaka (2002). Geographic information system-based regional transit feasibility analysis and simulation tool. *Transportation Research Record*, **1799**, 42-49.

Peng, Z., and K. J. Dueker (1995). Spatial data integration in route-level transit demand modeling. *Journal of the Urban and Regional Information Systems Association*, **7(1)**, 26-37.

Pickrell, D. H. (1992). A desire named streetcar. *Journal of the American Planning Association*, **58(1)**, 158-176.

Pike, D. H. and E. Vougioukas (1982). Camden's public transport accessibility model. *Traffic Engineering and Control*, **23(1)**, 15-22.

Polzin, S. E. and X. Chu (2003, 27 March). *NHTS early findings on public transportation travel trends*. Retrieved 23 Aug, 2004, from http://nhts.ornl.gov/2001/presentations/polzin/index.shtml

Polzin, S. E., R. M. Pendyala and S. Navari (2002). Development of time-of-day-based transit accessibility analysis tool. *Transportation Research Record*, **1799**, 35-41.

Pratt, R. H. and G. Park (2000). *Travel response to transportation system changes, interim handbook*: Transit Cooperative Research Program, Transportation Research Board.

Pulugurtha, S. S. and S. S. Nambisan (1999). Evaluating transit market potential and selecting locations of transit service facilities using GIS. *Journal of Public Transportation*, **2(4)**, 75-94.

Rastogi, R. and K. V. K. Rao (2003). Defining transit accessibility with environmental inputs. *Transportation Research Part D*, **8**, 383-396.

Ruppert, W. R. (1979, 9-12 July). *Levels of accessibility induced by urban public transport systems*. Paper presented at the Seminar M held at the PTRC Summer Annual Meeting, University of Warwick, England.

Sultana, S. (2002). Job/housing imbalance and commuting time in the Atlanta metropolitan area: Exploration of causes of longer commuting time. *Urban Geography*, **23(8)**, 728-749.

SG Associates, Inc., and Volpe National Transportation Systems Center (1998). *Guidelines for network representation of transit access: State-of-the-practice summary* (No. DOT-T- 99-05). Washington DC: US Department of Transportation, Federal Highway Administration.

Thakuriah, P., J. F. Ortega and P. S. Sriraj (2003). *Spatial data integration for low-income worker accessibility assessment: A case study of the Chicago metropolitan area*. Paper presented at the 2003 Mid-Continent Transportation Research Symposium, Ames, Iowa.

Thompson, G. L. (2001). New insights into the value of transit: Modeling inferences from Dade County. *Transportation Research Record*, **1753**, 52-58.

US Census Bureau (2004). *Centers of Population by Block Group*. Retrieved 23 Aug, 2004, from http://www.census.gov/geo/www/cenpop/blkgrp/bg_cenpop.html

Willier, T. E. (1947). Traffic and trade. *Traffic Quarterly*, **1(3)**, 211-220.

Zhao, F., M.-T. Li, L.-F. Chow, A. Gan and L. D. Shen (2002). *FSUTMS mode choice modeling: factors affect transit use and access* (No. NCTR 392-07, 416-03). Tampa, Florida: University of South Florida, National Center for Transit Research.

Zhao, F., L.-F. Chow, M.-T. Li, I. Ubaka and A. Gan (2003). Forecasting transit walk accessibility: Regression model alternative to buffer method. *Transportation Research Record*, **1835**, 34-41.

APPENDIX

Census Tract	Block Group	Total Population	Estimated Populations Served			Ratio	
			Area Ratio [A]	Simplified Parcel [B]	Proposed Parcel [C]	[A]/[C]	[B]/[C]
4400	4	1,597	142	10	10	14.27	1
4500	1	923	3	n/a	n/a	n/a	n/a
5200	1	1,121	116	n/a	n/a	n/a	n/a
5200	2	1,089	54	n/a	n/a	n/a	n/a
5200	5	1,050	704	396	391	1.80	1.01
5301	1	5,575	2,724	2,399	2,399	1.14	1.00
5301	2	816	811	816	816	0.99	1
5301	3	1,138	1,133	1,094	1,094	1.04	1
5301	4	980	890	639	639	1.39	1
5302	1	309	287	309	309	0.93	1
5302	2	39	36	39	39	0.93	1
5302	3	2,180	637	n/a	n/a	n/a	n/a
5802	2	2,016	3	n/a	n/a	n/a	n/a
6200	1	612	256	530	517	0.49	1.02
6200	2	695	367	685	680	0.54	1.01
6200	3	757	469	627	600	0.78	1.04
6200	4	841	480	469	456	1.05	1.03
6200	5	895	704	726	721	0.98	1.01
6400	1	824	819	673	660	1.24	1.02
6400	2	1,135	1,119	1,089	1,087	1.03	1.00
6400	3	627	597	596	572	1.04	1.04
6400	4	615	578	323	313	1.85	1.03
6500	2	658	496	458	456	1.09	1.00
6500	3	853	419	372	332	1.26	1.12
7000	2	1,155	301	265	218	1.38	1.22
7000	3	1,497	850	624	621	1.37	1.01
7000	5	1,043	1,043	876	874	1.19	1.00
7000	6	1,049	161	419	419	0.38	1
7100	1	890	379	264	264	1.44	1
7100	2	919	779	919	919	0.85	1
7200	1	495	278	11	11	26.08	1
7200	2	2,589	2,161	1,472	1,472	1.47	1
7300	1	809	338	120	108	3.12	1.11
7300	2	1,163	1,163	1,163	1,163	1	1
7300	3	246	153	n/a	n/a	n/a	n/a
7400	1	804	97	146	144	0.67	1.02
7400	2	790	790	756	755	1.05	1.00
7400	3	673	673	673	673	1	1
7400	4	1,671	1,671	1,671	1,671	1	1
7400	5	1,805	1,805	1,805	1,805	1	1
7400	6	1,291	1,229	1,077	1,075	1.14	1.00
7400	7	1,898	574	185	185	3.11	1

Census Tract	Block Group	Total Population	Estimated Populations Served			Ratio	
			Area Ratio [A]	Simplified Parcel [B]	Proposed Parcel [C]	[A]/[C]	[B]/[C]
7500	1	1,161	1,161	1,161	1,161	1	1
7500	2	1,071	1,071	1,069	1,063	1.01	1.01
7500	3	1,500	1,273	1,297	1,292	0.99	1.00
7500	4	890	882	890	890	0.99	1
7500	5	932	855	917	917	0.93	1
7600	1	557	557	557	557	1	1
7600	2	1,065	1,065	1,065	1,028	1.04	1.04
7600	3	849	849	849	849	1.00	1.00
7600	4	790	790	790	742	1.06	1.06
7700	1	505	472	422	401	1.18	1.05
7700	2	680	680	660	660	1.03	1.00
7700	3	885	718	655	636	1.13	1.03
7700	4	1,126	1,096	1,093	1,091	1.00	1.00
7700	5	794	794	786	784	1.01	1.00
7800	2	735	12	n/a	n/a	n/a	n/a
7800	3	784	276	123	112	2.45	1.10
7800	4	895	90	62	54	1.65	1.13
7800	5	704	150	84	77	1.96	1.09
7800	6	669	136	97	82	1.65	1.18
7800	7	790	96	78	68	1.41	1.15
7900	1	630	119	9	9	13.26	1.00
7900	2	861	164	74	74	2.21	1
7900	3	774	384	66	66	5.80	1
7900	4	784	784	773	722	1.09	1.07
7900	5	1,183	1,175	1,183	1,061	1.11	1.11
8001	1	767	767	767	767	1	1
8001	2	1,498	550	n/a	n/a	n/a	n/a
8001	3	1,145	722	776	776	0.93	1.00
8002	1	1,618	393	212	212	1.85	1.00
8100	1	2,431	2,129	2,431	941	2.26	2.58
8100	2	1,046	1,039	898	898	1.16	1.00
8200	1	333	333	333	333	1	1
8200	2	845	23	n/a	n/a	n/a	n/a
8200	3	1,697	1,454	1,078	1,073	1.35	1.00
8300	2	1,303	1,303	1,303	1,303	1	1
8300	3	579	439	490	490	0.90	1
8400	1	3,086	2,828	2,633	2,633	1.07	1
8400	2	752	332	500	500	0.66	1
8500	1	1,118	4	n/a	n/a	n/a	n/a
8500	3	3,720	562	n/a	n/a	n/a	n/a
8600	1	858	137	240	229	0.60	1.05
8600	2	2,135	121	n/a	n/a	n/a	n/a
8700	2	1,019	590	552	537	1.10	1.03
8700	4	1,237	37	11	11	3.30	1.00
8800	1	647	647	642	638	1.01	1.01

Census Tract	Block Group	Total Population	Estimated Populations Served			Ratio	
			Area Ratio [A]	Simplified Parcel [B]	Proposed Parcel [C]	[A]/[C]	[B]/[C]
8800	2	760	760	723	716	1.06	1.01
8800	3	755	755	735	731	1.03	1.01
8800	4	722	602	610	605	0.99	1.01
8800	5	622	455	364	357	1.27	1.02
8900	1	949	11	32	11	1	3.00
8900	2	592	207	60	40	5.20	1.50
8900	3	609	573	506	503	1.14	1.01
8900	4	865	857	693	678	1.26	1.02
8900	5	758	758	721	716	1.06	1.01
8900	6	823	823	805	794	1.04	1.01
9000	1	783	503	512	512	0.98	1
9000	2	506	28	2	2	13.98	1
9000	3	845	45	n/a	n/a	n/a	n/a
9100	1	1,131	144	60	n/a	n/a	n/a
9100	2	952	788	952	n/a	n/a	n/a
9200	1	925	863	925	n/a	n/a	n/a
9200	2	911	902	911	n/a	n/a	n/a
9300	1	960	1	n/a	n/a	n/a	n/a
9300	2	667	86	667	n/a	n/a	n/a
9400	2	863	24	211	211	0.11	1
9400	3	1,182	119	n/a	n/a	n/a	n/a
9500	1	940	100	5	5	19.45	1
9500	2	816	424	169	156	2.72	1.08
9500	4	787	3	n/a	n/a	n/a	n/a
9500	5	1,356	1,298	1,089	1,076	1.21	1.01
9500	6	1,243	1,228	1,121	1,094	1.12	1.02
10000	1	824	438	92	83	5.25	1.10
10000	2	1,130	31	2	2	13.11	1
10100	3	1,037	1	1	n/a	n/a	n/a
10100	4	1,259	110	12	9	12.53	1.33
Total		124,152	71,283	64,301	58,505	1.22	1.10
Percent		100%	57.4%	51.8%	47.1%		

Access to Destinations
D.M. Levinson and K.J. Krizek (editors)
© 2005 Elsevier Ltd. All rights reserved.

CHAPTER 12

PAVING NEW GROUND: A MARKOV CHAIN MODEL OF THE CHANGE IN TRANSPORTATION NETWORKS AND LAND USE

David Levinson and Wei Chen, University of Minnesota

INTRODUCTION

It has long been a mantra among planners that transportation policies and networks drive land use and it has also been a mantra among civil engineers that networks are built where the people are. Can it be that both are right?

Since the widespread introduction of paved roads early in the twentieth century, more and more roads, of higher and higher quality and capacity have been constructed. However, despite the growth in transportation networks, there has been even more growth in the demand for transportation networks. Vehicle kilometers traveled has outpaced lane kilometers in most cities (e.g. the Twin Cities of Minneapolis and St. Paul (Levinson and Karamalaputi, 2003)). This has led to significant increases in congestion, and over time, falling speeds on many links (TTI, 2002). This leads many in the engineering community to believe that when they are building roads, they are simply responding to existing needs.

Similarly with the growth of population and of cities, more and more land has been devoted to developed uses (commercial, employment, and residential) and less to "undeveloped" uses (agricultural and recreational). These trends are especially visible in growing metropolitan areas that have steadily added to their spatial extent. Of course, that development would be impossible with concurrent construction of infrastructure such as streets and highways. The phenomenon of induced demand, whereby an addition 1% roadways leads to some increase (typically 0.2%–0.8% (Parthasarathi *et al.*, 2003)) in travel demand leads many in the planning community to conclude that it is transportation capacity driving travel demand. While the length and number of trips per person may change (and has been increasing), transportation surely cannot be directly blamed for the increasing number of people that is the dominant part of the increase in travel demand.

In this paper we explore the inter-connectedness of the evolution of transportation networks and land use through the application of a Markov Chain model. This model investigates how individual cells, with both land use and transportation network attributes, change over time. We can see whether cells with more transportation network available are more likely to develop, and whether cells that are developed are more likely to attract additional highway investment. While this paper does not consider land use density directly, it does consider land use type, and as cells change type, we can conclude that some form of development is likely to be occurring.

The next section of this paper outlines the Markov Chain model. This is followed by a discussion of the Twin Cities data used in the study. The subsequent section develops the transition probability matrices used by the Markov Chain in our case. Those matrices are analyzed to understand the empirical regularities that appear in the data. They are then applied to both assess the predictive ability of the Markov Chain model (comparing what the model would predict with what actually happened), and then applying the model to forecast future changes in the Twin Cities area. The paper concludes with some suggestions for future research.

THE MARKOV CHAIN MODEL

It is fair to say that the co-evolution of urban highway network and land use is a complicated stochastic process; therefore, it is more appropriate to analyze the co-evolution with a probabilistic rather than a deterministic model. Among probabilistic processes, the Markov Chain Model, employed in this paper, provides a powerful tool for analyzing the system evolution through time series, and it has been applied in many fields of research.

The Markov Chain Model has a *discrete-time* version and a *continuous-time* version. In this study, we concentrate on the discrete-time version. A discrete-time Markov Chain Model describes the evolution of a process through a sequence of states S with equal time intervals, such as $S(0) \dashrightarrow S(1) \dashrightarrow S(2) \dashrightarrow \ldots\ldots S(t)$, where $S(t)$ indicates the state of the system at time t. $S(t)$ controls $S(t+1)$ through transition probability p_{ij}, where p_{ij} is given by

$$p_{ij} = P(S(t+1) = j \mid S(t) = i)$$

We can describe the transition probabilities in a more compact way by arraying them into a square matrix P, called the transition matrix. The transition matrix looks like this:

$$P = \begin{bmatrix} p_{00} & p_{01} & p_{02} & \cdots & p_{0n} \\ p_{10} & p_{11} & p_{12} & \cdots & p_{1n} \\ p_{20} & p_{21} & p_{22} & \cdots & p_{2n} \\ \vdots & \vdots & \vdots & \ddots & \vdots \\ p_{n0} & p_{n1} & p_{n2} & \cdots & p_{nn} \end{bmatrix}$$

Given the present state of the system $S(t)$, matrix P provides the probability to go in one step from state $S(t)$ to state $S(t+1)$, that is

$$S(t+1) = P \times S(t)$$

For a system with an initial state $S(0)$ and transition matrix P, the consecutive states in equal time intervals are estimated in the following succession:

$$S(1) = P \times S(0), \ S(2) = P \times S(1), \ S(3) = P \times S(2) \ \ldots \ldots$$

A generalized expression can be that given the current state $S(t)$, the kth power of the transition matrix P provides the conditional probability that after k steps' evolution the system becomes state $S(t+k)$, that is

$$S(t+k) = P^k \times S(t), \quad t, k = 0, 1, \ldots \ldots$$

This indicates that we can predict the future evolution of a system by determining the transition matrix P.

Of the few applications of Markov Chain Model in spatial social-economics and geography, land-use, and transportation research, Lever (1972) used a four-zone Markov Chain Model to predict the probable future distribution of manufacturing establishments. The transition matrix was estimated by counting the probability of the movements of firms from Zone i to Zone j,($i,j = 1,2,3,4$) throughout 1959 to 1969. Clark (1965) studied the movement of rental housing areas and divided the central city tracts into ten classes with $10 intervals in rents and each of the classes represented a state of the Markov Chain Model. Two decades of data, from 1940 (starting state) to 1950 (destination state) and from 1950 (starting state) to 1960 (destination state), were pooled in the same matrix to calculate the probability of the movements of tracts from class i to class j. Zhang and Li, (2005) used a two-dimensional Markov chain model to simulate multinomial land-cover classes and to estimate occurrence probability vectors for spatial uncertainty representation. Zhang, and Cho (2001) presented an evolutionary Markov chain Monte Carlo method to identify the functional structure of a target system that underlies the observed data. Janssens *et al.* (2005) developed and evaluated the implementation of an adapted Markov Chain modelling heuristic and simulation framework in the context of transportation research. Yin *et al.* (2003) simulated certain types of population dynamics using continuous-time finite-state Markov chains, and presented two case studies.

One was a process of drug delivery for which a closed-form solution of the forward equation can also be obtained; the other one simulated the birth, death, and growth of cell population dynamics using a Markovian model. Goulias (1999) used generalized mixed Markov latent class models to study the dynamics of travel patterns. The repeated nature of travelers' behavior "allows to distinguish units that over time change their behavior from those that are not and to uncover the underlying stochastic behavior generating the data... Travel pattern change is best explained by a single path of change with stationary day-to-day pattern transition probabilities that are different from their year-to-year counterparts."

In this paper, a Markov Chain Model is developed to analyze and predict the co-development of the highway network and land use. The sources of data and the Markov Chain Model are presented in the ensuing sections.

DATA

High-quality GIS land use and highway network maps for the Twin Cities Metropolitan Area from 1958 to 1990 were developed from paper maps (Table 1). Then a lattice layer is created which is composed of 188×188 m square cells. This lattice layer was transformed into polygon coverage which shares the same corridor system with the land use and highway network layers. The land use layer and highway layer were merged into the lattice layer, the cells of the lattice layer then contain the spatial information of land use and highways.

Note: Figures and Tables are located at the end of this chapter.

Figure 1 displays the gridded maps of land use for 1958, 1968, 1978, and 1990 wherein each cell is classified into one of five types of zones E, R, M, A, and W, based on land use:

- E (Employment zone), which contains Commercial areas, Industrial areas, Institutional and Office areas, and Airports;
- R (Residential zone), which contains both Single Family Residential and Multi-Family Residential areas;
- M (Mixed use), which contains both Employment and Residential areas;
- A (Agricultural and Recreational zones), which contains Agricultural, Recreational and also vacant areas; and
- W (Water zone), which is predominantly water covered.

Each of the cells is classified into one of the five types of zones based on the following rules: Firstly, for all the cells that are within or intersect Employment zones, they belong to E or M; a cell is classified as M if it intersects Employment zones and also contains Residential areas and a cell is classified as E if it is within or intersects Employment zones but does not contain Residential areas. Secondly, for the remaining cells, they belong to R if they are within or intersect Residential zones. Thirdly, for the cells not belonging to E, M, or R, they belong to

W if they are within or intersect Water zones. Finally, the cells belong to A if they do not belong to E, M, R, or W.

The highways are classified into two levels, the upper level highways composed of Interstates and Divided Highways and the lower level highways composed of Undivided Highways and County Highways. Each of the cells has one of the four attributes: U (the cell only contains upper level highways), L (the cell only contains lower level highways), B (the cell contains both upper and lower level highways), and N (the cell does not contain either upper or lower level highways).

Combining the 5 land use and 4 highway classifications, we can get the 20 types of cells: EU, EL, EB, EN, MU, ML, MB, MN, RU, RL, RB, RN, WU, WL, WB, WN, AU, AL, AB, and AN, which represent the 20 classes of the Markov Chain Model.

TRANSITION MATRIX

A Markov Chain Model moves from one state to the next controlled by the transition matrix. In this study Years 1958, 1968, 1978 and 1990 are the successive states[1]. A transition probability matrix can be derived from each pair of the successive states.

$$S(1958) \rightarrow S(1968) \rightarrow S(1978) \rightarrow S(1990) \ldots\ldots$$

For instance, let $P_{EU,EU}$ denote the probability that a cell being EU in one state continues to be EU in the next state, and let $P_{EU,MB}$ denote the probability that a cell being EU in one state changes to be MB in the next state. Using this procedure, we can get 20×20 transition probabilities that are assembled into a transition matrix P.

$$P = \begin{bmatrix} P_{EU,EU} & P_{EU,MU} & P_{EU,RU} & P_{EU,WU} & P_{EU,AU} & P_{EU,EL} & P_{EU,ML} & P_{EU,RL} & \cdots & P_{EU,MN} & P_{EU,RN} & P_{EU,WN} & P_{EU,AN} \\ P_{MU,EU} & P_{MU,MU} & P_{MU,RU} & P_{MU,WU} & P_{MU,AU} & P_{MU,EL} & P_{MU,ML} & P_{MU,RL} & \cdots & P_{MU,MN} & P_{MU,RN} & P_{MU,WN} & P_{MU,AN} \\ P_{RU,EU} & P_{RU,MU} & P_{RU,RU} & P_{RU,WU} & P_{RU,AU} & P_{RU,EL} & P_{RU,ML} & P_{RU,RL} & \cdots & P_{RU,MN} & P_{RU,RN} & P_{RU,WN} & P_{RU,AN} \\ P_{WU,EU} & P_{WU,MU} & P_{WU,RU} & P_{WU,WU} & P_{WU,AU} & P_{WU,EL} & P_{WU,ML} & P_{WU,RL} & \cdots & P_{WU,MN} & P_{WU,RN} & P_{WU,WN} & P_{WU,AN} \\ P_{AU,EU} & P_{AU,MU} & P_{AU,RU} & P_{AU,WU} & P_{AU,AU} & P_{AU,EL} & P_{AU,ML} & P_{AU,RL} & \cdots & P_{AU,MN} & P_{AU,RN} & P_{AU,WN} & P_{AU,AN} \\ P_{EL,EU} & P_{EL,MU} & P_{EL,RU} & P_{EL,WU} & P_{EL,AU} & P_{EL,EL} & P_{EL,ML} & P_{EL,RL} & \cdots & P_{EL,MN} & P_{EL,RN} & P_{EL,WN} & P_{EL,AN} \\ P_{ML,EU} & P_{ML,MU} & P_{ML,RU} & P_{ML,WU} & P_{ML,AU} & P_{ML,EL} & P_{ML,ML} & P_{ML,RL} & \cdots & P_{ML,MN} & P_{ML,RN} & P_{ML,WN} & P_{ML,AN} \\ P_{WL,EU} & P_{WL,MU} & P_{WL,RU} & P_{WL,WU} & P_{WL,AU} & P_{WL,EL} & P_{WL,ML} & P_{WL,RL} & \cdots & P_{WL,MN} & P_{WL,RN} & P_{WL,WN} & P_{WL,AN} \\ \vdots & \vdots & \vdots & \vdots & \vdots & \vdots & \vdots & \vdots & \ddots & \vdots & \vdots & \vdots & \vdots \\ P_{MN,EU} & P_{MN,MU} & P_{MN,RU} & P_{MN,WU} & P_{MN,AU} & P_{MN,EL} & P_{MN,ML} & P_{MN,RL} & \cdots & P_{MN,MN} & P_{MN,RN} & P_{MN,WN} & P_{MN,AN} \\ P_{RN,EU} & P_{RN,MU} & P_{RN,RU} & P_{RN,WU} & P_{RN,AU} & P_{RN,EL} & P_{RN,ML} & P_{RN,RL} & \cdots & P_{RN,MN} & P_{RN,RN} & P_{RN,WN} & P_{RN,AN} \\ P_{WN,EU} & P_{WN,MU} & P_{WN,RU} & P_{WN,WU} & P_{WN,AU} & P_{WN,EL} & P_{WN,ML} & P_{WN,RL} & \cdots & P_{WN,MN} & P_{WN,RN} & P_{WN,WN} & P_{WN,AN} \\ P_{AN,EU} & P_{AN,MU} & P_{AN,RU} & P_{AN,WU} & P_{AN,AU} & P_{AN,EL} & P_{AN,ML} & P_{AN,RL} & \cdots & P_{AN,MN} & P_{AN,RN} & P_{AN,WN} & P_{AN,AN} \end{bmatrix}$$

The changes of the cells between different classes are summarized in three two-way tables for the 1958–1968 period, 1968–1978 period, and 1978–1990 period respectively (Table 2). A transition matrix can be derived from each of the two-way tables and each of the transition probabilities is calculated by its corresponding cell entry divided by the row total, for example,

$$P_{EU,MB} = \frac{EU,MB}{EU,EU + EU,MU + EU,RU + EU,WU + EU,AU + EU,EL + \cdots + EU,WN + EU,AN}$$

Clearly, the sums of the rows of a transition matrix are always equal to 1.

For the 1958 – 1968 period, 1968 - 1978 period, and 1978 - 1990 period respectively, we get three transition matrices $P_{1958-1968}$, $P_{1968-1978}$ and $P_{1978-1990}$ (Appendix: Table 3). The highlighted cells indicate the maximum of each row.

From Tables 2 and 3, we can derive some important evolution tendencies of highways and land use.

The percentage of land use distribution in different zones is shown in Figure 2. From 1958 to 1990, there was a significant increase (from 9% to 30%) of the proportion occupied by Zone E and M (Employment zones and the mixed Employment and Residential Zones), and also a significant increase (from 21% to 43%) of the proportion occupied by Zone R (Residential zones). Meanwhile, there was a significant decrease (from 62% to 20%) of the proportion occupied by Zone A (Agricultural and Recreational zones). These changes display a clear tendency toward urbanization over the last four decades of the twentieth century.

Accompanying the urbanization of land use, upper level highways (Interstates and Divided Highways) experienced a tremendous growth from 1958 to 1990, which was measured by the number of cells containing the upper level highways (Figure 3). Meanwhile, the proportions of the upper level highways within Zone E, M, and R increased significantly (from 43% to 83%), while the proportions of the upper level highways within Zone A decreased significantly (from 55% to 14%) (Figure 4). This was caused by the expansion of Employment and Residential zones and the contraction of Agricultural and Recreational zones (Figure 2), and also caused by transportation and land use interactions. Transportation and land use interactions can be simplified as follows: land development generates travel demand, which in turn leads to the development of new transportation infrastructure in the urbanized areas; the developed transportation infrastructure improves accessibility and mobility, which attracts further land development.

Compared with upper level highways, the lower level highways had moderate development and seemed to be close to a saturation state after 1978 (Figure 3). Furthermore, the proportions of lower level highways within Zone E, M, and R increased significantly (from 47% to 86%) while decreasing significantly within Zone A (from 51% to 12%) (Figure 4), similar to the upper level highways. This was partially caused by the expansion of

Employment and Residential zones and the contraction of Agricultural and Recreational zones (Figure 2), and partially due to the fact that lower level highways are the connecting links among urban settlement and the upper level highways, so their growth tends to accompany the development of Zone E, M, and R and the upper level highways. The share of cells with lower level highways also declines if those roads are upgraded.

Next, we will discuss the effects of transportation on land development. Figure 5 shows the development of Agricultural and Recreational zones. It is obvious that highways do affect the development of Agricultural and Recreational zones. Agricultural and Recreational zones without highways (AN) have the lowest probability to convert to urbanized land use (E, M, and R), while Agricultural and Recreational zones that contain both upper and lower level highways (AB) have the highest probability to convert to urbanized land use. These facts indicate that the construction of highways contributes to the urbanization of land development. Figure 6 shows the development of urbanized land use (E, R, M). Compared with Agricultural, the development of urbanized zones is less influenced by highways. Independent of whether the zone contains highways, Employment zones are most likely to continue being Employment zones, and next most likely to change to the Mixed zones. Highways do not obviously influence the development of Mixed zones either. Independent of their highways, Mixed zones are most likely to keep their land use attributes unchanged, and next most likely to change to Employment or Residential zones. Residential zones, however, are more likely to change to Mixed zones if they contain highways. Residential zones without highways (RN) have the highest probability of remaining unchanged, while Residential zones that contain both upper and lower level highways (RB) have the highest probability of converting to Mixed use zones.

The effect of land use on transportation growth is an equally important topic, addressed in Levinson and Chen (2004), which demonstrates that an area's land use attributes and population density level do have significant relationships with the area's likelihood of highway development.

PREDICTION

Section four presented the Markov Chain Model and transition matrices, some important evolution rules of highways and land use were summarized. Markov Chain Models also allow us to predict the future development of a system. To do this, we use one state as well as the transition matrix obtained from this state and its immediately preceding state to predict the next state, such as, $\widehat{S}(t) = P_{t-2,t-1} \times S(t-1)$ and $\widehat{S}(t+1) = P_{t-1,t} \times S(t)$. Furthermore, if we assume $P_{t-2,t-1} \approx P_{t-1,t}$ we obtain $\widehat{S}(t+1) = P^2_{t-2,t-1} \times S(t-1)$.

Table 4 presents the observed and predicted *S(1978)* and *S(1990)*, where the predicted *S(1978)* and *S(1990)* are obtained through $\hat{S}(1978) = P_{1958\text{-}1968} \times S(1968)$, $\hat{S}(1990) = P^2_{1958\text{-}1968} \times S(1968)$, and $\hat{S}(1990) = P_{1968\text{-}1978} \times S(1978)$.

Chi-square tests of goodness-of-fit are conducted for the three pairs of predicted and observed percentage distribution in Table 4, the null hypothesis is that the differences between the predicted and observed distribution is due to chance only. All the four *p* values for the calculated chi-squares are much larger than 0.1, so the null hypothesis is not rejected, that means the difference between predicted and observed distribution under the null hypothesis is the result of chance, and therefore negligible. The predicted distribution is acceptable.

The quality of the predictions using Markov Chain Model depends on the constancy of the transition matrices. We can get good prediction results if the successive transition matrices are approximately constant, for example, the predicted distributions of *S(1978)* and *S(1990)* are close to their observed distributions, which indicates that the transition matrix of 1958-1968 period is similar to that of 1968-1978 period, and the transition matrix of 1968-1978 period is similar to that of 1978-1990 period. Since the transition matrix of 1978-1990 period is the most recent one we have in the database, we use this transition matrix to predict the future development of the system as below, assuming that this transition matrix is approximately constant for these years.

$$\hat{S}(2000) = P_{78\text{-}90} \times S(90), \ \hat{S}(2010) = P_{78\text{-}90}^2 \times S(90) \ ... \ \hat{S}(2050) = P_{78\text{-}90}^6 \times S(90).$$

Another commonly used prediction method is to pool the data of different periods into the same two-way table and derive a single transition matrix from the table, and use this transition matrix to predict the ensuing states, such as,

$$\left. \begin{array}{l} S(58) \rightarrow S(68) \\ S(68) \rightarrow S(78) \\ S(78) \rightarrow S(90) \end{array} \right\} \rightarrow \cdots\cdots S(t)$$

The matrix obtained from the pooled data may be rational for some other cases, but we do not think it would work better than the matrix of the 1978-1990 period in predicting the future development of the urban system. The growth pattern of the urban system in the 1978-1990 period should be more similar to the future development than that in the 1958-1968 and 1968-1978 periods. Clearly the closer the prediction the more accurate the results (assuming, as here, the transition matrices are based on a substantially long period of time that smoothes out the development cycle).

Figure 7 presents the predicted development of land use and highways in the next half century base on the transition matrix of the 1978-1990 period. It is expected that, from 1990 to 2050, the Employment zones keep stationary, while the Mixed use zones increase from 16% to 22%

and Residential zones increase from 43% to 48%. The Agricultural and Recreational zones decrease from 20% to 9%. The predicted results indicate that if the development pattern of 1978-1990 period is continued, we would expect to have a highly urbanized region in 2050, where more than 80% of land is used for commercial, industrial, institutional, office and residential purposes and only less than 10% of land in the currently designed metro area is left for agricultural and recreational purposes. Meanwhile, it is expected that the upper level highways continue to increase from 1990 to 2050 (measured by the number of the cells that contain the highways), and in 2050 about 15 percent of the cells contain upper level highways; the lower level highways, however, seem to have reached the stationary state and remain in about 16 percentage of cells. It is also expected that (Figure 8), in 2050, 86 percentage of the upper level highways and 91 percentage of the lower level highways are within the urbanized zones (Zones E, M, and R).

CONCLUSION

This paper employs a Markov Chain Model to analyze the spatial co-evolution of transportation and land use. A transition matrix records the interaction between transportation and land use and it is used to predict the future development of transportation and land use.

The study is based on the highway network and land use of the Twin Cities Metropolitan Area from 1958 to 1990. Through the Markov Chain Model we find that highways do affect the development of Agricultural and Recreational zones. The Agricultural and Recreational zones that contain highways are much more likely to convert to Employment and Residential zones than the Agricultural and Recreational zones without highways. Compared with Agricultural and Recreational zones, the development of urbanized zones is less influenced by highways. Employment zones are least likely to change their land use attributes. Residential zones, however, are more likely to change to Mixed use zones if they contain highways.

The prediction function of the Markov Chain Model has been performed and Chi-square tests tell that the predicted distributions of S(1978) and S(1990) are close to their observed distributions, which indicates that the Markov Chain Model works well to predict the future system development at least for the next decade. We also use the transition matrix of the 1978-1990 period to predict the future development of the system in the next half century assuming that this transition matrix is approximately constant during these years, and the predicted results show that if the future development of the system follows the same growth pattern of 1978-1990 period, we will expect to see a highly urbanized Metropolitan Area in the next half century.

Markov Chain Model allows us to describe, analyze and predict the development of complicated dynamic systems. Further improvements of the Markov Chain Model presented in this paper will result in a greater understanding of the dynamics of urban highway networks and land use.

FIGURES

Figure 1. The gridded land use zones

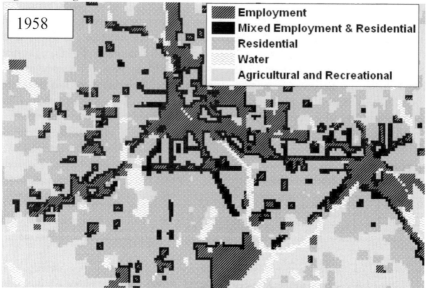

Figure 1. (cont'd)

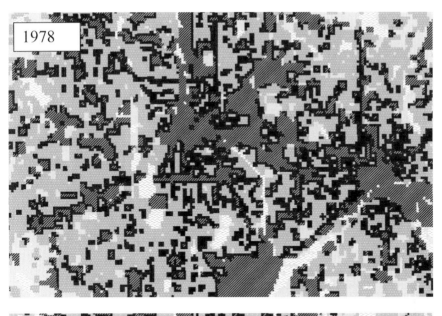

Figure 2. The Percentage of Land use Distribution

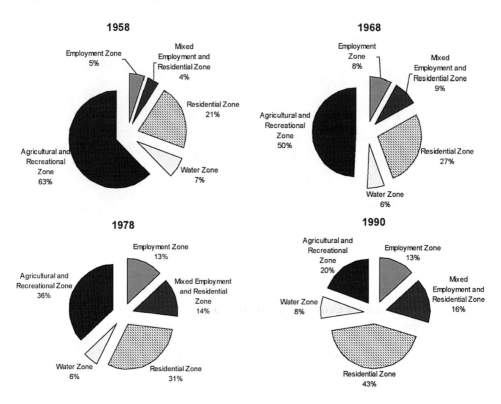

Figure 3. The growth of upper and lower level highways.

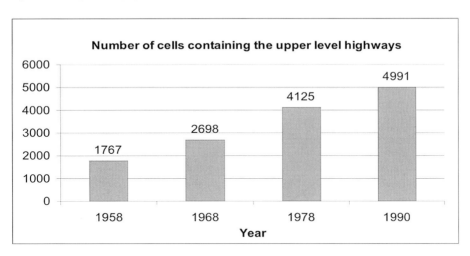

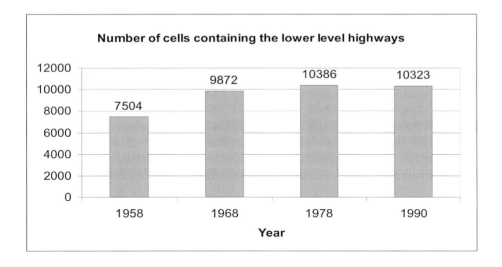

Figure 4. Distribution of upper and lower level highways by zone.

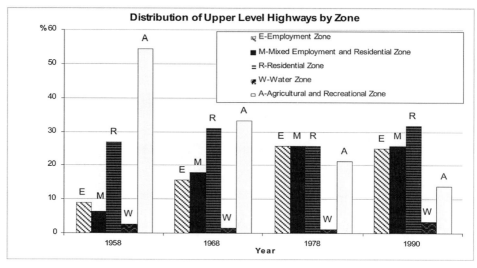

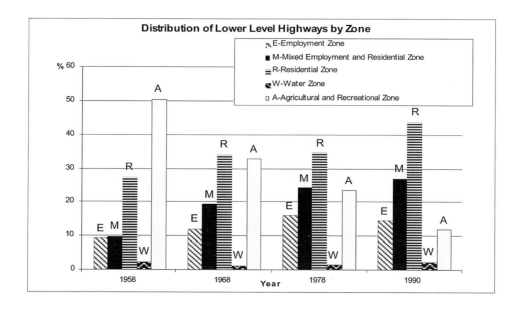

Figure 5. The development of Agricultural and Recreational zones.

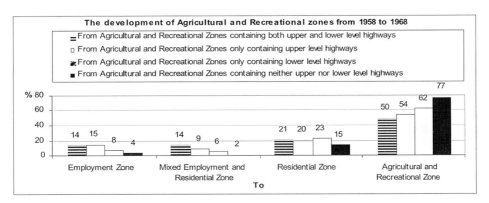

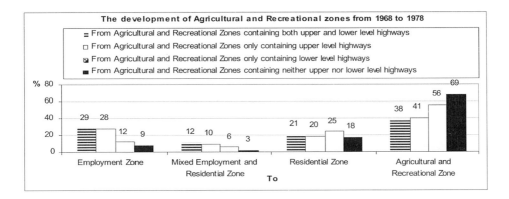

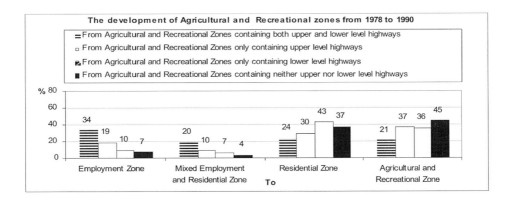

Figure 6. The development of urbanized land use – Employment zones, Residential zones, and the mixed Employment and Residential zones.

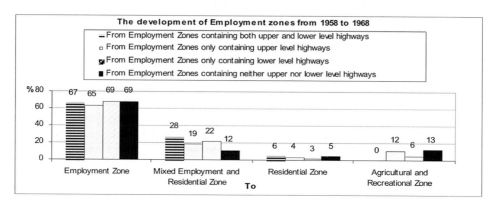

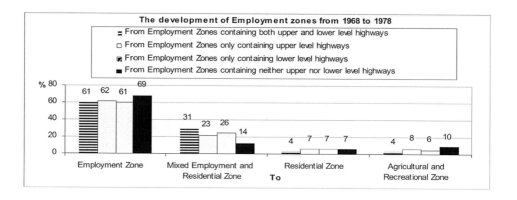

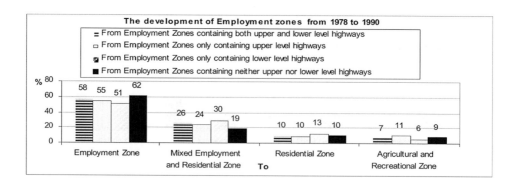

Figure 6. (cont'd)

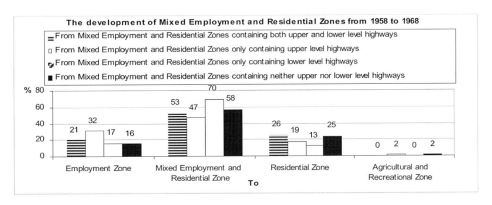

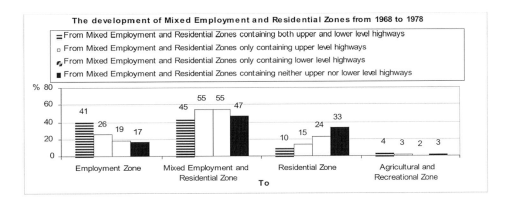

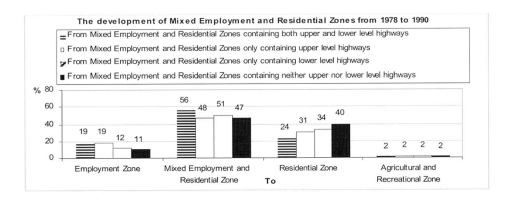

Figure 6. (cont'd)

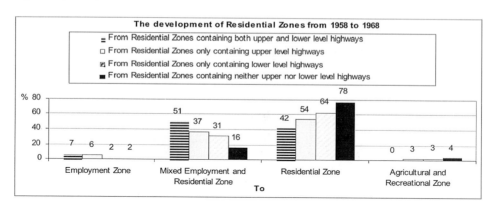

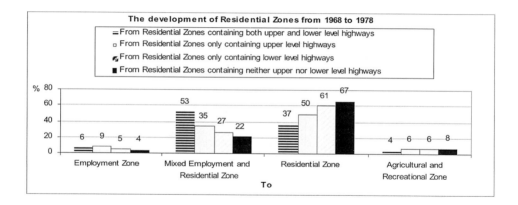

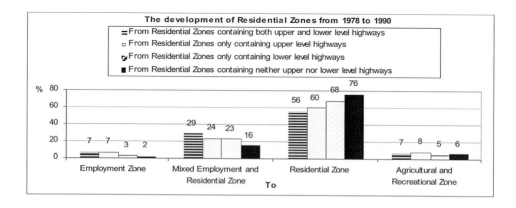

Figure 7. The development of land use and highways.

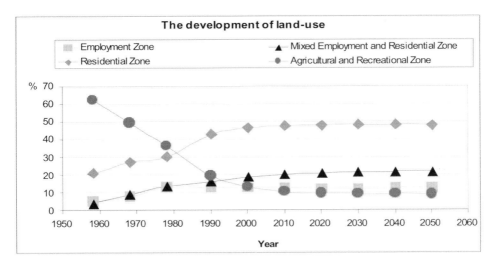

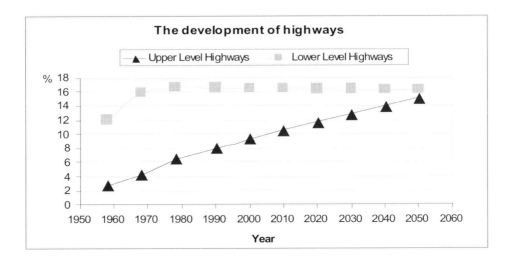

Figure 8. Distribution of upper and lower level highways by zone.

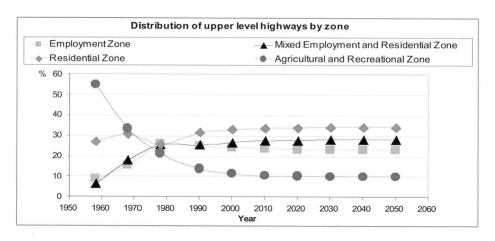

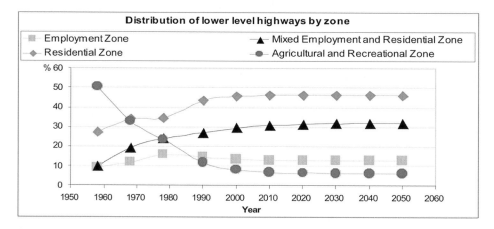

TABLES

Table 1. GIS data summary.

GIS Map	Source
Twin Cities Metropolitan Area Land use Distribution 1958, 1968, 1978 and 1990.	Generalized Land Use maps (paper copy) for 1958, 1968 and 1978 issued by Twin Cities Metropolitan Council. (* 1990 land use map can be downloaded at http://www.datafinder.org/).
Twin Cities Metropolitan Area Highway Network 1968, 1971, 1975, 1978, 1981, 1985, and1990.	Minnesota Official Transportation Maps, issued by Minnesota Department of Transportation *(* before 1978, it was called 'Minnesota Official Highway Maps' and 'Minnesota Department of Highway').*

Table 2. The changes of the cells between different classes.

1958 (rows) → 1968 (columns)

1958 \ 1968	EU	MU	RU	WU	AU	EL	ML	RL	WL	AL	EB	MB	RB	WB	AB	EN	MN	RN	WN	AN	row sum
EU	78	18	6	0	16	5	7	0	0	0	4	2	0	0	1	4	0	0	0	0	141
MU	22	41	15	0	2	2	2	1	0	0	6	2	1	0	0	0	0	1	0	0	95
RU	26	151	206	1	9	0	1	6	0	0	0	6	19	0	3	1	3	3	0	0	435
WU	8	2	9	16	3	0	0	0	0	0	1	0	0	0	1	0	0	0	0	2	42
AU	127	71	167	8	434	0	3	0	0	0	6	4	9	0	27	2	3	4	0	14	879
EL	11	6	0	0	5	426	140	17	0	28	16	2	0	1	0	20	4	1	1	7	685
ML	2	15	5	0	0	115	465	81	0	3	3	15	3	0	0	3	15	5	0	0	730
RL	2	21	52	0	4	31	571	1150	5	53	2	14	22	0	3	0	27	58	0	3	2018
WL	2	0	4	2	0	10	6	53	54	37	0	1	1	0	1	0	0	1	9	3	184
AL	18	20	27	0	42	259	179	791	25	2179	4	2	13	0	17	18	8	31	0	79	3712
EB	3	1	0	0	0	1	1	0	0	0	8	3	1	0	0	0	0	0	0	0	18
MB	1	0	1	0	0	0	0	0	0	0	3	10	4	0	0	0	0	0	0	0	19
RB	0	3	2	0	0	1	0	1	0	0	2	20	16	0	0	0	0	0	0	0	45
WB	0	0	2	1	0	0	0	0	0	0	0	0	0	0	0	0	0	0	0	0	3
AB	2	3	3	0	6	1	0	0	0	1	10	9	16	0	37	0	1	0	0	1	90
EN	25	4	0	2	12	103	32	17	5	23	4	0	0	0	2	1517	248	111	91	281	2477
MN	2	9	14	0	0	25	115	48	2	4	1	1	4	0	0	229	801	329	5	28	1617
RN	1	26	123	0	8	20	223	680	8	33	0	2	17	0	1	161	1454	7312	274	390	10733
WN	4	0	10	7	11	17	3	35	18	16	0	0	0	2	1	98	15	471	2979	432	4119
AN	17	7	61	2	240	104	75	376	8	760	2	0	4	0	19	1400	576	4661	439	25351	34102
column sum	351	398	707	39	792	1120	1823	3256	125	3137	72	93	130	3	113	3453	3155	12988	3800	26589	62144

1978 (rows) → 1990 (columns)

1978 \ 1990	EU	MU	RU	WU	AU	EL	ML	RL	WL	AL	EB	MB	RB	WB	AB	EN	MN	RN	WN	AN	row sum
EU	458	201	80	41	89	0	0	0	0	0	9	4	1	0	0	2	1	0	0	1	887
MU	158	393	262	9	21	0	3	0	0	0	3	8	4	0	0	0	3	0	0	0	864
RU	62	212	532	20	75	0	0	3	0	0	2	3	4	0	0	0	2	1	0	0	916
WU	10	2	5	16	8	0	0	0	0	0	0	0	0	0	0	0	0	0	0	0	41
AU	153	74	239	30	293	0	0	0	0	0	0	2	0	0	3	0	0	0	0	0	794
EL	41	13	6	0	9	628	396	164	27	71	10	15	5	0	2	70	16	10	1	2	1486
ML	17	21	7	0	1	239	1123	751	11	47	6	10	7	0	0	23	35	25	0	1	2324
RL	9	16	24	0	2	90	727	2178	76	168	1	13	15	0	2	8	29	79	6	6	3449
WL	2	0	0	1	0	15	23	47	28	6	0	0	2	0	0	6	1	4	9	0	144
AL	14	2	10	2	11	199	144	962	82	810	7	5	11	0	10	10	9	40	13	19	2360
EB	21	8	1	0	0	0	0	0	0	0	81	38	16	2	12	0	0	0	0	0	179
MB	2	13	5	0	0	0	0	1	0	0	35	97	41	3	3	0	1	0	0	0	201
RB	0	8	12	0	0	0	0	3	0	0	11	36	68	3	11	0	0	2	0	0	154
WB	0	0	0	0	0	0	0	0	0	0	0	1	2	1	0	0	0	0	0	0	4
AB	11	4	2	1	0	0	0	0	0	0	18	13	18	0	18	0	0	0	0	0	85
EN	37	11	10	3	3	92	21	9	1	11	3	2	1	0	1	3235	995	535	236	474	5680
MN	16	16	21	1	2	10	47	32	1	2	0	1	0	0	0	520	2371	2003	45	87	5175
RN	6	25	77	8	15	5	52	90	5	6	0	0	4	0	2	291	2086	10391	573	862	14498
WN	2	0	0	9	4	2	0	1	4	0	0	0	1	3	1	196	37	608	2228	286	3382
AN	53	17	90	18	89	35	21	86	8	41	1	1	3	0	3	1341	701	7107	1350	8556	19521
column sum	1072	1036	1383	159	622	1315	2557	4327	243	1162	187	249	203	12	68	5702	6287	20805	4461	10294	62144

1968 (rows) → 1978 (columns)

1968 \ 1978	EU	MU	RU	WU	AU	EL	ML	RL	WL	AL	EB	MB	RB	WB	AB	EN	MN	RN	WN	AN	row sum
EU	211	77	23	1	28	0	0	0	0	0	5	4	0	0	0	2	0	0	0	0	351
MU	97	212	57	3	13	0	0	0	0	0	4	5	1	0	0	3	2	1	0	0	398
RU	59	234	336	8	44	0	1	0	0	0	1	9	4	0	1	0	1	9	0	0	707
WU	12	2	2	12	11	0	0	0	0	0	0	0	0	0	0	0	0	0	0	0	39
AU	218	77	148	7	317	0	0	1	0	0	5	3	4	0	6	2	0	2	1	1	792
EL	84	28	6	2	3	527	226	68	18	66	28	18	4	0	2	30	10	0	0	0	1120
ML	32	61	35	0	4	284	884	383	15	27	20	28	10	0	2	9	21	8	0	0	1823
RL	16	63	109	1	13	141	768	1777	49	179	2	11	37	1	8	7	30	36	0	8	3256
WL	6	0	0	0	0	11	17	35	30	24	0	0	0	0	0	0	0	0	1	1	125
AL	39	15	21	1	26	326	169	727	22	1682	5	3	10	0	14	13	5	13	1	45	3137
EB	6	3	1	0	0	0	1	0	0	0	38	18	2	0	3	0	0	0	0	0	72
MB	5	9	4	0	1	1	4	0	0	0	32	28	5	1	3	0	0	0	0	0	93
RB	1	11	5	0	4	0	0	0	0	0	7	57	41	2	1	0	0	1	0	0	130
WB	0	1	0	0	0	0	0	0	0	0	0	2	0	0	0	0	0	0	0	0	3
AB	4	3	2	0	4	1	0	0	0	1	28	10	22	0	38	0	0	0	0	0	113
EN	33	7	3	0	8	85	19	11	0	14	2	1	0	0	0	2201	448	219	92	310	3453
MN	7	17	17	0	2	24	56	47	0	5	0	4	1	0	0	502	1393	983	20	77	3155
RN	8	31	94	2	30	23	136	235	3	22	2	2	8	0	0	458	2568	8074	366	926	12988
WN	1	0	0	4	7	0	0	3	4	3	0	0	0	0	0	277	44	595	2459	403	3800
AN	48	13	53	0	279	63	43	162	3	337	0	0	3	0	7	2176	653	4557	442	17750	26589
column sum	887	864	916	41	794	1486	2324	3449	144	2360	179	201	154	4	85	5680	5175	14498	3382	19521	62144

Table 3. Transition matrices $P_{1958-1968}$, $P_{1968-1978}$ and $P_{1978-1990}$.

$P_{1978-1990}$	EU	MU	RU	WU	AU	EL	ML	RL	WL	AL	EB	MB	RB	WB	AB	EN	MN	RN	WN	AN	TOTAL
EU	51.63	22.66	9.02	4.62	10.03	0.00	0.00	0.00	0.00	0.00	1.01	0.45	0.11	0.00	0.00	0.23	0.11	0.00	0.00	0.11	100
MU	18.29	45.49	30.32	1.04	2.43	0.00	0.35	0.00	0.00	0.00	0.35	0.93	0.46	0.00	0.00	0.00	0.35	0.00	0.00	0.00	100
RU	6.77	23.14	58.08	2.18	8.19	0.00	0.00	0.33	0.00	0.00	0.22	0.33	0.44	0.00	0.00	0.00	0.22	0.11	0.00	0.00	100
WU	24.39	4.88	12.20	39.02	19.51	0.00	0.00	0.00	0.00	0.00	0.00	0.00	0.00	0.00	0.00	0.00	0.00	0.00	0.00	0.00	100
AU	19.27	9.32	30.10	3.78	36.90	0.00	0.00	0.00	0.00	0.00	0.25	0.00	0.00	0.38	0.00	0.00	0.00	0.00	0.00	0.00	100
EL	2.76	0.87	0.40	0.00	0.61	42.26	26.65	11.04	1.82	4.78	0.67	1.01	0.34	0.00	0.13	4.71	1.08	0.67	0.07	0.13	100
ML	0.73	0.90	0.30	0.00	0.04	10.28	48.32	32.31	0.47	2.02	0.26	0.43	0.30	0.00	0.00	0.99	1.51	1.08	0.00	0.04	100
RL	0.26	0.46	0.70	0.00	0.06	2.61	21.08	63.15	2.20	4.87	0.03	0.38	0.43	0.00	0.06	0.23	0.84	2.29	0.17	0.17	100
WL	1.39	0.00	0.00	0.69	0.00	10.42	15.97	32.64	19.44	4.17	0.00	0.00	1.39	0.00	0.00	4.17	0.69	2.78	6.25	0.00	100
AL	0.59	0.08	0.42	0.08	0.47	8.43	6.10	40.76	3.47	34.32	0.30	0.21	0.47	0.00	0.42	0.42	0.38	1.69	0.55	0.81	100
EB	11.73	4.47	0.56	0.00	0.00	0.00	0.00	0.00	0.00	0.00	45.25	21.23	8.94	1.12	6.70	0.00	0.00	0.00	0.00	0.00	100
MB	1.00	6.47	2.49	0.00	0.00	0.00	0.00	0.50	0.00	0.00	17.41	46.26	20.40	1.49	1.49	0.00	0.50	0.00	0.00	0.00	100
RB	0.00	5.19	7.79	0.00	0.00	0.00	0.00	1.95	0.00	0.00	7.14	23.38	44.16	1.95	7.14	0.00	0.00	1.30	0.00	0.00	100
WB	0.00	0.00	0.00	0.00	0.00	0.00	0.00	0.00	0.00	0.00	0.00	25.00	50.00	25.00	0.00	0.00	0.00	0.00	0.00	0.00	100
AB	12.94	4.71	2.35	1.18	0.00	0.00	0.00	0.00	0.00	0.00	21.18	15.29	21.18	0.00	21.18	0.00	0.00	0.00	0.00	0.00	100
EN	0.65	0.19	0.18	0.05	0.05	1.62	0.37	0.16	0.02	0.19	0.05	0.04	0.02	0.00	0.02	56.95	17.52	9.42	4.15	8.35	100
MN	0.31	0.31	0.41	0.02	0.04	0.19	0.91	0.62	0.02	0.04	0.00	0.02	0.00	0.00	0.00	10.05	45.82	38.71	0.87	1.68	100
RN	0.04	0.17	0.53	0.06	0.10	0.03	0.36	0.62	0.03	0.04	0.00	0.00	0.03	0.00	0.01	2.01	14.39	71.67	3.95	5.95	100
WN	0.06	0.00	0.00	0.27	0.12	0.06	0.00	0.03	0.12	0.00	0.00	0.00	0.03	0.09	0.03	5.80	1.09	17.98	65.88	8.46	100
AN	0.27	0.09	0.46	0.09	0.46	0.18	0.11	0.44	0.04	0.21	0.01	0.01	0.02	0.00	0.02	6.87	3.59	36.41	6.92	43.83	100

$P_{1968-1978}$	EU	MU	RU	WU	AU	EL	ML	RL	WL	AL	EB	MB	RB	WB	AB	EN	MN	RN	WN	AN	TOTAL
EU	60.11	21.94	6.55	0.28	7.98	0.00	0.00	0.00	0.00	0.00	1.42	1.14	0.00	0.00	0.00	0.57	0.00	0.00	0.00	0.00	100
MU	24.37	53.27	14.32	0.75	3.27	0.00	0.00	0.00	0.00	0.00	1.01	1.26	0.25	0.00	0.00	0.75	0.50	0.25	0.00	0.00	100
RU	8.35	33.10	47.52	1.13	6.22	0.00	0.14	0.00	0.00	0.00	0.14	1.27	0.57	0.00	0.14	0.00	0.14	1.27	0.00	0.00	100
WU	30.77	5.13	5.13	30.77	28.21	0.00	0.00	0.00	0.00	0.00	0.00	0.00	0.00	0.00	0.00	0.00	0.00	0.00	0.00	0.00	100
AU	27.53	9.72	18.69	0.88	40.03	0.00	0.13	0.00	0.00	0.00	0.63	0.38	0.51	0.00	0.76	0.25	0.00	0.25	0.13	0.13	100
EL	7.50	2.50	0.54	0.18	0.27	47.05	20.18	6.07	1.61	5.89	2.50	1.61	0.36	0.00	0.18	2.68	0.89	0.00	0.00	0.00	100
ML	1.76	3.35	1.92	0.00	0.22	15.58	48.49	21.01	0.82	1.48	1.10	1.54	0.55	0.00	0.11	0.49	1.15	0.44	0.00	0.00	100
RL	0.49	1.93	3.35	0.03	0.40	4.33	23.59	54.58	1.50	5.50	0.06	0.34	1.14	0.03	0.25	0.21	0.92	1.11	0.00	0.25	100
WL	4.80	0.00	0.00	0.00	0.00	8.80	13.60	28.00	24.00	19.20	0.00	0.00	0.00	0.00	0.00	0.00	0.00	0.00	0.80	0.80	100
AL	1.24	0.48	0.67	0.03	0.83	10.39	5.39	23.18	0.70	53.62	0.16	0.10	0.32	0.00	0.45	0.41	0.16	0.41	0.03	1.43	100
EB	8.33	4.17	1.39	0.00	0.00	0.00	1.39	0.00	0.00	0.00	52.78	25.00	2.78	0.00	4.17	0.00	0.00	0.00	0.00	0.00	100
MB	5.38	9.68	4.30	0.00	1.08	1.08	4.30	0.00	0.00	0.00	34.41	30.11	5.38	1.08	3.23	0.00	0.00	0.00	0.00	0.00	100
RB	0.77	8.46	3.85	0.00	3.08	0.00	0.00	0.00	0.00	0.00	5.38	43.85	31.54	1.54	0.77	0.00	0.00	0.77	0.00	0.00	100
WB	0.00	33.33	0.00	0.00	0.00	0.00	0.00	0.00	0.00	0.00	0.00	0.00	0.00	66.67	0.00	0.00	0.00	0.00	0.00	0.00	100
AB	3.54	2.65	1.77	0.00	3.54	0.88	0.00	0.00	0.00	0.88	24.78	8.85	19.47	0.00	33.63	0.00	0.00	0.00	0.00	0.00	100
EN	0.96	0.20	0.09	0.00	0.23	2.46	0.55	0.32	0.00	0.41	0.06	0.03	0.00	0.00	0.00	63.74	12.97	6.34	2.66	8.98	100
MN	0.22	0.54	0.54	0.00	0.06	0.76	1.77	1.49	0.00	0.16	0.00	0.13	0.03	0.00	0.00	15.91	44.15	31.16	0.63	2.44	100
RN	0.06	0.24	0.72	0.02	0.23	0.18	1.05	1.81	0.02	0.17	0.02	0.02	0.06	0.00	0.00	3.53	19.77	62.17	2.82	7.13	100
WN	0.03	0.00	0.00	0.11	0.18	0.00	0.00	0.08	0.11	0.08	0.00	0.00	0.00	0.00	0.00	7.29	1.16	15.66	64.71	10.61	100
AN	0.18	0.05	0.20	0.00	1.05	0.24	0.16	0.61	0.01	1.27	0.00	0.00	0.01	0.00	0.03	8.18	2.46	17.14	1.66	66.76	100

$P_{1958-1968}$	EU	MU	RU	WU	AU	EL	ML	RL	WL	AL	EB	MB	RB	WB	AB	EN	MN	RN	WN	AN	TOTAL
EU	55.32	12.77	4.26	0.00	11.35	3.55	4.96	0.00	0.00	0.00	2.84	1.42	0.00	0.00	0.71	2.84	0.00	0.00	0.00	0.00	100
MU	23.16	43.16	15.79	0.00	2.11	2.11	2.11	1.05	0.00	0.00	6.32	2.11	1.05	0.00	0.00	0.00	0.00	1.05	0.00	0.00	100
RU	5.98	34.71	47.36	0.23	2.07	0.00	0.23	1.38	0.00	0.00	0.00	1.38	4.37	0.00	0.69	0.23	0.69	0.69	0.00	0.00	100
WU	19.05	4.76	21.43	38.10	7.14	0.00	0.00	0.00	0.00	0.00	2.38	0.00	0.00	0.00	2.38	0.00	0.00	0.00	4.76	0.00	100
AU	14.45	8.08	19.00	0.91	49.37	0.00	0.34	0.00	0.00	0.00	0.68	0.46	1.02	0.00	3.07	0.23	0.34	0.46	0.00	1.59	100
EL	1.61	0.88	0.00	0.00	0.73	62.19	20.44	2.48	0.00	4.09	2.34	0.29	0.00	0.15	0.00	2.92	0.58	0.15	0.15	1.02	100
ML	0.27	2.05	0.68	0.00	0.00	15.75	71.10	11.10	0.00	0.41	0.41	2.05	0.41	0.00	0.00	0.41	2.05	0.68	0.00	0.00	100
RL	0.10	1.04	2.58	0.00	0.20	1.54	28.30	56.99	0.25	2.63	0.10	0.69	1.09	0.00	0.15	0.00	1.34	2.87	0.00	0.15	100
WL	1.09	0.00	0.00	2.17	1.09	0.00	5.43	3.26	28.80	29.35	20.11	0.00	0.54	0.54	0.54	0.00	0.00	0.54	4.89	1.63	100
AL	0.48	0.54	0.73	0.00	1.13	6.98	4.82	21.31	0.67	58.70	0.11	0.05	0.35	0.00	0.46	0.48	0.22	0.84	0.00	2.13	100
EB	16.67	5.56	0.00	0.00	0.00	5.56	5.56	0.00	0.00	0.00	44.44	16.67	5.56	0.00	0.00	0.00	0.00	0.00	0.00	0.00	100
MB	5.26	5.26	5.26	0.00	0.00	0.00	0.00	0.00	0.00	0.00	15.79	52.63	21.05	0.00	0.00	0.00	0.00	0.00	0.00	0.00	100
RB	0.00	6.67	4.44	0.00	0.00	2.22	0.00	2.22	0.00	0.00	4.44	44.44	35.56	0.00	0.00	0.00	0.00	0.00	0.00	0.00	100
WB	0.00	0.00	66.67	33.33	0.00	0.00	0.00	0.00	0.00	0.00	0.00	0.00	0.00	0.00	0.00	0.00	0.00	0.00	0.00	0.00	100
AB	2.22	3.33	3.33	0.00	6.67	1.11	0.00	0.00	0.00	1.11	11.11	10.00	17.78	0.00	41.11	0.00	1.11	0.00	0.00	1.11	100
EN	1.01	0.16	0.00	0.08	0.48	4.16	1.29	0.69	0.20	0.93	0.16	0.00	0.00	0.00	0.08	61.24	10.01	4.48	3.67	11.34	100
MN	0.12	0.56	0.87	0.00	0.00	1.55	7.11	2.97	0.12	0.25	0.06	0.06	0.25	0.00	0.00	14.16	49.54	20.35	0.31	1.73	100
RN	0.01	0.24	1.15	0.00	0.07	0.19	2.08	6.34	0.07	0.31	0.00	0.02	0.16	0.00	0.01	1.50	13.55	66.13	2.55	5.63	100
WN	0.10	0.00	0.24	0.17	0.27	0.41	0.07	0.85	0.44	0.39	0.00	0.00	0.05	0.00	0.02	2.38	0.36	11.43	72.32	10.49	100
AN	0.05	0.02	0.18	0.01	0.70	0.30	0.22	1.10	0.02	2.23	0.01	0.00	0.01	0.00	0.06	4.11	1.69	13.67	1.29	74.34	100

REFERENCES

Anderson, T. W. and L. A. Goodman (1957). Statistical inference about Markov chains. *Annals of Mathematical Statistics*, **28(1)**, 89-110.

Brown, L. A. (1970). On the use of Markov chains in movement research. *Economic geography*, **46**, Issue supplement. Commission on quantitative methods, 393-403.

Clark, W. A. V. (1965). Markov chain analysis in geography: an application to the movement of rental housing areas. *Annals of the Association of American Geographers*, **55(2)**, 351-359.

Goodman, Leo A. (1962). Statistical Methods for Analyzing Processes of Change. *American Journal of Sociology*, **1968(1)**, 57-1978.

Goulias, K. (1999). Longitudinal analysis of activity and travel pattern dynamics using generalized mixed Markov latent class models. *Transportation Research Part B*, **33**, 535-557.

Janssens, D., G. Wets, T. Brijs, and K. Vanhoof (2005). The development of an adapted Markov chain modeling heuristic and simulation framework in the context of transportation research. *Expert Systems with Applications*, **28**, 105-117.

Kelley, A.C. and L. W. Welss (1969). Markov processes and economic analysis: the case of migration. *Econometrica*, **37(2)**, 280-297.

Krishnan, P. (1977). A Markov chain approximation of conjugal history. *Mathematical models of sociology*, Issue Editor P. Krishnan, Great Britain, 127-133.

Lever, W.F. (1972). The intra-urban movement of manufacturing: a Markov approach. *Transactions of the institute of British geographers*, **56**, 21-38.

Levinson, D. and W. Chen (2004). *Area-Based Models of New Highway Route Growth*. Presented at World Conference on Transport Research, Istanbul.

Levinson, D. and R. Karamalaputi (2003). Induced Supply: A Model of Highway Network Expansion at the Microscopic Level. *Journal of Transport Economics and Policy*, **37(3)**, 297-318.

Parthasarathi, P., D. Levinson and R. Karamalaputi (2003). Induced Demand: A Microscopic Perspective. *Urban Studies*, **40(7)**, 1335-1353.

Singer, B. and S. Spilerman (1976). The representation of social processes by Markov models. *American Journal of Sociology*, **82(1)**, 1-54.

Stafford, James (1977). Urban growth as an absorbing Markov process. *Mathematical Models of Sociology*, Issue Editor P. Krishnan, Great Britain, 135-142.

Texas A&M University, Texas Transportation Institute, *2002 Urban Mobility Report*.

Yin, K., H. Yang, P. Daoutidis, and G. Yin (2003). Simulation of population dynamics using continuous-time finite-state Markov chains. *Computers and Chemical Engineering*, **27**, 235-249.

Zhang, B.-T. and D.-Y. Cho (2001). System identification using evolutionary Markov chain Monte Carlo. *Journal of Systems Architecture*, **47(7)**, 587-599.

Zhang, C. and W. Li (2005). Markov Chain Modeling of Multinomial Land-Cover Classes. *GIScience and Remote Sensing*, **42(1)**, 1-18.

NOTES

[1] Since Year 1988 land-use map is not available, Year 1990 map is used as a substitute. Years 1958, 1968, 1978 and 1990 are considered the time points of equal intervals.

Access to Destinations
D.M. Levinson and K.J. Krizek (editors)
© 2005 Elsevier Ltd. All rights reserved.

267

CHAPTER 13

ACCESSIBILITY AND FREIGHT: TRANSPORTATION AND LAND USE— EXPLORING SPATIAL-TEMPORAL DIMENSIONS

Clarence Woudsma, University of Waterloo
John F. Jensen, University of Calgary

INTRODUCTION

Considering transportation's influence on urban development patterns, the central question concerns how accessibility influences land use development and the decisions made by locators (firms or individuals). As an integral part of the interrelationship between transportation (T) and land use (LU), accessibility is an obvious focus of much research. However, there are still important questions about its characteristics and influence. This is particularly true if we think beyond people movement and personal activities to the movement of goods (freight) and land use classes associated with goods distribution activity or Distribution-Logistics-Warehousing (DLW) (Woudsma, 2001). The central aim of this paper is to explore T/LU relationships and accessibility from a freight perspective. DLW falls within the context of logistics and supply chain management, a recent evolution of the goods production-distribution process. The importance of understanding the urban T/LU problem within the context of these dimensions is increasing with the growing recognition that the DLW sector has a strong influence on regional prosperity and is a powerful factor affecting regional structural change and urban land use (Hesse, 2002). Canadian statistics reveal that "more than 90% of the goods moved within Canada depend on truck transportation, either solely or as part of an intermodal shipment" (Industry Canada, 2000). Trucks carry nearly all of the commodities from warehouses—or the more appropriately termed distribution centres - to points of consumption of retail sale in metropolitan regions (Wegmann *et al.*, 1995). In Calgary, Alberta, transportation and warehousing grew 20% between 1990 and 2001 making it one of the top two fastest growing economic sectors (The City of Calgary, 2002a; The City of Calgary, 2002c).

Given that spatial interaction is an important geographical dimension of logistics, this research focuses on issues related to mobility and accessibility. Mobility refers to "the ability and knowledge to travel from one location to another in a reasonable amount of time and for acceptable costs" (Meyer & Miller, 2001, p 95). Accessibility is broadly defined as an "aggregate measure of the size and closeness of activity opportunities of a given type to a particular location" (Meyer & Miller, 2001, p 336). Travel occurs when mobility and accessibility combine with purpose (e.g. transport of goods to market). Thus, as the demand for speed, time sensitivity, and delivery frequency increases (Andel, 1998; Hesse, 2002; McCann & Sheppard, 2003) and considering i) a growing sensitivity to proximity and ii) that trucks are the dominant distribution mode, it follows that the distribution of urban DLW land use types should be influenced by issues such as impedances induced by the urban transportation system. This notion is widely acknowledged in the seminal (Weber, 1909; Alonso, 1964), contemporary (e.g., Hesse & Rodrique, 2004), and integrated (e.g., Wegener, 1994) T/LU modeling (Hunt *et al.*, 1999; McCann & Sheppard, 2003) literature, but has yet to be fully demonstrated empirically. Woudsma (2001), and others (e.g., Ortúzar & Willumsen, 1996), suggest that this is partly due to a lack of reliable data on firms and urban goods movements.

Specifically, this work explores and quantifies the relationship between DLW land use development patterns and transportation system performance expressed in terms of traffic congestion and accessibility. Traffic congestion is viewed as a serious problem with respect to the physical distribution of goods in urban areas (Ogden, 1991; Morris *et al.*, 1998; Gorys & Hausmanis, 1999; Golob & Regan, 2001). The potentially important influence of traffic congestion on DLW land use distribution prompted the formulation of the research question: *Does transportation system performance have a quantifiable influence on the timing and spatial character of DLW land use development in Calgary, Alberta?*

BACKGROUND

Noting the obvious complexity of city systems, it is not surprising that limitations exist with respect to our understanding of the elements of urban spatial structure, particularly from the perspective of transportation's influence on land use. This is the case for several reasons. First, although major land use theories incorporate transportation costs as having a significant impact on location, this has been typically limited to composite costs related to distance (e.g. distance from the central business district in a monocentric city or distance from inputs of production) or a composite inter-zonal cost (e.g., de la Barra, 1989). Second, urban planning efforts with respect to these elements have tended to occur separately and with a distinct bias towards households and automobiles as opposed to firm location and the movement of goods (Woudsma, 2001). Finally, integration of these elements into models (Hunt *et al.*, 1999) involves complex relationships that are difficult to untangle (Pickrell, 1999).

Furthermore, although it is generally accepted that transportation influences land use patterns, our understanding is incomplete with respect to the specifics and strength of the relationship.

For example, there has been an emphasis on examining the impacts of new infrastructure as opposed to levels of traffic or performance on the existing system. Additionally, some have argued that the ubiquity of reasonable access in modern cities minimizes the influence of transportation (e.g., Giuliano, 1995), while others claim that as congestion worsens the influence on development patterns increases (e.g., Cervero & Landis, 1996).

Thus, the central argument considered here is that system-related transportation costs, specifically traffic congestion, influence the distribution of Distribution-Logistics-Warehouse (DLW) land use types. Theoretically, it is expected that locations with better mobility and accessibility are more attractive for DLW development and will be developed faster. Although good accessibility does not guarantee that land use will follow (Davis, 1999)—other factors (e.g. economy, politics, services, and personal choice) also play into the equation. However, for the general reasons cited above, it has yet to be fully demonstrated empirically that urban DLW land use distribution does in fact respond to these costs.

Theoretical Context

The two main theoretical dimensions associated with a) transportation and land use and b) location theory share a common thread of explaining the role of transportation costs on location choice. Classic geographic theories advanced by Von Thunen and later Alonso employed the notion of bid-rent where the price or rent of land or building stock is influenced by distance or transportation costs, giving rise to distinct urban forms such as concentric rings with higher value land uses at the core and lower value land uses at the periphery (Torrens, 2000). More modern theories, relevant to this research are based on neo-classical economic theory with the transportation land use relationship reflected in the profit-maximizing behaviour of firms or the utility-maximizing behaviour of individuals (Meyer & Miller, 2001) Firms (or individuals) trade off accessibility and their willingness to pay for space; for example, industrial firms with large land demands and limited need for access to population would tend to locate at the periphery of urban areas, keeping land costs low, enhancing the profitability of their operation (Boarnet & Crane, 2001). In general, locations with lower transport costs should be more attractive for development and, therefore, will be developed faster (Wegner & Fürst, 1999).

Clearly, theory postulates that transportation cost is a component of DLW development. However, it is important to note that there is disagreement as to its importance as an influencing factor. Some argue that the influence of transport is declining (Giuliano, 1989), and in fact, contemporary location theory places less emphasis on transportation and more on the influence of agglomeration and associated industry linkages, such as clustering, in explaining locating behaviour (Anas *et al.*, 1998). On the other hand, recent theoretical debates have highlighted the fact that transportation cost is becoming more important than ever in the globalized economy with its emphasis on flexible production and use of supply chain logistics (McCann & Sheppard, 2003). Further, recent empirical efforts employing discrete choice methods to examine the decision behaviour of firms has highlighted the fact that transportation costs do matter, but vary, for example, depending on the context—e.g. a new entrant versus a relocation (Leitham *et al*, 2000; Holguin-Veras *et al*., 2005). Given these mixed views, there is a need for a more comprehensive understanding of the response of sectors like distribution and logistics to transportation-related cost.

Transportation and Land Use

In a effort to understand the T/LU relationship, a large body of work deals with models that simulate the interaction between the transportation and land use systems (Wegener, 1994; Miller *et al.*, 1999; Wegner & Fürst, 1999; Hunt *et al.*, 1999). These models are useful in that they are able to reasonably replicate actual land use and transportation patterns; they provide a tool to forecast future urban patterns; and they facilitate controlled testing of theories and practices. Thus, they represent a potential means to address the research question here; however, these models have some limitations. For example, a review (Hunt *et al.*, 1999) identified several shortcomings including, i) temporal resolution, ii) type of transportation model, and iii) availability of data.

Table 1. Urban change processes
Source: adapted from (Wegner & Fürst, 1999:43; Shaw & Xin, 2003)

Urban change process	Examples
Very slow change	*Networks* (e.g., transport networks): are the most permanent elements of cities
Slow change	*Land use*: distribution is often stable, changes are incremental *Workplaces* (e.g., warehouses, office buildings, shopping centers): exist much longer than the firms or institutions that occupy them *Housing*: exist longer that the households that live in it
Fast change	*Employment*: refers to firms that open, close, expand, or relocate *Population*: refers to households that form, grow, decline, dissolve, or relocate
Intermediate change	*Goods transport*: adjusts quickly to changes in demand *Travel*: adjusts quickly to changes in traffic conditions

Temporal resolution refers to the timing of model feedback mechanisms. The T/LU relationship is generally expressed as a feedback cycle of trips generated on a transportation network by a land use model versus accessibility provided to the land use model by the transportation network. T/LU cycle completion time is recognized as having an important bearing on the relationship, particularly with respect to the timing of the various components (Meyer & Miller, 2001:129). As such, a lag is normally invoked between the transportation sub-model and the land use model. Although this lag is typically applied through feedback of the transportation model output into the next period's land use model (Abraham & Hunt, 1999), there are limitations in our understanding of the magnitude of the lag. It is widely recognized that the various model components operate on different time scales, but to date quantification of this timing is limited to broad categories (e.g., Bourne, 1982; Wegner & Fürst, 1999) such as those illustrated in Table 1. This implies that, although a lag is typically incorporated into most integrated models, there is insufficient evidence to suggest a standardized or empirically validated lag period (typically measured in years) that differs depending on the land use type and process under consideration.

Overall, although these models are increasing in complexity, they continue to be limited by the simplicity of theories linking firm location decisions to transportation costs. Furthermore, these models tend to be data intensive and the difficulty of obtaining detailed historical data on land use patterns poses a serious challenge (Still, 1996; Davis, 1999; Pickrell, 1999; Miller *et al.*, 2003).

Mobility and Accessibility

Accessibility and Mobility represent the focal points for the interaction between the land use and transportation systems. Typically, accessibility is associated with a particular place within the urban system and mobility is associated with the actors in the urban system. An individual can possess high mobility (auto owner vs transit dependent) which influences their ability to reach places, and a location can have high accessibility (central location with many modal interconnections) which influences its activity level and type. These ideas also apply when we consider goods movement and non-residential locations, and can be understood in the context of "logistical friction" (Hesse & Rodrique, 2004). According to Hesse and Rodrique, logistical friction represents the cumulative pressure that is exerted on the supply chain (Table 2). This friction is expressed in the form of system-related transport costs, such as distance, time, access to roads and congestion. It is not clear, however, if or how these frictional components are internalized in the decisions of the marketplace - for example, if firms in fact respond to urban traffic congestion (Anas *et al.*, 1998). Nor is there a clear understanding about the nature and timing of the response by those firms that are influenced by negative mobility and accessibility issues. Furthermore, it is also unclear as to the point at which these issues become a problem. For example, Giuliano (1989) indicates that firms seek to achieve some minimum accessibility threshold and that all comparably suitable sites are acceptable unless this threshold is breached.

Table 2. Logistical friction

Source: (Hesse & Rodrique, in press)

Impedance Factor	Assessment Measures
Transport/Logistics Costs	Distance, Time, Composition, Transshipment, Decomposition
Supply Chain Complexity	Number of suppliers, Number of distribution centers, Number of parts/variety of components
Transactional Environment	Competition, (sub-) contracting, inter-firm relationships, power issues, (de-) regulation
Physical Environment	Infrastructure supply, Road bottlenecks and congestion, Urban density, Urban adjustments

In order to assess the impact of friction related to mobility and accessibility on DLW land use distribution, it is necessary to measure it. This is a difficult challenge because this friction exhibits considerable spatial variability (Weber & Kwan, 2002). Although there is little agreement regarding what constitutes a good measure, it is basically understood that most accessibility-type measures consist of two elements: transportation and activity (Handy & Niemeier, 1997). The transportation element represents transportation costs related to travel, while the activity element captures the attractiveness of a location as a trip destination. The focus of this research is on the transportation element, specifically daily traffic congestion (as opposed to non-recurrent congestion—e.g. due to traffic accidents). Traffic congestion is defined here as "travel time or delay in excess of that normally incurred under light or free-flow conditions" (Lomax *et al.*, 1997).

Friction related to intraurban transportation has typically been measured by travel distance or travel time. Time-based measures are a preferred means of assessing network impedance, but the challenge of determining them in a complex urban road network is considerable and approaches are varied. Nevertheless, there is some agreement that this variable is best measured through a network from an origin to a destination as a time-based measure (Lomax *et al.*, 1997; Larsen, 2001) preferably using congested speeds to measure shortest paths (Martin & McGuckin, 1998); although, other measures (e.g. space-time measures) are also suitable. An accessibility surface (Weber & Kwan, 2002) is an effective means to express such a variable (MacKinnon & Lau, 1973).

SPATIAL DEPENDENCE

A variety of methods have been used to quantify the T/LU relationship including, among others (see reviews by, Vessali, 1996; Badoe & Miller, 2000), ordinary regression (e.g., Sivitanidou, 1996; Boarnet, 1996; Miller & Ibrahim, 1998; Davis, 1999; Kawamura, 2001), hedonic models (e.g., Bowes, 2001) and four stage models (including gravity models) (e.g., Hirschman & Henderson, 1990; Miller & Demetsky, 1998). Where conventional statistical methods have been applied, there is an underlying assumption that the data is statistically independent and identically distributed (iid) (Anselin, 1995: 266). However, given the likelihood of clustering of DLW land use types (Shukla & Waddell, 1991; Ellison & Glaeser, 1997), it was expected that spatial autocorrelation would be present in the dependent variable (DLW land use development). The iid assumption does not hold in the presence of spatial autocorrelation.

For zone-based data, similar to the data structure used in this research, spatial autocorrelation refers to a situation in which the value of an attribute in one zone depends on the values of the attribute in neighbouring zones (Fotheringham *et al.*, 2000) The consequence of spatial autocorrelation for standard regression is the presence of remaining spatial dependence in the residuals even after structural similarities of neighbouring zones have been controlled for. This is contrary to the assumption of independent error terms. Considering this condition, analytical methods selected for this research are influenced by the recent work of Haider and Miller (2000) and others (Anselin, 1988; Baller *et al.*, 2001; Anselin, 2002) who advocate the use of spatially autoregressive models (SAR) as a preferred method in the presence of spatial autocorrelation.

In this specific case, a spatial lag model is considered to be the most appropriate approach to modeling spatial dependence in zone-based data (for a thorough discussion, see Anselin, 1988; Fotheringham *et al.*, 2000: 166-171; Baller *et al.*, 2001; Anselin, 2002). The spatial lag model stipulates the effect of neighbouring zones through the inclusion of an extra explanatory variable in the standard regression model (e.g., mean of the dependent variable for adjacent zones) expressed as

$$y = X\beta + \rho Wy + \varepsilon \tag{1}$$

where y is the dependent variable, X is the independent variable and related parameter β, W is an adjacency or spatial weights matrix for the zones, is a regression coefficient or spatial autoregressive parameter for the adjacency variable and ε is an error term . Equation 1 expressed in reduced form

$$y = (I - \rho W)^{-1} X\beta + (I - \rho W)^{-1} \varepsilon \tag{2}$$

demonstrates that this model captures the influence of unmeasured independent variables in the error term as well as the effect of neighbouring zones in the dependent variable. This type of model is appropriate under the assumption that activity in a zone will somewhat mimic that of its neighbouring zones (Fotheringham *et al.*, 2000: 166-171), which is expected here.

STUDY AREA

Calgary, Alberta was selected as the study area for several important reasons. First, the key spatial data inputs required for this research are in place (Staley & Sheldrake, 2001; Sheldrake,

2001; Staley, 2002) and include i) transportation system flows represented by average weekday traffic (AWDT) on Calgary's major streets and expressways for the period 1964-2001 and ii) land use in the form of approximately 270,000 individual land use parcels and their attributes including actual use codes and building year of construction. Second, Calgary is a young, dynamic community that is experiencing considerable economic growth. In particular, Calgary is strategically located as a major distribution hub for western Canadian markets and the DLW sector represents one of the major growth areas where firms like WalMart and Sears have recently established new regional distribution centers. Third, Calgary has a well-developed urban planning process that is facing significant challenges in the areas of transportation and land use planning. Transportation planners in Calgary are currently engaged in a program to advance Calgary's regional transportation model to include commercial travel (The City of Calgary, 2001). Recent trends in industrial development have prompted city planners to more closely evaluate the quality and quantity of available land (The City of Calgary, 2002c), and, in addition, transportation infrastructure falling behind the speed of growth was the number one issue for Calgarians in the 2001 Customer Satisfaction Survey (The City of Calgary 2001 Annual Report (2002b)). The report also stresses Calgary's intent to combine land use planning initiatives and transportation operations strategies. It is apparent from these points that research of this nature in Calgary is both appropriate (data is available) and timely (the city is experiencing transportation and land use challenges).

METHODOLOGY

Research methods (Figure 1) involve comprehensive variable development designed to establish an empirical link between transportation system performance and DLW land use development. With this aim in mind, transportation variables are created to represent network impedance and accessibility, while land use variables are created to represent DLW development and available land. All variables are developed based on a regular 855 m^2 sample grid. This grid size was established based on the extent of the City of Calgary official boundaries in 2003 and the need to keep the number of study cells (zones) reasonable. Part of the challenge was trying to integrate the fine detail of the parcel level with the more general concept of location accessibility. That is, it was decided that established a measure of accessibility for each parcel would have been both time consuming, and in the context of this research, unnecessary. The use of grid cells of this size facilitated both the use of parcel level detail and the creation of accessibility variables. Miller *et al.* (2003) suggest that grid cells may provide a suitable means to avoid aggregation bias problems. Variable development is followed by modeling procedures designed to estimate the effects on DLW development of independent variables (e.g, network performance, available land and serviced land).

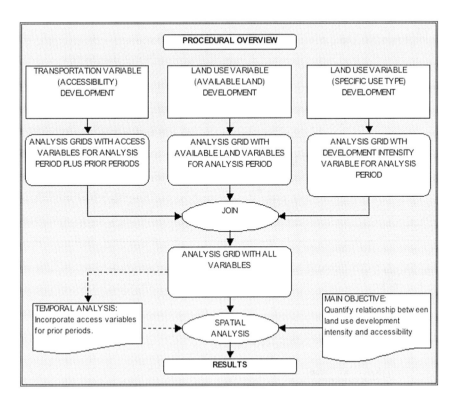

Figure 1. Procedural overview

Transportation Variables

Transportation data for Calgary includes average weekday traffic volumes (AWDT) on major roads for the years 1964 to 2000 provided by the City of Calgary in the form of paper flow maps. Data from the flow maps is manually incorporated into a GIS database and merged with a road network output from *The Calgary Transportation Model* (The City of Calgary, 1991; Sheldrake, 2001; Jensen, 2004) to produce a subset road network of only major roads with traffic volume attributes (Figure 2) (Staley, 2002). The transportation model is an a.m. peak hour model of the City of Calgary based on a single carriageway road network subset and implemented using the EMME/2 system. Key model attributes include, number of lanes, capacity, posted speed limit, signalization and link length. To facilitate temporal analysis, the subset network is also coded to reflect historical presence/absence of roads based on existing paper road maps for the City of Calgary. Thus, it is possible to represent both historical traffic volume and network structure. Variables created from this base data set include analysis hour volumes, network impedance and accessibility.

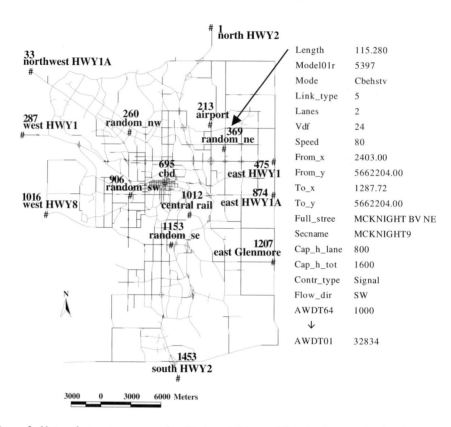

Figure 2: Network structure, example of link variables and Destinations nodes for shortest path analysis

Traffic analyses typically focus on 15-minute or hourly flows and frequently on the peak hour of traffic for the peak direction because this represents the highest capacity requirements (Transportation Research Board, 2000). Although, there are those that suggest other times of day should also be considered—particularly with respect to truck movements (Woudsma, 2001). Given that *The Calgary Transportation Model* is based on a.m. travel (The City of Calgary, 1991), AWDT volumes are disaggregated into a.m. peak hour volumes using conversion factors developed directly from City of Calgary year 2001 six hour 15-minute traffic counts collected for the a.m. peak period (7 to 9 am). The conversion factor is applied to each respective link AWDT volume to extract an appropriate directional analysis hour volume.

These directional analysis hour volumes are then used as input to produce a time-based friction or impedance variable representing congested link travel time. A common means to calculate congested link travel time is found in a formulation developed by the U.S. Bureau of

Public Roads (BPR) to estimate link travel times as a function of the volume/capacity ratio (Martin & McGuckin, 1998). In this formulation, impedance is a measure of the congested travel time on link given the traffic volume on that specific link with respect to capacity. Thus, an impedance variable is developed to represent the basic influence of increasing traffic volume without the complexity of estimating delays due to queuing of traffic.

The resulting impedance variable provides a cost input for subsequent network analysis. Network analysis involves the connection of all sample grid centroids to the nearest node on each individual historical network. Then, shortest path procedures are implemented. Shortest path procedures follow an iterative process of finding the shortest distance from an origin node to a destination node and all intermediate nodes on the path (e.g., Heywood, Cornelius, & Carver, 1998; Miller & Shaw, 2001; Chang, 2002).

All grid centroids represent origins. Destinations are defined based on criteria described by Hesse (2002). According to Hesse, firms can be classified according to three locational attitudes: i) the pure cost minimizer (e.g., large freight forwarders or large retail chains) locates at strategic places near major transport corridors; ii) the fast and flexible respondent (e.g., wholesale distributors, parcel services) prefers proximity to customers; and iii) the inter-modal operator (e.g., container shippers, major freight carriers) is drawn to large transport modal interfaces such as airports and rail yards. Considering this classification scheme, destination nodes are selected to include major entry/exit points to the city, the central business district, the airport, a major central rail terminal and four random points (Figure 2).

Final inputs include i) the road network in 5-year increments with connected grid centroids representing origins and destinations, ii) congested link travel time representing impedance, and iii) free flow travel time representing impedance. Shortest paths are iteratively run from each grid centroid to the destination points for the years 1985, 1990, 1995, and 2000 first using free flow travel time and then congested travel time as the impedance value. Given the scope of this research, it was decided not to pursue the earlier time frames at this stage. The resulting cumulative shortest path values are attributes that represent the cost of travel from a grid centroid to a given destination.

Figure 3 illustrates the resulting independent variable(s) *AMPH* , which represents accessibility based on congested travel time during the AM peak hour, 2000. This value is the shortest path travel time from a given grid centroid to a pre-selected destination point. The result is a set of variables (referred to hereafter as Set 1), one created for each chosen destination point and an additional variable representing the average for all destination points. The average surface variable represents overall accessibility for each grid cell to all points in the city. In all cases, the pattern is intuitive—the most accessible locations are central and in terms of access to features of interest, basically, the further you are away from the feature (airport), the longer your travel time to get there. In order to try and tease out the influence of system performance alone—exclude the influence of distance—a new set of variables (hereafter Set 2) were introduced.

The independent variable(s) *NAMH* (prefix NAMH—Normalized Congestion AM peak hour) represents accessibility as the percent change between free flow and congested travel time (*AMPH*) from a given grid centroid to a pre-selected destination points (see Lomax *et al.*, 1997 for a discussion about congestion measures). This set of variables reflects the difference in travel time due to congestion and in a sense, isolates the influence of congestion from travel distance. Figure 4 displays a number of example surfaces based on this new set of variables. The pattern is obviously different from those in Figure 3, with the congestion influence spatially varying across the city, following the underlying network structure somewhat. This is most dramatically realized in Figure 5, which captures the exposure to congestion as an average of getting to all destination points. The peaks indicate areas where users would be exposed to greater congestion influences in trying to access other destinations. The question pursued here is whether, as a variable reflecting accessibility, this has an influence on patterns of development.

Land Use Variables

Land use data for Calgary includes a comprehensive registered land parcel rolling database (*pardat*) containing approximately 270,000 parcel records including attributes such as parcel address, geographic coordinates, zone (e.g. I-2 industrial), use type (e.g. multi bay warehouse) and year of construction. Use type consists of approximately 415 separate codes to describe individual parcel land use. Year represents initial year of development and is documented from the year 1900 to present. Given that this is a rolling database, land use type and year only reflect the original date of development and any changes are lost. Therefore, transportation influence on land use is considered from the perspective of new growth as opposed to changes to existing land use.

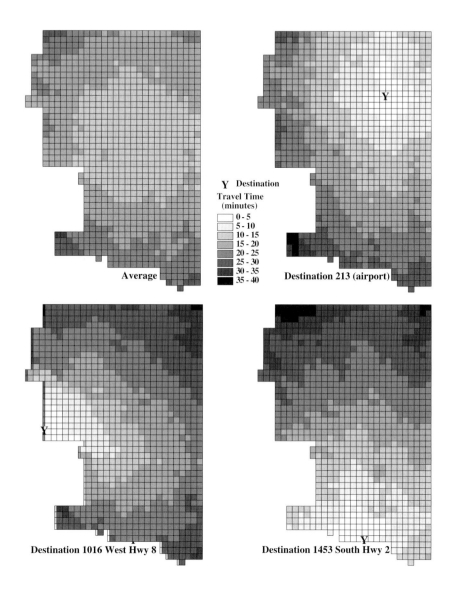

Figure 3: 2000 Accessibility variable maps (Set 1 AMPH)

The four maps illustrate cumulative congested travel time (accessibility) to the airport, west highway 8 and south highway 2 as well as the average accessibility to all destinations. Lighter colours indicate better accessibility or lower travel time. Note the distinct imprint of distance in the form of concentric rings.

Figure 4: 2000 Accessibility variable maps (Set 2 NAMH)

The four maps illustrate the influence of congestion alone on travel (accessibility measure) to the airport, west highway 8 and south highway 2 as well as the average congestion influence for all destinations. Lighter colours indicate better accessibility or lower congestion influence. Note that the distance imprint identified in Figure 3 is not apparent.

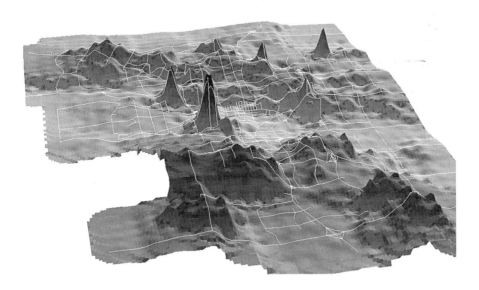

Figure 5: Pseudo 3D surface representing year 2000 congestion in Calgary

This is a Pseudo 3D impression of congestion influence exclusive of distance (percent change between free flow and congested link travel time). Greater elevation (darker colours) represents increasing congestion influence (value range is 1.3 to 10.7% increase over free flow). This means that the most congested locations (highest elevation) experience on average a 10.7% increase in travel time between free flow and congested travel to any of the sample destinations used in this research.

Additional data includes parcel polygons, building footprints, a vector representation of water mains and a general land use classification. This is by no means a comprehensive set of land use variables, with, for example, key variables like land value missing. Attempts were made to incorporate the all important influence of land value, however, it was not possible to obtain property values at the level of spatial and temporal detail used in this study. It will be addressed in subsequent research efforts. The parcel polygons represent all land parcels within the Calgary city limits and when joined with *pardat* provide area, use type and year of development attributes for all land parcels (Figure 6). Water mains provide a proxy for historical municipal services or serviced land in that they include a year of construction. The land use classification provides a more aggregate representation of Calgary land use.

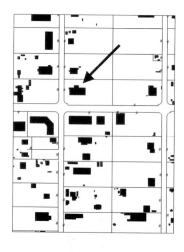

PARDAT POINT	
Parcel id	860546875
Address	9028 44 ST SE
Xcoord	2121.19
Ycoord	5648509.05
Zone	I-4
Type	WAREHOUS
Year	1979

FOOTPRINT POLYGON	
Area	1957.45
Xcoord	2172.01
Ycoord	5648499.88

PARCEL POLYGON	
Parcel id	860546875
Area	19143.98
Xcoord	2119.67
Ycoord	5648509.05

Figure 6: Parcel data detail

This figure illustrates the parcel data set. On the left is a graphic showing parcels (parcel polygon) with building footprints (footprint polygon) and a point representing pardat output (pardat point). The tables represent the attributes associated with the features of the parcel indicated with an arrow.

Variables developed from this base data set include i) a dependent variable represented as m^2 of DLW development per sample grid cell per time period, ii) land available for DLW development, also represented as m^2 per sample grid cell per time period, and iii) a variable representing serviced land, produced from the water main database. Thus, for each sample grid cell, it is possible to identify i) DLW development, ii) land available for development and iii) presence of city services.

For the purposes of this research, the dependent variable, DLW development is created using a query of use codes designed to extract only DLW-related parcels (e.g., multi-bay warehouse and warehouse with internal office space). Given that there is a scarcity of DLW land use development in any given year, a five-year time block (1996-2000) is arbitrarily selected to ensure that sufficient development is captured to facilitate meaningful statistical analysis. This parcel query is summarized on total developed DLW parcel area per grid cell (m^2 per sample grid cell per time period) (Figure 7).

Available land is determined based on an intersection of i) the sample grid ii) the Calgary Land Classification (year 2001) and ii) the merged pardat/parcel polygon database. Inclusion of land available for development is essential because DLW development can only occur *where land is*

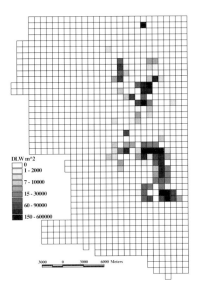

Figure 7: DLW development per grid cell (1996–2000)

available. The result is a parcel polygon data set that includes sample grid identifier, major land use classes and pardat attributes (year of construction and use type). This intersection is summarized on sample grid cell in order to produce area amounts per grid cell representing available land in the form of total and proportional areas for the target time period. Figure 8 illustrates sample grid cells with land available for non-residential development in the 1996-2001 time period. Area amounts are calculated to express land available as of the end of the year 1995. Thus, for each sample grid cell, it is possible to identify for each land use development class i) total area, ii) developed area, and iii) vacant area or available land (Figure 8).

The presence of water mains provides a proxy for serviced land (e.g., water, power, sewage). Serviced land is established based on an intersection of the Calgary watermain network and the sample grid. Sample grid cells are coded according to the presence or absence of water mains based on year of construction present in the watermain dataset. The result is a proxy representation of serviced land.

Spatial Modeling

Spatial modeling is performed using SpaceStat (Anselin, 2001) and is primarily based on procedures prescribed by Anselin (1988; 1998). Quantification of the empirical link established above in the form of comprehensive transportation and land use variables begins with exploratory data analysis (EDA) followed by exploratory spatial data analysis (ESDA) (for a review, see Anselin, 1998), which focuses on spatial autocorrelation of the dependent variable (e.g., Anselin, 1996) and heterogeneity (e.g., spatial regimes). This exploratory analysis sets up modeling procedures designed to estimate the effects on DLW development of the independent variables

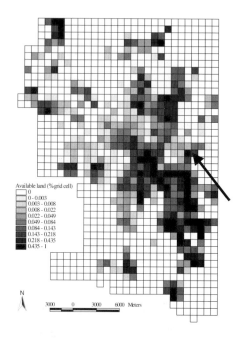

Grid cell area	731025.0000
Total non-res	68765.6485
Total res	213995.8912
Total urban reserve	230641.5873
Total other	39209.5515
Total all types	552612.6785
Unavailable land	178412.3215
Proportion non-res	0.0941
Proportion res	0.2927
Proportion urban	0.3155
Proportion other	0.0536
Proportion all types	0.7559
Proportion unavailable	0.2441
Developed non-res	12643.3303
Developed res	192968.6956
Developed urban	178840.4220
Developed other	4.7912
Developed all types	384457.2391
Vacant non-res	56122.3182
Vacant res	21027.1956
Vacant urban reserve	51801.1653
Vacant other	39204.7603
Vacant all types	168155.4394
%Vacant non-res	7.68
%Vacant res	2.88
%Vacant urban reserve	7.09
%Vacant other	5.36
%Vacant all types	23.00

Figure 8: Sample grid—available non-residential land (period 1996-2001)

An example attribute table for the grid cell indicated by the arrow is presented to the right of the sample grid. Note that 23% of the total grid cell is vacant or available for development (all use types) while only 7.68% (56,122 m²) of the grid cell is available for non-residential development.

(e.g., network performance, available land and serviced land) with adjustments for spatial dependence and heterogeneity. Then, a spatial autoregressive model that best addresses spatial dependence in the data is selected to quantify the relationship. Finally, modeling efforts are expanded to include accessibility from several time periods as a means to capture a temporal lag.

A spatial lag model, estimated using a maximum likelihood approach, is selected based on results obtained from EDA, ESDA and the OLS diagnostics. Model inputs include i) DLW development (years 1996-2000) as the dependent variable, ii) accessibility (both including (Set 1) and excluding distance (Set 2) for the am peak hour (e.g year 1995), available land for the period 1996-2000 and serviced land as of the year 2000 as the independent variables, iii) a sample grid representing the east region of the City of Calgary and including only those cells with DLW development potential based on year 2001 land use, and iv) two corresponding weights matrices based on i) distance and ii) nearest neigbour. The analysis was restricted to the East side of the city (point iii) in order to minimize the influence of a high number of cells with no DLW development in the analysis (the West half of the City). This decision, driven by computational demands, also makes intuitive sense given that zoning restrictions would typically prohibit large DLW developments in those primarily residential areas. However, it is important to keep in mind that our accessibility calculations do in fact reflect the entire city transportation system.

Given these inputs, an exhaustive and systematic set of model runs is carried out considering all possible variable and weights matrix combinations. Variable combinations found in the best model runs are then re-run using similar combinations of accessibility variables for the years 1980, 1985 and 1990 in order to explore the temporal component of the problem. The results of these temporal runs are examined for the presence of significant accessibility variables.

RESULTS AND DISCUSSION

Figure 9 is a useful starting point for an exploration into the nature of the relationship between the transportation system and patterns of development for the DLW sector. The last 20 years can be characterized for Calgary as an era of rapid growth, approaching 1,000,000 population, coupled with minimal expansion of major roadways and expressways. The challenge is to determine what kind of influence this steep rise in congestion is having—in particular on those firms involved in warehousing and distribution.

The results in Table 3 represent the top four spatial model runs ranked according to the Akaike Information Criterion (AIC) (Anselin, 1988; 1992). The model with the lowest AIC is best. The dependent variable in all cases is DLW development (square root of total parcel development m^2 per grid cell for the period 1996-2000). The variables, which consistently were positively associated with DLW development, were 1) the amount of available land and 2) whether or not a grid cell was serviced. This result was not surprising, but reaffirmed our confidence in the variable specifications. In relation to the question of accessibility, the results in this table suggest that *there is a significant inverse relationship* between DLW land development and transportation system performance in Calgary. That is, even if a grid cell had a significant amount of land available for warehouse development, the amount of actual development that occurred was negatively impacted by the influence of congestion—reduction in accessibility - related to that grid cell. In terms of the locations (e.g. getting to "the airport"), Table 3 indicates they are in the North central and eastern parts of the city. Furthermore, there are three Set 2 variables (congestion isolated) and one model with a Set 1 (congested travel time) variable location. In terms of the strength of that influence, in some cases (examining the results on a cell by cell basis) there is up to a 28% penalty or reduction in development related to accessibility. The locations (destinations) associated with the significant access variables make sense given, for example, that highway 2 is the major North-South corridor in the Province of Alberta and a designated NAFTA corridor as well. Therefore, distribution firms would see good access to this highway as an important consideration.

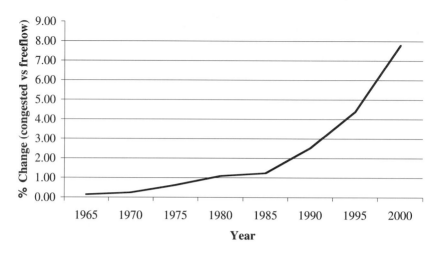

Figure 9: Influence of congestion on travel in Calgary over time

The graph represents the change over time in the average difference between congested and free flow travel times for all major links in the City of Calgary's road network. It is meant to reflect the growth in traffic outstripping the supply of capacity.

However, as Table 4 suggests, the influence of accessibility is spatially uneven with respect to the locations of interest, and depending on the specification of the accessibility variable (Set 1 or Set 2). This conclusion is supported by accessibility surfaces developed from time-based measures and consistent with findings in other contexts (e.g., Weber & Kwan, 2002). *When distance is considered*, (Set 1) accessibility based on congested travel time to a given location presents the characteristic pattern of decreasing accessibility with increasing distance from that location. This characteristic pattern suggests that proximity to a destination has implications for accessibility regardless of traffic, which is clearly in line with classical theory (Weber, 1929). On the other hand, the spatial variability of accessibility when expressed based on *congestion exclusive of distance* (Set 2) indicates that *congestion alone* is a factor influencing DLW development patterns. This may reaffirm the emphasis on travel time as part of research efforts to understand accessibility, but also serves to underscore the importance of accounting for spatial influences (distance) when examining transportation system performance. Time based measures are obviously influenced by distance. In this case, we've tried to illustrate that in terms of system performance, it is important to control for this influence.

In addition to spatial variability, the T/LU relationship in this case also exhibits a distinct *temporal lag*. The notion of a lag in the T/LU relationship is well recognized in the literature (e.g., Bourne, 1982; Wegner & Fürst, 1999); however, it is only broadly defined. To begin with, Figure 10 illustrates the temporal relationship between the statistical significance (left hand vertical axis with lower bars (below 0.05) indicating significance at the 95% level) of the Set 1 variables associated with the major North and East exit/entrance points and the airport and the congestion curve from Figure 9 (right hand vertical axis indicating % above free flow). Clearly, the pattern is fairly consistent through time and is indicative of the importance of proximity or distance alone. Regardless of the year, proximity to these key locations is a significant explanatory component of the pattern of DLW land use development.

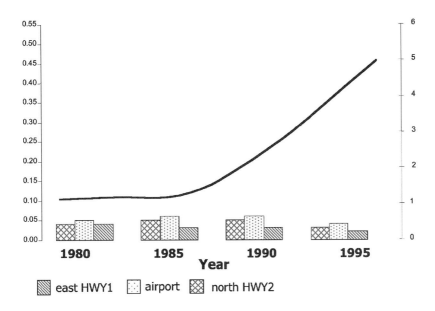

Figure 10: Set 1 Example—Variable Significance vs. Increasing Congestion Influence

Table 3: Results—top 4 spatial model runs ranked according to the Akaike Information Criterion

Variable	Beta coefficient	Standard error	z-value	Probability
SPATIAL LAG	0.529	0.071	7.422	0.000
CONSTANT	-22.941	14.156	14.156	0.105
NAMH1	-669.114	256.955	-2.604	0.009
PVAC_WHS	266.925	25.997	10.268	0.000
WATER_95	26.200	11.870	2.207	0.027
SPATIAL LAG	0.535	0.071	7.577	0.000
CONSTANT	-20.018	15.264	-1.312	0.190
NAMH213	-396.651	163.074	-2.432	0.015
PVAC_WHS	269.788	26.092	10.340	0.000
WATER_95	34.902	12.026	2.902	0.004
SPATIAL LAG	0.571	0.068	8.383	0.000
CONSTANT	-12.751	19.289	-0.661	0.509
AMPH1	-0.030	0.014	-2.094	0.036
PVAC_WHS	268.635	26.119	10.285	0.000
WATER_95	23.520	12.202	1.928	0.054
SPATIAL LAG	0.552	0.069	8.003	0.000
CONSTANT	-27.302	14.547	-1.877	0.061
NAMH260	-595.619	303.582	-1.962	0.050
PVAC_WHS	261.612	26.101	10.023	0.000
WATER_95	27.718	11.902	2.329	0.020

Model specifications:
    Spatial Lag Model – maximum likelihood estimation
    Data set = east region, grid cells with DLW development potential

Summary statistics:
    Number of observations = 314
    AIC = 3709.96, 3711.17, 3712.34, 3712.89
    $R2$ = 0.426, 0.423, 0.417, 0.418
    SIG-SQ = 7471.81, 7495.42, 7488.51, 7519.87

Diagnostics:
    Heteroskedasticitiy, Breusch-Pagan[b] = 295.709**, 290.629**, 295.947**, 281.678**
    Likelihood ratio test on spatial lag dependence[a] = 45.336**, 46.362**, 57.372**, 51.290**
    Lagrange multiplier test on spatial error dependence[a] = 0.095, 0.037, 0.005, 0.439

Weights matrix:
    NNWR1_12 = 12 nearest neighbours, row standardized

Dependent variable:
    square root of total parcel development (m^2) per grid cell, years 1996-200

Independent variable:
    NAMH1 = % change between free flow and congested travel time to north Hwy 2
    NAMH213 = % change between free flow and congested travel time to the airport
    AMPH1 = congested travel time to north Hwy 2
    NAMH260 = % change between free flow and congested travel time to random NW point
    PVAC_WHS = % grid cell area available for development, years 1996-2000
    WATER_95 = serviced versus non-serviced grid cell

** $p \le .01$ (two-tailed test)
[a] distributed as ^{2} with 1 degree of freedom, [b] distributed as ^{2} with 3 degrees of freedom

Table 4: Accessibility Variable Results

AIC	Variable	Beta coefficient	Standard error	z-value	Probability
Set 2 variables = % change between free flow & congested travel time to destination					
3709.96	HWY2N	-669.114	256.955	-2.604	0.009
3711.17	AIRPORT	-396.651	163.074	-2.432	0.015
3712.89	RANDNW	-595.619	303.582	-1.962	0.050
3713.50	HWY2S	727.104	398.992	1.822	0.068
Set 1 variables = congested travel time to destination					
3712.34	HWY2N	-0.030	0.014	-2.094	0.036
3713.25	HWY2S	0.024	0.013	1.885	0.060
3713.28	HWY8W	0.041	0.022	1.871	0.061
3713.46	RANDSE	0.041	0.022	1.818	0.069
Model specifications: Spatial Lag Model – maximum likelihood estimation					
Summary statistics: Number of observations = 314					
Weights matrix: NNWR1_12 = 12 nearest neighbours, row standardized					
Dependent variable: WALL_SQR = square root of total parcel development (m²/grid cell), 1996-200					

Figure 11 is the same time period and locations, but reflects the Set 2 variables—those in which distance is isolated and congestion is most prominent. In this case, the temporal pattern is quite different and indicates that a lag falls in the 5 to 10 year range for Calgary. In other words, current DLW land development patterns demonstrate a significant relationship with traffic conditions at least five years in the past and in some cases up to 10 years. Note that the height of the bars in 1980 and 1985 are well above 0.1, the cut off indicating significance at a 90% level. As one moves forward in time, and as congestion in the City increased overall, the statistical influence of accessibility related to level of development was realized. This result implies that accessibility patterns based on current traffic conditions could be indicative of DLW land use patterns five to ten years hence, an empirically verified notion that should be considered by both planners and predictive modelers. However, the question remains as to whether this is truly a lagged relationship or is instead reflecting the rapid growth of a mid-sized city in the 1970's that is now pushing 1,000,000 population and experiencing the growth pains of a true "big city". An extension of this project is currently being pursued in which other combinations of time periods (i.e. different 5 year development blocks) are being tested for the existence of this lag.

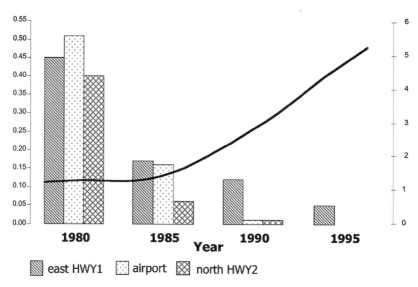

Figure 11: Set 2 Example—Variable Significance vs. Increasing Congestion Influence

CONCLUSION

This paper has presented arguments that identify a lack of understanding with respect to urban DLW land development, particularly as it relates to transportation system performance and the influence of accessibility. This lack of understanding has arisen because the problem is complex, data is difficult to collect and much is yet unknown about the interrelationship between these two key elements of urban form.

With this problem in mind, a simplified spatial-temporal approach was applied to answer the research question: *Does transportation system performance have a quantifiable, although temporally lagged, influence on the timing and spatial character of DLW land use development in Calgary, Alberta?* The approach was driven by the fact that DLW land use types in Calgary are not randomly distributed in space. The presence of spatial dependence in the data has methodological implications in that analyses are likely to produce misleading results unless these interactions are taken into account.

Nevertheless, results presented here showing increasing congestion in Calgary coupled with indications of a significant relationship between DLW land development and transportation system performance confirm the importance of continued exploration of the urban T/LU problem. To this end, this research offers a methodology with which to examine and measure the T/LU relationship. Application of this methodology produced a number of useful outputs including i) congestion-based accessibility surfaces for several important destinations in the

City of Calgary, ii) confirmation of a significant relationship between DLW land use development and transportation system performance, iii) the identification of a specific temporal lag in the T/LU relationship. Although this research imparts important information with respect to the T/LU relationship, several issues relevant to this and future research bear further consideration in order to enhance the credibility of the interpretations presented herein. A common concern with respect to T/LU studies is the issue of aggregation (both spatial and temporal). Although there is no agreement in the literature with respect to spatial aggregation of urban data, the most common method of aggregation is the census tract or transportation zone. This research assumed a regular grid based on comments by Miller *et al.* (2003); however, future studies should explore the implications of using varying spatial representations such as buildings, parcels or finer grid cells. In addition, considering that accessibility in Calgary is spatially uneven, future investigations should focus on the impact of local versus global transportation system performance. A similar scale issue extends into the temporal realm given that this research incorporated DLW land development as a five year time block and temporal lag investigations were conduced in five year increments. It would be useful to introduce a finer temporal scale.

Several variable development issues also follow this line of thought. On the transportation side of the problem, analyses were conducted based on typical daily traffic, which excludes the congestion effect of non-recurrent traffic (e.g., construction, accidents and cuing). Inclusion of non-recurrent traffic may paint an entirely different picture. Although consideration of this aspect of the congestion problem was not possible in this research, it certainly merits examination in future research. Additionally, analyses were conducted based on the AM peak hour of traffic. Woudsma (2001), and others, indicate that the AM peak hour is not necessarily the most important time for DLW travel, which suggests that future investigations should be conducted including accessibility variables based on other times of day.

On the land use side of the problem, analyses were conducted based on an aggregation of DLW use types into one single class while controlling for available and serviced land. Clearly different use types within the DLW sector (e.g., multibay warehouse versus regional distribution centre or intermodal shipper truck/rail versus truck/air) are likely to exhibit different responses with respect to the transportation system and access (see, Hesse, 2002). Further work is required to examine the various DLW land use classes separately.

REFERENCES

Abraham, J., & Hunt, J. (1999). Firm location in the MEPLAN Model of Sacramento. In *Proceedings of the 79th Annual Meeting of the Transportation Research Board* Washington, DC.

Alonso, W. (1964). *Location and land use*. Cambridge, M.A.: Harvard University Press.

Anas, A., R. Arnott, and K. A. Small (1998). Urban spatial structure. *Journal of Economic Literature,* **36(3)**, 1426-1464.

Andel, T. (1998). What every warehouse should know about transportation. *Transportation & Distribution,* **39(5)**, 89.

Anselin, L. (1988). *Spatial econometrics: Methods and models.* Kluwer Academic Publishers, Dordrecht.

Anselin, L. (1992). *SpaceStat tutorial.* Regional Research Institute, West Virginia University, Morgantown, WV.

Anselin, L. (1995). *SpaceStat version 1.80 user's guide.* Regional Research Institute, West Virginia University, Morgantown, WV.

Anselin, L. (1996). The Moran scatterplot as an ESDA tool to assess local instability in spatial association. In: *Spatial Analytical Perspectives on GIS: GIS Data IV* (M. Fischer, H. J. Scholten, & D. Unwin, eds.), pp. 111-125. Taylor & Francis, London.

Anselin, L. (1998). Exploratory spatial data analysis in a geocomputational environment. In: *Geocomputation: A Primer* (P. A. Longley, S. M. Brooks, R. McDonnell, & B. Macmillan, eds.). John Wiley & Sons Ltd.

Anselin, L. (2001). *SpaceStat, a software program for the analysis of spatial data, version 1.91.* Regional Research Institute, West Virginia University, Morgantown, WV.

Anselin, L. (2002). Under the hood: Issues in the specification and interpretation of spatial regression models. *Agricultural Economics,* **17**(3), 347-267.

Badoe, D. A., and E. J. Miller (2000). Transportation-land-use interaction: Empirical findings in North America, and implications for modeling. *Transportation Research D,* **5(4)**, 235-263.

Baller, R. D., L. Anselin, S. F. Messner, G. Deane, and D. F. Hawdins (2001). Structural covariates of U.S. county homicide rates: Incorporating spatial effects. *Criminology,* **39(3)**, 561-590.

Boarnet, M., and R. Crane (2001). The influence of land use on travel behavior: specification and estimation strategies. *Transportation Research A,* **35**, 823-845.

Boarnet, M. G. (1996). The direct and indirect economic effects of transportation infrastructure. *The University of California Transportation Center, University of California at Berkeley, Working Paper UCTC No. 340,* 1-26.

Bourne, L. S. (1982). Urban spatial structure: An introductory essay on concepts and criteria. In: *Internal Structure of the City: Readings on Urban Form, Growth, and Policy* (L. S. Bourne, ed.), Second Ed., pp. 28-45, Oxford University Press, New York.

Bowes, D. R. (2001). Identifying the impacts of rail transit stations on residential property values. *Journal of Urban Economics,* **50**, 1-25.

Cervero, R., and J. Landis (1996). Why the transportation-land use connection is still important. *TR News,* **187**, 9-11.

Chang, K. (2002). *Introduction to geographic information systems.* Boston: McGraw Hill.

The City of Calgary (1991). *1991 Calgary regional transportation model primer.*

The City of Calgary (2001). *New directions for the Calgary Regional Transportation Model.* The City of Calgary Planning and Transportation Forecasting.

The City of Calgary (2002a). *Calgary's economic times: May issue.* City of Calgary: Customer Services & Communications, Creative Resources.

The City of Calgary (2002b). *2001 annual report.*

The City of Calgary (2002c). *Calgary industrial growth strategy: Part 1.*

Davis, J. S. (1999). Land use impacts of transportation: A guidebook. *NCHRP Report 423A* . Washington D.C.: National Academy Press.

de la Barra, T. (1989). *Integrated land use and transport modeling: Decision chains and hierarchies.* Cambridge Urban and Architectural Series, no. 12. Cambridge University Press, Cambridge, U.K.:.

Ellison, G., & E. L. Glaeser (1997). Geographic concentration in U.S. manufacturing industries: A dartboard approach. *The Journal of Political Economy, 105(5)*, 889.

Fotheringham, A. S., C. Brunsdon, and M. Charlton (2000). *Quantitative geography: perspectives on spatial data analysis.* Sage Publications, London.

Giuliano, G. (1989). New directions for understanding transportation and land use. *Environment and Planning A, 21*, 145-159.

Giuliano, G. (1995). The weakening transportation-land use connection. *Access, 6,* 2-10.

Golob, T. F., & A. C. Regan (2001). Impacts of highway congestion on freight operations: perceptions of trucking industry managers. *Transportation Research, 35A(7)*, 577-599.

Goodchild, M. F. (1986). *Spatial Autocorrelation.* CATMOG 47. Geo Books, Norwich, U. K.

Gorys, J., & I. Hausmanis (1999). A strategic overview of goods movement in the greater Toronto area. *Transportation Quarterly, 53*(2), 101.

Griffith, D. A. (1987). *Spatial autocorrelation: A primer.* Washington: Association of American Geographers.

Haider, M., and E. J. Miller (2000). Effects of transportation infrastructure and location on residential land values: Application of spatial-autoregressive techniques. *Transportation Research Record, 1722*, 1-8.

Handy, S. L., & D. A. Niemeier (1997). Measuring accessibility: An exploration of issues and alternatives. *Environment and Planning A, 29*, 1175-1194.

Hesse, M. (2002). Changes in goods distribution and the city: Physical distribution as an indicator of urban and regional development. In: *Monitoring Cities: International Perspectives* (W. K. D. Davies and I. J. Townshend, eds.), pp. 357-372. Graphcom Printers Ltd, Calgary.

Hesse, M., & J. Rodrique (2004). The transport geography of logistics and freight distribution. *The Journal of Transport Geography, 12(3)*, 171-184.

Heywood, I., S. Cornelius, and S. Carver (1998). *An introduction to geographical information systems.* Prentice Hall, Upper Saddle River, NJ.

Hirschman, I., and M. Henderson (1990). Methodology for assessing local land use impacts of highways. *Transportation Research Record, 1274*.

Holguin-Veras, J., N. Xu, H. S. Levinson, C. E. McKnight, R. D. Weiner, R. D. Paaswell, K. Ozbay, and D. Ozmen-Ertekin (2005). An Analysis Of The Geographic Patterns of Firm Relocations To New Jersey (1990-1999) and Its Relationship to Transportation, *TRB 2005 Annual Conference Proceedings*, CD-ROM.

Hunt, J., D. Kriger, and E. Miller (1999). Current operational urban land-use transport modeling frameworks. *In Proceedings of the 79th Annual Meeting of the Transportation Research Board* Washington, DC.

Industry Canada (2000). *Logistics and supply chain management: Overview and prospects.* Ottawa, Ontario: Service Industries and Capital Projects Branch, Industry Canada.

Jensen, J. F. N. (2004). *Urban development and transportation: A spatial-temporal analysis.* Masters Thesis, University of Calgary, Calgary, Alberta.

Kawamura, K. (2001). Empirical examination of relationship between firm location and transport facilities. *Transportation Research Record,* **1747**, 97-103.

Larsen, M. (2001). Direction for freight performance measures. In: *Performance Measures to Improve Transportation Systems and Agency Operations, Transportation Research Board Conference Proceedings,* **26**, pp 143-146. National Academy Press, Washington, DC.

Leitham, S., R. W. McQuaid, and J. D. Nelson (2000). The influence of transport on industrial location choice: a stated preference experiment, *Transportation Research Part A* **34**, 515-535

Lomax, T., S. Turner, G. Shunk, H. S. Levinson, R. H. Pratt, P. N. Bay, and G. B. Douglas (1997). Quantifying congestion: Volume 1, final report. *National Cooperative Highway Research Program (NCHRP) Report 398* . Washington D.C.: National Academy Press.

MacKinnon, R. D., and R. Lau (1973). Measuring accessibility change. In: *The Form of Cities in Central Canada: Selected Papers* (L. S. Bourne, & R. D. J. W. MacKinnon, eds.), pp. 120-137. University of Toronto Press, Toronto.

Martin, W. A., and N. A. McGuckin (1998). Travel estimation techniques for urban planning. *National Cooperative Highway Research Program (NCHRP) Report 365* . National Academy Press, Washington DC.

McCann, P., and S. Sheppard (2003). The rise, fall and rise again of industrial location theory. *Regional Studies,* **37**(6&7), 649-663.

Meyer, M. D., and E. J. Miller (2001). *Urban transportation planning: a decision-oriented approach: Second edition.* Boston : McGraw Hill.

Miller, E. J., J. D. Hunt, J. E. Abraham, and P. A. Salvini (2004). Microsimulating urban systems. *Computers, Environment and Urban Systems,* **28**:1 ,9-44

Miller, E. J., and A. Ibrahim (1998). Urban form and vehicular travel: some empirical findings. *Transportation Research Record,* **1617**, 18.

Miller, E. J., D. S. Kriger, and J. D. Hunt (1999). Integrated urban models for simulation of transit and land-use policies: Guidelines for implementation and use. *Transit Cooperative Research Program (TCRP) Report 48* . Washington D.C.: National Academy Press.

Miller, H. J., and S.-L. Shaw (2001). *Geographic information systems for transportation: Principles and applications (spatial information systems).* Oxford University Press.

Miller, J. S., and M. J. Demetsky (1998). *Using historical data to measure transportation infrastructure constraints on land use.* (Report No. FHWA/VTRC 98-R32). Virginia Transportation Research Council, Charlottesville, VA.

Moore, T., and P. Thorsnes (1994). The transportation/land use connection: A framework for practical policy. *American Planning Association Planning Advisory Service Report, 448/449.*

Morris, A. G., A. L. Kornhauser, and M. J. Kay (1998). Urban freight mobility: collection of data on time, costs, and barriers related to moving product into the central business district. *Transportation Research Record,* **1613**, 27.

Odland, J. (1988). *Spatial Autocorrelation*. Newbury Park: Sage publications.

Ogden, K. W. (1991). Truck movement and access in urban areas. *Journal of Transportation Engineering, 117*(1), 71-90.

Ortúzar, J. d. D., & Willumsen, L. G. (1996). *Modeling transport*. Second ed.,). Chichester: John Wiley & Sons.

Pickrell, D. (1999). Transportation and land use. In José A. Gómez-Ibáñez, William B. Tye, & Clifford Winston (Editors), *Essays in Transportation Economics and Policy: A Handbook in Honor of John R. Meyer* . Brookings Institution Press, Washington, DC.

Shaw, S. L., and X. Xin (2003). Integrated land use and transportation interaction: A temporal GIS exploratory data analysis approach. *Journal of Transport Geography, 11,* 103-115.

Sheldrake, M. (2001). *Migration of traffic volume flow map from CADD to GIS.* University of Calgary, Department of Geography (unpublished report).

Shukla, V., and P. Waddell (1991). Firm location and land use in discrete urban space. *Regional Science and Urban Economics, 21*, 225-253.

Sivitanidou, R. (1996). Warehouse and distribution facilities and community attributes: An empirical study. *Environment and Planning A, 28*, 1261-1278.

Staley, R. (2002). *Traffic flow GIS database development.* University of Calgary (unpublished report).

Staley, R., and M. Sheldrake (2001). *Flow map progress report.* University of Calgary, Department of Geography (unpublished report).

Still, B. (1996). The importance of transport impacts on land use in strategic planning. *Traffic Engineering & Control, 37*(10), 564.

TD Bank Financial Group (2003). *The Calgary-Edmonton corridor: Take action now to ensure tiger's roar doesn't fade.* TD Economics Special Report, April 22.

Torrens, P. M. (2000). How land use transportation models work. *Centre for Advanced Spatial Analysis, University College, London, Working Paper Series, 20*, 1-75.

Transportation Research Board (2000). *Highway capacity manual - HCM 2000 (Metric Units).*

Vessali, K. V. (1996). Land use impacts of rapid transit: A review of the empirical literature. *Berkeley Planning Journal, 11*, 71-105.

Weber, A. (1909). Uber den Standort der Industrien. Translated by Friedrech C. J.(1929) *Alfred Weber's Theory of the Location of industries* . University of Chicago Press, Chicago.

Weber, A. (1929). *Theory of the location of industries*. University of Chicago Press, Chicago.

Weber, J., and M.-P. Kwan (2002). Bringing time back in: A study on the influence of travel time variations and facility opening hours on individual accessibility. *The Professional Geographer, 54*(2), 226-240.

Wegener, M. (1994). Operational urban models: State of the art. *Journal of the American Planning Association, 60*, 17-29.

CHAPTER 14

ACCESSIBILITY IN THE LUCI2 URBAN SIMULATION MODEL AND THE IMPORTANCE OF ACCESSIBILITY FOR URBAN DEVELOPMENT

John R. Ottensmann, Indiana University-Purdue University

INTRODUCTION

The *luci2 Urban Simulation Model* forecasts changes in employment and the conversion of nonurban land to residential and employment-related uses. The model and its initial implementation for Central Indiana have several distinctive aspects: It covers a large area including 44 counties and 8 Metropolitan Statistical Areas. It easily allows the creation and comparison of future development scenarios reflecting a wide range of policy choices and assumptions affecting future development. And it is designed for use directly by citizens and policy makers, requires no special expertise to use, and is freely-distributed to interested users on the web.

Accessibilities to both employment and population have long been considered to play a significant role in shaping patterns of urban development and have been used in urban simulation models. The *luci2* model provides the opportunity to examine the importance of accessibility both by considering the estimation of the equations for the model and by the comparison of scenarios created by the model using alternative assumptions regarding the relative importance of accessibility.

This paper describes the structure of the *luci2 Urban Simulation Model* and the roles played by accessibility to population and employment in the model. The estimation of the model equations using historical data provides evidence of the importance of accessibility for urban development and shows that changes in accessibility can also be significant predictors of urban change. Use of the model to generate scenarios using higher and lower accessibility coefficients directly demonstrates the effects of accessibility on the simulated patterns of urban development.

ACCESSIBILITY IN URBAN SIMULATION MODELS

Nearly half a century ago, Hansen (1959) and Guttenberg (1960) argued that accessibility to employment plays a central role in shaping patterns of urban development. Likewise, Huff (1963) and Hansen and Lakshmanan (1965) used accessibility to population in models predicting the location of retail activity.

The measure of accessibility used most frequently in urban simulation models is a generalized form of the measure originally proposed by Hansen. This is the sum over all zones of the number of opportunities in each zone multiplied by some function of the cost of travel to the zone, $\sum O_j f(C_{ij})$. The cost function will generally include one or more empirically-determined parameters that reflect the extent to which interaction decreases as a function of distance. Spatial interaction (gravity) models, frequently used in urban simulation models, incorporate Hansen accessibility in their structure (Fotheringham and O'Kelly, 1989).

But Hansen-type measures are only one form for measuring accessibility that can be used in urban simulation models. In addition to this measure, Handy and Neiemeier (1997) identify two other forms for the measurement of accessibility. Cumulative opportunities measures specify the number of opportunities that can be reached within some given distance or travel time. A measure of accessibility based upon random utility theory employs the utilities associated with the set of opportunities available to an individual as estimated using a multinomial logistic choice model (the logsum).

Even simpler measures of accessibility exist and have been used in simulation models. If opportunities are assumed to be concentrated in a single center, such as employment in the central business district (CBD), distance or travel time to that center can be used as a measure of accessibility. This is the approach used in the theoretical monocentric models developed in urban economics (Mills, 1972; Muth, 1969). If multiple centers exist, this approach can be extended by using distances or travel times to each of those centers.

An additional simple approach can be employed by assuming that certain major transportation facilities provide faster travel and therefore more convenient access to opportunities throughout the region. In this case, distances or travel times to such transportation facilities, such as distances to the nearest freeway interchange, can be used as a measure of accessibility.

Some of the earliest models developed to forecast land use employed the simpler measures of accessibility. The land use forecasts developed for the Chicago Area Transportation Study in the 1950s used distance to the CBD (Creighton, 1970). The Niagara Frontier model for the

Buffalo area (Lathrop, *et al.*, 1965) employed travel time to the CBD in an intervening opportunities model. The North Carolina model developed by Chapin and Weiss (1962) used travel distance to the CBD along with a measure of accessibility to multiple employment centers that was the sum of the distances to those centers multiplied by their employment (in effect, a simplified type of Hansen accessibility).

Most of the urban simulation models developed after these initial efforts have employed some form of Hansen accessibility. Interestingly, however, some newer approaches to land use modeling in the past decade have returned to using the simple measures. Landis has developed a series of simulation models within geographic information systems using discrete choice models. In the California Urban Futures II model, accessibility is included by using distances to two CBDs and distances to the nearest freeway interchange and BART station (Landis and Zhang, 1998a). The California Urban and Biodiversity Model likewise uses distances to the nearest major highway and CBD (Landis, 2001). In his most recent modeling effort involving the creation of multiple models for regions in California, the number of jobs located within a travel time of 90 minutes is used as an accessibility measure along with distance to the nearest freeway interchange (Landis and Reilly, 2003).

Some new approaches use cellular automata for urban simulation. Such models, when they include measures of accessibility, employ the simple measures. Silva and Clarke (2002) developed the SLEUTH model (slope, land use, exclusion, urban extent, transportation, and hillshade), which used a simple measure of accessibility, whether the grid cell is within a specified distance from (within the buffer around) major roads. Pijanowski's (Pijanowski, *et al.*, 2002) Land Use Transformation Model used distances from major roads and highways. Li and Yeh (2002) and Chen, *et al.* (2002), in models of Chinese urban areas, used both distances to multiple centers and distances to the nearest access points for multiple transportation facilities.

In the seminal Model of Metropolis, which greatly influenced much of the work on urban simulation that followed, Lowry (1964) used the basic Hansen measures of accessibility. Accessibility to employment was the primary predictor of the location of residential activity, and accessibility to population was used for predicting the location of local-service employment. Many subsequent urban simulation models have been developed using the basic concepts employed in the Lowry model. For a review of some early work following Lowry, see Goldner (1971).

Many of the models that followed explicitly employed spatial interaction models. Calibration of these models included determination of the value(s) of the accessibility coefficient(s) in the cost function. The most highly developed and widely used (in the United States) of these models is Putman's DRAM/EMPAL (and now, METROPILIUS) model (Putman, 1983; Putman and Chan, 2001). When used within the Integrated Transportation Land-Use Package, updated travel times from the travel demand models can be used in each simulation round.

In a review of urban simulation models, Wegener (1994) observed that most models were now grounded on the application of random utility theory and are implemented using discrete choice models. Separate measures of accessibility are included in these models as factors contributing to the levels of utility of the location choices being modeled. Only a few of the more widely used current models can be discussed here. Major comparative reviews of urban simulation models have been recently completed and provide information on many of the models that have been developed (Agarwal, *et al.*, 2000; David Simmonds Consultancy, 1999; Southworth, 1995; U.S. Environmental Protection Agency, 2000).

Two of the more widely used models using discrete choice models are MEPLAN (Hunt and Simmonds, 1993) and Tranus (de la Barra, 1989, 2001). The developers of these models originally worked together and the models share many characteristics. They use an expanded form of the Lowry framework, with multiple residential and business activities represented in a spatially-disaggregated matrix, to which input-output methods are applied. Spatial allocation is accomplished using multinomial logit models. Hansen-type measures of accessibility are used in these models. Both MEPLAN and Tranus are integrated land use-transportation models, including the modeling of travel behavior. This allows the land use model to use updated measures of accessibility for each simulation period using the travel times produced by the transportation models.

UrbanSim (Waddell, 2000, 2001, 2002) uses discrete choice models to model the decisions made by actors involved with land use, including businesses, households, and developers. UrbanSim is also an integrated land-use transportation model. Once again, measures of accessibility are used in the choice models. However, UrbanSim uses measures of accessibility based upon the measures of utility associated with travel between pairs of zones produced in the modal choice model, with the accessibity for each zone being the logsum of the utilities for each of the other zones.

Current work is extending these types of models. The PECAS model was developed by the Oregon Department of Transportation for their second-generation TLUMIP program (Abraham and Hunt, 2002; Hunt, *et al.*, 2001). This model combines the multiple sector structure of MEPLAN and Tranus with the microsimulation of UrbanSim, more completely representing the operation of the land market. The degree of disaggregation by the characteristics of actors is so great that the number of possible combinations far exceeds the total number of actors in the area being modeled, so Monte Carlo methods must be used to make selections in the simulation process. The PECAS model is now also being implemented in Sacramento.

The continued use of measures of accessibility in urban simulation models suggests the importance of accessibility for predicting land use change. In only a limited number of cases, however, have model developers presented results that allow consideration of the role played by accessibility and the relative importance of accessibility compared with other predictors of land use change. Putman and Ducca (1978) present the results from the calibration of the

DRAM model for 11 urban areas in North America. The trip function parameters (there are two cost function parameters in the model) show significant variations across income quartiles and urban areas. Putman's only observation is that coefficients tend to be lower as one moves from east to west. Because DRAM is a spatial interaction model, it is not possible to compare the relative importance of accessibility with other predictors of land use change.

For the California Urban Futures II model, Landis and Zhang (1998b) present the results by county for multinomial logit models estimated to predict various types of land use conversion. The accessibility measures in the model are distances to San Francisco and San Jose and distances to the nearest freeway interchange and BART station. The coefficients for these variables were statistically significant in about two thirds of the cases. In quite a few cases, the coefficients estimated were positive, indicating increased probabilities of conversion with distance, opposite to what had been expected. Because unstandardized regression coefficients were provided, the relative importance of accessibility compared with the other predictors in the models cannot be determined.

More recently, Landis and Reilly (2003) report on models estimated to predict conversion from undeveloped to developed land in four regions of California. The accessibility measures for these models included distance to the nearest freeway interchange and the number of jobs within 90 minutes travel time. Coefficients for the jobs measure were statistically significant for all four regions; coefficients for the freeway interchange distance were significant in all but the South San Joaquin Valley region. Landis presents standardized coefficients, so it is possible to compare the relative importance of the accessibility measures to the other variables in the models (which included 11 predictors). Distance to freeways was the single most important predictor in three of the regions. In the one region where that measure was not significant, the number of jobs within 90 minutes was the third most important predictor. Number of jobs was also above average in importance in the other regions.

Waddell (2000) presents results estimated for the residential components of UrbanSim for the Eugene-Springfield, Oregon area. Included in the model for household bid-price functions are accessibility to employment and retail activity (both logsums of utility measures) as well as travel time to the CBD. Waddell reports that "in almost all cases" the coefficients were significant and had the expected signs, but he does not indicate which were significant. The coefficients were positive for the two accessibility measures and for time to the CBD. Waddell suggests the positive coefficient for travel time to the CBD may be accounting for negative externalities associated with proximity to the center, with the accessibility measures accounting for the desire for convenient access to those opportunities. Since the unstandardized coefficients are presented, it is not possible to assess the relative importance of the accessibility measures.

THE LUCI2 URBAN SIMULATION MODEL

The *luci2 Urban Simulation Model* is the successor to the *LUCI: Land Use in Central Indiana Model* (Ottensmann, 2003).[1] The *luci2* model separately simulates the conversion of nonurban land to residential and employment-related land uses and includes models to forecast employment change for multiple industry groups. The original *LUCI* model simulates conversion only to the single category of urban use and uses simple shift-share methods for the forecasting of employment change. In addition, the *luci2* model has been developed so that it can be used to implement models of varying complexity for different areas, while the *LUCI* model was limited to Central Indiana.

The simulation model can best be characterized as using random utility theory, with aggregate discrete choice models for the decisions to convert available nonurban land to residential and employment-related uses. The models predict the choices of developers leading to the conversion of land from nonurban to residential and employment-related uses. The assumption is that the decisions by developers incorporate their assessments of the demand by consumers for residential land use and firms for employment-related land use.

This basic approach to modeling land use change is perhaps closest to that used by Landis in the California Urban Futures II and California Urban and Biodiversity Analysis models (Landis, 2001; Landis and Zhang, 1998a, 1998b). The major difference in the methods being used here as compared with those employed by Landis and most others using choice models for urban simulation is that they have used the more familiar discrete logit models, while *luci2* employs aggregate logit models predicting the proportions of the land in each simulation unit converted to urban uses. (For comparison of discrete and aggregated logit models, see Wrigley, 1985.) Models of the following form are estimated using historical data on land use change:

$$\mathrm{logit}(p_i) = \log\left(\frac{p_i}{1-p_i}\right) = \beta_0 + \sum_k \beta_k X_{ik}$$

where p_i is the proportion of available nonurban land converted to residential or employment-related use in each simulation zone i, and X_{ik} is a set of k predictors for each of the simulation zones, including measures of accessibility. Additional models are estimated to predict the net population density for the simulation zones and the net employment density for the employment zones. The specification of these models—the variables used as predictors of the probability of land use change and land use densities—can vary for different implementations of the model. The detailed specification of the models included in the current *Central Indiana Implementation* are provided below in the section discussing the estimation of accessibility in the model.

The *luci2* model also predicts changes in employment in the employment zones for major industry groups. The first industry group is local-service employment; the remaining industry groups can be specified in implementing the model. Models are estimated using historical data on employment change by industry group for the employment zones. Predictors can include population, employment, and land use characteristics of the zone along with, again, measures of accessibility.

Simulation of urban development in the model is driven by an exogenous forecast of population growth for the entire region. Total employment in the region is assumed to grow at the same rate as the population. This growth in population and employment must then be accommodated by the conversion of land to residential and employment-related uses.

The simulation proceeds in a series of simulation periods (5 year periods for the current implementation) to the final target year. In each simulation period, employment growth is forecasted for the employment zones, and the amounts of land converted to urban uses are forecasted for the simulation zones. Values are then updated for the zones and the process is repeated on a round-by-round basis for each subsequent simulation period.

The first step in each simulation period is the prediction of employment by industry for each of the employment zones using the models that have been estimated. These predictions are then adjusted so that the rate of total employment growth for the region equals the rate of population growth and so that the proportion of total employment in the local-service employment sector remains constant. (The model simulation process is shown in Figure 1.)

The next stage is the simulation of employment-related development for the simulation zones. (This precedes the allocation of residential development, as it is assumed that employment-related uses will be able to outbid residential uses.) The amount of new employment-related land use required within each employment zone is predicted using the employment density model, and the probability of conversion of land to employment-related uses is predicted using the aggregate logit model. Because employment-related development is expected to be clustered within the larger employment zones and not be uniformly distributed across the grid cells in those zones, constraints are placed on the allocation of such development to the grid cells. In general, the allocation of new employment-related uses to the simulation zones is limited to some specified maximum percentage of the cells within each employment zone, and the amount of such uses allocated to a simulation zone must exceed a specified minimum.

The final step is the simulation of residential development. The logit model predicts the probability of development for each simulation zone and the population density model predicts the density of new development for each zone. This produces a tentative allocation that will result in development that will accommodate more or less population growth than has been exogenously specified. The predicted probabilities of residential development are then iteratively adjusted up or down (using a power transform) until the predicted amounts of residential development at the predicted densities accommodates the exogenously specified population growth.

Figure 1.luci2 Simulation Process

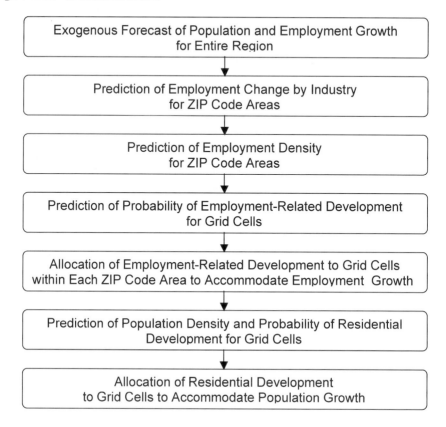

As mentioned above, a primary purpose of the model is to allow the users to generate alternative scenarios reflecting differing policy choices and assumptions regarding future urban development. The model includes a wide variety of options for doing so. Some of the more important include changing the densities of residential development from those predicted by the model; considering the effects of major transportation improvements (distances to major facilities are included as predictors in various models); altering policies regarding the expansion and/or requiring of water and sewer utility service; restricting development on agricultural land and in certain sensitive areas; imposition of urban growth boundaries; and, of greatest relevance to this paper, changing the relative importance of accessibility to employment for residential development. For all of these scenarios, the emphasis is on providing the ability to identify the qualitative changes that would be associated with scenario choices.

At the conclusion of each simulation, the model displays the results for the new scenario in the form of maps and provides summary tabular results, including the amounts of land converted to urban uses, including estimates of the amounts of agricultural and sensitive lands converted. For both the maps and the tables, the results for the new scenario are displayed side-by-side with a comparison scenario to facilitate the understanding of the implications of the choices made in creating the new scenario.

THE CENTRAL INDIANA IMPLEMENTATION

The *luci2* model has been developed as a general-purpose urban simulation model so that the program can be used to develop models for different areas. The initial implementation of the model—the subject for further consideration in this paper—is for the 44-county Bureau of Economic Analysis region for Indianapolis. The area included in the model is approximately the central third of Indiana and encompasses eight Metropolitan Statistical Areas. The map in Figure 2 shows the area covered by the *Central Indiana Implementation*.

The land use change equations in the model were estimated and the model has been implemented using land use data derived from LandSat Thematic Mapper satellite imagery for 1987, 1993, and 2000. The images were first classified by land cover on a pixel-by-pixel basis, including classes for high- and low-intensity development. This left many pixels within urban areas classified as forest (areas of heavy tree canopy) and herbaceous (grassland, e.g., backyards). Using current land use data for Marion County (Indianapolis), a reclassification procedure was developed using information on the land cover in the neighborhood of each pixel and ancillary data, including the pattern of the current road network and census block data for population and housing units. This produced a classification of the pixels as nonurban, urban residential, and urban employment-related (commercial, industrial, and employment-intensive special uses). Classification accuracy for Marion County exceeded 85 percent. This reclassification procedure was applied to the land cover data for the entire Central Indiana region to estimate land use at the pixel level. These data were then aggregated to a set of 17,369 1.61 kilometer (km) square grid cells (1-mile square grid cells) that are used for the estimation and as the simulation zones in the model. Urban parkland and publicly-owned nonurban land are designated as not being available for urban development.

The Indiana Business Research Center provided employment data for major industry groups by ZIP code for 1995 and 2000 tabulated from the workforce (ES-202) data collected by the Indiana Department of Workforce Development. Data from *ZIP Code Business Patterns* were then used to estimate values that were missing because of disclosure rules. For the estimation of the models to forecast employment, the data were further aggregated into four major industry categories. Local-service employment includes retail, service, public administration, and construction employment. These classes showed similar patterns of change when estimating the forecasting models. The three additional industry categories are manufacturing; finance, insurance, and real estate; and a category combining transportation and

Figure 2. Indiana, Central Indiana, and Included Metropolitan Areas

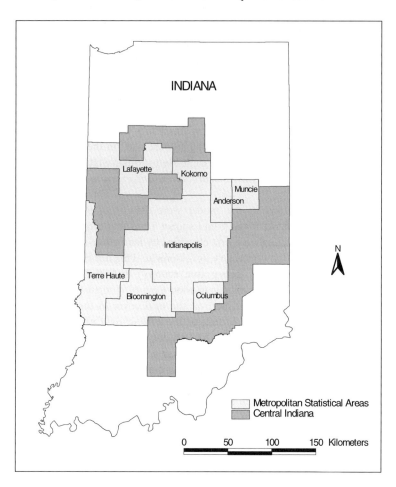

communications and wholesale trade. The employment zones used for the estimation and for the simulation model are the 320 ZIP code areas within the Central Indiana region.

The other major data collection effort for the model involved determining the areas currently provided with water and sewer utility service. This required contacting all of the water and sewer utility providers in the region to obtain maps of the areas in which they provided service. These data were then converted to the system of grid cells, with a cell being considered to be provided with the utility service if any portion of the cell was currently being served (so these are dummy variables).

ESTIMATING THE ROLE OF ACCESSIBILITY

The equations for the simulation model were estimated using the historical land use and employment data described above, along with additional sources of data. Measures of accessibility to employment, employment change, population, and population change are included as predictors in many of the models. In most cases, accessibility (including in several instances, changes in accessibility) proved to be the most important factor in accounting for the variation in the variable being predicted. This section describes the estimation of the models and provides the results.

The measures of accessibility used as predictors in the models are calculated using the standard Hansen equation:

$$A_i = \sum_j O_j f(C_{ij})$$

where A_i is the accessibility at point i, O_j is the number of opportunities at point j, and $f(C_{ij})$ is a function of the generalized travel cost from i to j. A negative-exponential cost function of distance is used:

$$f(C_{ij}) = e^{-\beta d_{ij}}$$

where β is an empirically-determined accessibility coefficient and d_{ij} is the distance either from the center of the simulation zone (grid cell) i to the centroid for the ZIP code j when calculating accessibilities for the simulation zones or from ZIP code centroid i to ZIP code centroid j when calculating accessibilities for the ZIP codes. Distance was used rather than travel time because suitable travel time data were not available for the entire region. The negative-exponential form for the cost function was used because it performed better in the models than the alternative inverse-power form.

For the models involving prediction of residential development, employment in the ZIP code areas is the measure of opportunities. Accessibilities calculated for the models used in the prediction of employment change and development use population in the ZIP code areas as the measure of opportunities.

In addition, measures of change in accessibility to employment and population are used. Change in accessibility is

$$\Delta A_i^{12} = A_i^2 - A_i^2 = \sum_j O_j^2 f(d_{ij}) - \sum_j O_j^1 f(d_{ij}) = \sum_j \Delta O_j^{12} f(d_{ij})$$

where ΔA_i^{12} is the change in accessibility from time *1* to time *2*; A_i^1 and A_i^2 are accessibilities at times *1* and *2*; O_j^1 and O_j^2 are opportunities at *j* at times *1* and *2*; and ΔO_j^{12} is the change in opportunities from time *1* to time *2*. Thus, change in accessibility to opportunities is equivalent to accessibility to the change in opportunities.

The values of the accessibility measures depend upon the value of the accessibility coefficient β, which determines the rate at which accessibility declines with distance from the opportunities. Estimating models incorporating such measures of accessibility requires the estimation of the value for that coefficient. All of the models were estimated using standard multiple regression methods, which do not allow for the calculation of the accessibility values from the distance and opportunity data and for the estimation of the value of the accessibility coefficient. Therefore, an iterative procedure was followed for the estimation of the models and the accessibility coefficient.

A simple program was developed to calculate the set of accessibility values for any accessibility coefficient and to estimate iteratively the best-fit value for the accessibility coefficient for predicting the desired dependent variable. (This is essentially the simple regression of the accessibility variable on the dependent variable.) The program can output the calculated set of accessibility values for any accessibility coefficient. These accessibility variables were then employed in the estimation of the regression models using standard statistical software (SPSS).

The procedure for estimating the models using accessibility proceeded iteratively. The accessibility program was used to create the accessibility variable with the accessibility coefficient that best predicted the dependent variable. Additional accessibility variables were also created with lower and higher accessibility coefficients. These were brought into the dataset being used for the model estimation. Model development began with the accessibility variable using the best-fit accessibility coefficient. As additional predictors were added to the model, it was no longer necessarily the case that this remained the best-fit accessibility coefficient. So the model was re-estimated using alternative accessibility variables calculated using the lower and higher accessibility coefficients, finding the accessibility variable that produced the best prediction (highest R^2). (As necessary, new accessibility variables were calculated using even lower or higher accessibility coefficients or intermediate accessibility coefficients.) When the (almost) final model had been developed, additional accessibility variables were created with accessibility coefficients in the vicinity of the current best-fit accessibility coefficient. The process was continued to determine the best-fit accessibility coefficient to two significant digits. Note that in the final stages of this process, small changes in the accessibility coefficient did not result in changes to the R^2 value at the precision at which it was reported. The best fit was determined by finding the accessibility variable with the accessibility coefficient that produced the smallest standard error of estimate. Changes in the second significant digit of the accessibility coefficient often produced changes only in the fourth or fifth significant digit of the standard error of estimate. So it seems reasonable to conclude that the estimation of the best-fit accessibility coefficient to two significant digits is sufficient and the most that can be supported by the data and the models.

Probability of residential development model

The equation predicting the probability that available nonurban land in a grid cell will be converted to residential use is the single most important equation in the *luci2* model. As described above, this is an aggregated logit model:

$$\text{logit}(p_i) = \log\left(\frac{p_i}{1 - p_i}\right) = \beta_0 + \sum_k \beta_k X_{ik}$$

The dependent variable is the proportion p_i of the available nonurban land at the start that was converted to residential use from 1995 to 2000. (This is derived from the proportion of land converted from 1993 to 2000, converted from a 7-year rate to a 5-year rate for consistency with the 5-year simulation period in the model.) Following Wrigley (1985), such an aggregated logistic model should be estimated using weighted least squares (WLS), with the weight being as follows:

$$w_i = n_i p_i (1 - p_i)$$

where n_i is the number of observations in grid cell i that have been aggregated to produce the proportion. In this case, n_i is the number of pixels of available nonurban land at the start that forms the denominator of the proportion.[2]

Accessibility to employment is expected to be a significant predictor of the location of new development because of the importance of the journey to work. Two measures of accessibility are included as predictors in the model: accessibility to employment by ZIP code in 1995 and change in accessibility to employment from 1995 to 2000. Accessibility to employment is included using a natural log transformation. Change in accessibility to employment is in its natural form. The other predictors included in the model are as follows:

- Whether a grid cell is provided with water utility service or sewer utility service

- Whether the center of a grid cell is less than 1.61 km (1 mile) from a freeway interchange or is 1.61 to 4.83 km (1 to 3 miles) from a freeway interchange and whether the center is less than 1.61 km or is 1.61 to 4.83 km from a four-lane highway that is not a freeway (This is another component of accessibility and is intended to capture the faster speeds and lower travel times that such facilities make possible, since distances rather than travel times had to be used for the calculation of the accessibility measures; these are not highly correlated with the accessibility measures.)

- The natural logarithm of the proportion of land in the 3 by 3 grid-cell neighborhood of the grid cell that was residential at the start and its square (This captures the logistic trend in urban development over time that is frequently observed in developing areas: new development in an undeveloped area starts out slowly. The rate of development accelerates as the area becomes more developed. Finally, the rate of development slows as the area becomes more fully developed.)

- The logit of the proportion of available nonurban land in the grid cell converted to residential use in the preceding period, again, a 5-year rate, estimated from the observed changes from 1985 to 1993 (This is a persistence term.)

- The overall total battery score for students taking the Indiana Statewide Test of Educational Progress (ISTEP) in 2000 for the school district

The results for the estimation of the model to predict the logit of the proportion of available land converted to residential use are presented in Table 1. The model has an R^2 of 0.398. All but one of the variables included in the model are statistically significant. The standardized regression coefficients (beta coefficients) are most relevant for assessing the relative contribution of the predictors in accounting for the variation in the dependent variable. In this model the variable proportion residential in the 3 by 3 neighborhood is entered in both the log form and the square of the log form. These variables are naturally highly correlated and have opposite signs and correspondingly very large beta coefficients. The net contribution of this variable in accounting for the variation in the dependent variable is the sum of these beta coefficients, 0.220. So this is still the variable that accounts for the greatest proportion of the variation, but this only slightly exceeds the beta coefficient of 0.209 for the log accessibility to employment variable. But in addition, change in accessibility to employment has a lower, but still highly significant beta coefficient of 0.074. So taken together, the two accessibility measures play the most important role in predicting the conversion of land to residential use.

The accessibility coefficients estimated for the two measures differ considerably. The accessibility coefficient for accessibility to employment is 0.099 and the accessibility coefficient for change in accessibility to employment is over twice as large, 0.22. (The accessibility coefficients using the negative-exponential form are dependent upon the units in which distance is measured, kilometers here.) So the effect of employment change on residential development drops off much more quickly with distance. Employment change has its greatest effect on residential development in areas closer to where that change is taking place, while overall employment levels have a more far-reaching, broader effect.

Table 1. Logit Proportion Land Converted to Residential Use, 1995-2000

Independent Variable	Regression Coefficient	Beta Coefficient	t value	
Log accessibility to employment[a]	0.352	0.209	19.36	**
Change accessibility to employment (thousands)[b]	0.116	0.074	6.66	**
Sewer service available	0.094	0.025	3.04	*
Water service available	0.280	0.072	9.05	**
Within 1.61 km of freeway interchange	0.096	0.014	2.16	**
1.61-4.83 km from freeway interchange	0.044	0.011	1.54	
Within 1.61 km of other four-lane highway	0.159	0.037	5.62	**
1.61-4.83 km from other four-lane highway	0.152	0.039	5.70	**
Log prop land residential in 3x3 neighborhood	8.251	0.592	24.53	**
Square log prop land residential in 3x3 n'hood	-10.917	-0.372	17.55	**
Logit proportion converted residential 1988-93	0.056	0.076	9.19	**
IN State Test of Educational Progress score	0.028	0.070	8.52	**
Constant	-8.950		37.47	**
R-squared	0.398			
N	17,052			

[a]Accessibility coefficient 0.099

[b]Accessibility coefficient 0.22

*Significant at the 0.01 level

**Significant at the 0.001 level

Population density model

The second equation used in the simulation of the conversion of nonurban land to residential use predicts the population density for residential development. The dependent variable used in estimating the model is the natural log of residential population density in 2000, obtained by dividing the 2000 population estimated for the grid cells from the census block data by the amount of residential land in the grid cells in 2000. The expectation that accessibility to employment would be a significant, positive predictor of population density arises directly from the standard urban economic model. Higher accessibility to employment would be associated with higher demand for land for residential use, resulting in higher land prices, which then translate into higher population densities for residential development.

Results for the model estimation are shown in Table 2. R^2 for the model is 0.417, with all four of the predictors highly statistically significant. The number of cases used for the estimation was much lower than the total number of grid cells because the majority of cells had no residential land so densities could not be calculated. Log accessibility to employment was the most important predictor in the model with a beta coefficient of 0.384, somewhat more important than the availability of sewer utility service, which was the second most important predictor of population density. The log of the distance to the nearest interstate highway interchange and the log of the statewide education test scores were the remaining predictors.

Table 2. Log Residential Population Density, 2000

Independent Variable	Regression Coefficient	Beta Coefficient	t value	
Log accessibility to employment[a]	0.209	0.384	29.56	**
Log distance to nearest freeway interchange	-0.042	-0.046	4.09	**
Sewer service available	0.591	0.312	27.09	**
Log IN State Test of Educational Progress score	-1.436	-0.127	-13.16	**
Constant	11.517		25.26	**
R-squared	0.417			
N	6,337			

[a]Accessibility coefficient 0.30

**Significant at the 0.001 level

The accessibility coefficient for employment accessibility in this model was 0.30, three times as great as the accessibility coefficient for employment accessibility in the previous model predicting the probability of residential development. So the effect of employment locations on population densities appears to decline more rapidly with distance than for the probability of development.

Probability of employment-related development model

The simulation of the conversion of available nonurban land to employment-related land use is similar to the simulation of conversion to residential use. The most important equation predicts the probability of employment-related development. It has been estimated using as the dependent variable the proportion of the available nonurban land at the start converted to employment-related use from 1995 to 2000. In this instance, log of accessibility to population is included as a predictor. Businesses presumably value such accessibility both for access to employees and, for businesses serving residents of the area, for access to markets.

Results for the estimation of this model are presented in Table 3. R^2 is 0.503 and all eight of the predictors are highly statistically significant. The log accessibility to population variable is easily the most important in accounting for the variation in employment-related development, with the beta coefficient of 0.398 being twice as great as the next largest, for the availability of sewer service. Other predictors included in the model are the availability of water utility service, dummy variables for the distances to the highway transportation infrastructure, both interstate highway interchanges and other four-lane highways, and the amount of employment-related land use in the grid cell at the start of the period, a simple measure of tendencies toward agglomeration.

Table 3. Logit Proportion Land Converted to Employment-Related Use, 1995-2000

Independent Variable	Regression Coefficient	Beta Coefficient	t value	
Log accessibility to population[a]	0.784	0.398	53.28	**
Sewer service available	0.899	0.198	26.30	**
Water service available	0.317	0.068	9.31	**
Within 1.61 km of freeway interchange	0.651	0.110	18.79	**
1.61-4.83 km from freeway interchange	0.118	0.027	4.53	**
Within 1.61 km of other four-lane highway	0.697	0.161	24.45	**
1.61-4.83 km from other four-lane highway	0.499	0.117	17.35	**
Employment-related land use in 1993	2.073	0.112	18.95	**
Constant	-12.315		96.57	**
R-squared	0.503			
N	17,066			

[a]Accessibility coefficient 0.21

**Significant at the 0.001 level

The accessibility coefficient for accessibility to population is 0.21. This falls within the range of the accessibility coefficients estimated for the previous two models. So despite the switch from residential to employment development and from accessibility to employment to accessibility to population, the magnitudes of the accessibility coefficients are comparable.

Employment density model

Since employment data for the model are only available for ZIP codes, the equation to predict the density of employment was estimated using data for those areas. In the simulation, employment density is predicted for the ZIP codes, and all grid cells within each ZIP code are assumed to have employment-related development occur at that density. The dependent variable for the model that was estimated was actually the inverse of density, the amount of employment-related land use in a ZIP code area divided by the number of employees in that ZIP code, in thousands. This form performed better in the estimation of the model.

Accessibility to population might be expected to affect the density of new employment-related development in several ways. Higher population demand might be expected to create more employment-related development in an area, more competition for land, higher prices, and higher densities for employment-related use. Also, higher population densities might be expected to be directly associated with higher land prices, following the standard urban economic model, leading to higher densities for employment-related uses as well.

A very simple model predicts employment-related land use per thousand employees, with the results shown in Table 4. R^2 is 0.430 with both predictors statistically significant. In this case, the most significant predictor was not accessibility to population. Rather, the best predictor was a composite measure, the percentage of housing units in the ZIP code built before 1940 times the percentage of the land in that ZIP code urban. This measure seeks to identify the older, built up urban areas with the highest employment densities. Accessibility to population still accounted for significant variation in the dependent variable, with a beta coefficient of -0.267. The estimated regression coefficients are negative, as expected, with higher accessibilities being associated with lower amounts of land being used per employee (and thus higher employment densities).

Table 4. Employment-Related Land Use per Thousand Employees, 2000

Independent Variable	Regression Coefficient	Beta Coefficient	t value	
Accessibility to population (thousands)[a]	-0.003	-0.267	5.12	**
Percent housing units built before 1940 times percent land urban	-2.992	-0.473	9.07	**
Constant	-1.359		18.58	**
R-squared	0.430			
N	297			

[a]Accessibility coefficient 0.13

**Significant at the 0.001 level

The accessibility coefficient for the accessibility to population measure was 0.13. Once again, this falls within the range of the accessibility coefficients estimated for the previous models.

Change in local-service employment model

The model also includes equations to predict changes in employment for the four industry groups. Accessibility plays a major role in the prediction of change in the local-service employment, which includes retail trade, services, public administration, and construction employment. As described above, accessibility has been employed in the prediction of retail activity for at least four decades. It is expected that accessibility to population would be important for the location of changes in local-service employment, as such firms would seek to locate proximately to their markets.

Changes in employment in the local-service industries from 1995 to 2000 by ZIP code were used to estimate the model. The results are presented in Table 5. R^2 is 0.668, and the three predictors are statistically significant. Change in accessibility to population by ZIP code from 1995 to 2000 is by far the most important predictor in the model, with a beta coefficient of 0.385. Growth in local service employment is occurring in proximity to areas of population growth. The two other predictors in the model are the total population residing within the ZIP code areas and the change in the proportion of land urban in those areas in the prior period.

Table 5. Change in Local-Service Employment, 1995-2000

Independent Variable	Regression Coefficient	Beta Coefficient	t value	
Change in accessibility to population[a]	0.127	0.385	6.32	**
Population, 1995 (thousands)	9.692	0.167	3.28	*
Change in proportion land urban, 1988-1993	215.501	0.227	3.34	**
Constant	-44.509		1.27	
R-squared	0.668			
N	315			

[a]Accessibility coefficient 0.28

*Significant at the 0.01 level

**Significant at the 0.001 level

The accessibility coefficient for the change in accessibility to population is 0.28, which once again falls within the range of accessibility coefficients established by the earlier models. It is near the high end of this range, suggesting that firms find it important to locate in relatively close proximity to the growing markets.

EFFECTS OF CHANGE IN ACCESSIBILITY IN LUCI2

The *luci2* model has been developed to allow users to generate alternative scenarios reflecting different policy choices and assumptions about conditions that may affect future urban development. One of the options provided in the model allows changes in the relative importance future residential locators will place on accessibility to employment. Therefore, the model can be used to simulate the effects of changes in the importance of accessibility on patterns of residential development.

Broad changes in socioeconomic conditions might be expected to alter the importance that future residential locators might place on accessibility to employment in making their choices of residences. For example, major increases in energy prices, causing increased costs for commuting, might increase the importance people place on proximity to employment opportunities. Alternatively, widespread adoption of telecommuting could have the opposite effect, reducing the importance of accessibility.[3]

The *luci2* model implements changes in the relative importance of accessibility to employment by changing the values of the accessibility coefficients for the accessibility to employment measures included in the models to predict the probability of residential development and population density. The current implementation allows the user to vary the coefficients from one-half the estimated accessibility coefficients for the maximum reduction in the importance of accessibility to twice the estimated coefficients for the maximum increase in the importance of accessibility.[4]

When the *luci2* model is started, the user first sees the results of the Current Trends scenario, which is a simulation of urban development based upon the assumptions that development will continue to occur as it has in the recent past. Population in the region is assumed to grow at the rate of growth in the 1990s. Utilities are extended to serve new areas as growth occurs. No new transportation improvements are assumed to be implemented. No new restrictions are placed on the development of agricultural or sensitive lands. The parameters estimated for the various models are used and assumed to remain constant.

Taking this scenario as the starting point, two additional scenarios have been created varying only the relative importance of accessibility to employment. The first assumes the minimum allowable importance of accessibility, the halving of the accessibility coefficients. The second assumes the maximum importance of accessibility, the doubling of the accessibility coefficients.

The maps in Figure 2 show the simulated patterns of residential development for these two scenarios, mapping the simulated change in the percentage of land in residential use. As one would expect, new residential development is concentrated around the peripheries of the major urban areas in the region. (The "holes" in the areas of development are, of course, areas that had been developed prior to the starting year of the simulation in 2000, where additional development cannot occur.) The differences in the simulated patterns of development for the two scenarios are significant and as expected. With lower accessibility coefficients, development is much more spread out, with the areas of significant new development extending out farther from the existing urban areas (more urban sprawl). Using the higher accessibility coefficients, new residential development is much more tightly clustered around the existing urban areas. The pattern of residential development simulated for the Current Trends scenario is between these two extremes.

Figure 3. Simulated Residential Development Using Alternative Accessibility Coefficients, 2000-2025 (Change in Percent Land Residential by Grid Cell)

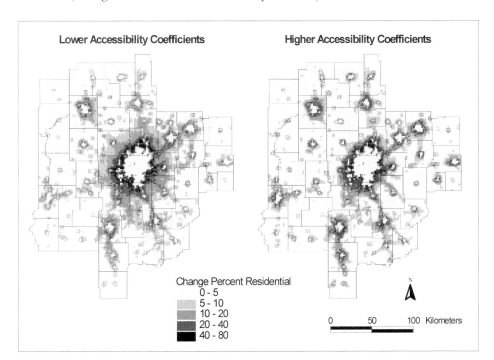

In addition to providing mapped output, the *luci2* model also reports a variety of summary statistics for each of the scenarios created. Table 6 presents a few of these results for the lower accessibility coefficient, Current Trends, and higher accessibility coefficient scenarios. The population density for new residential development for the scenario using the higher accessibility coefficients is 619 persons per square kilometer (km^2), significantly higher than the density for the scenario using the lower coefficients, 532 per km^2. While change to the accessibility coefficient for accessibility to employment in the population density model accounts for some of the difference, the majority results from the changes in the probability of residential development model. With the scenario for higher relative importance of accessibility, more of the new residential development occurs closer to the existing urban centers. These areas have higher accessibilities to employment and are more likely to be provided with sewer utility service. Therefore, the population densities predicted for new development in these areas are higher. Conversely, the greater amounts of development occurring in more remote areas with the lower importance of accessibility scenario are predicted to occur at lower densities because of reduced accessibility to employment and greater likelihood of the absence of sewer service.

Table 6. Simulation Results for Alternative Accessibility Coefficients, 2000-2025

	Lower Accessibility Coefficients (0.5 times estimated)	Accessibility Coefficients Estimated (Current trends)	Higher Accessibility Coefficients (2 times estimated)
Residential population density			
Density at start (per sq km)	1,022	1,022	1,022
Density of new development (per sq km)	532	560	619
Density at end (per sq km)	766	784	816
Percentage change in density	-25.05%	-23.29%	-20.16%
Residential land			
Residential land at start (sq km)	2,981	2,981	2,981
Land converted to residential use (sq km)	2,222	2,111	1,911
Residential land at end (sq km)	5,203	5,092	4,892
Percentage change in residential land	74.54%	70.82%	64.11%
Journey to work			
Mean journey to work at start (km)	14.95	14.95	14.95
Mean journey to work at end (km)	19.57	18.11	16.25
Percentage change in mean journey to work	30.90%	21.14%	8.70%

Since the densities of new residential development are higher for the scenario assuming greater importance of accessibility, it follows that the amounts of land converted from nonurban to residential use will be lower. The second section of Table 6 illustrates this. For the scenario with the higher importance of accessibility, 1,911 km² of nonurban land are converted to residential uses. With the scenario for reduced importance of accessibility, 2,222 km² of land are converted. This is an increase in land consumption for residential development of 311 km² or 16 percent.

The model also estimates the length of the mean journey to work at the beginning and end of the simulation for each scenario. Doing this involves many judgments and assumptions that are described below, so these results should only be considered to be suggestive. A standard, double-constrained spatial interaction model is used with data on population and employment by ZIP code and the distances between the ZIP code centroids to predict the journey-to-work pattern:

$$T_{ij} = k_i l_j O_i D_j e^{-\beta(d_{ij}+\delta)}$$

where T_{ij} are the trips from ZIP code i to ZIP code j, O_i are the origins (employees by place of residence) in i, D_j are the destinations (employees by place of work) in j, and d_{ij} are the distances from i to j. The parameters k_i and l_j are the balancing factors used to make the total predicted origins and destinations equal to the actual totals. The parameter β is the distance-function parameter associated with rate of decline of interaction with distance. The parameter

δ is a fixed amount added to each distance to reflect terminal times for trips. Intrazonal distances d_{ii} are set to one-half the distance to the nearest ZIP code centroid.

Parameters for the model were first estimated using 2000 data, finding parameter values that most closely replicated the shape of the trip-length distribution for the region from the 2000 census. (This distribution was for travel time in minutes, not distances, but was the only data available for the entire area.) The estimated value for the accessibility coefficient β was 0.148. When initially used to predict the journey to work for the simulation results in the model, this model did not result in predictions of increases in the journey to work comparable to the increase observed from 1990 to 2000 in the census, from 20.52 minutes to 22.60 minutes, a 10.1 percent increase. This led to consideration that the accessibility coefficient for the journey-to-work model may be declining over time. So the model was re-estimated using 1990 data, finding the value for the accessibility coefficient that resulted in a prediction of the mean trip length for 1990 that was lower than the prediction of the trip length for 2000 by the proportion observed in the census.[5] The model then assumes the continued decline of the accessibility coefficient when calculating the final mean journey-to-work length.[6]

For scenarios involving greater or lesser importance of accessibility to employment for residential development, it seems reasonable to assume that the value of the accessibility coefficient for the journey-to-work model should likewise be increased or decreased. Increasing or decreasing the coefficient at the same rate as the accessibility to employment coefficients produces unrealistically low or high predictions of the mean journey to work for increases and decreases in the importance of accessibility to employment. But changes in the accessibility to employment coefficients affect only the simulated distribution of new residential development, since the model does not simulate changes to the locations of existing residential uses or employment. So a lesser change in the journey-to-work accessibility coefficient seemed warranted. A relatively arbitrary decision was made to change this coefficient by the proportional change specified for the accessibility to employment coefficient multiplied by the proportion of the final population that is new population growth since the start of the simulation.

With these explanations and caveats regarding the prediction of the mean journey to work in the model, consider the predictions for the scenarios involving decreased and increased importance of accessibility to employment, shown in the final section of Table 6. For the scenario with the lower accessibility coefficients, the increase in the mean journey to work distance is about 31 percent. With the increased importance of accessibility to employment, the increase is only 9 percent. The Current Trends scenario has an intermediate predicted increase in the mean journey to work of 21 percent (which is not inconsistent with the observed increase of the mean journey to work for the region from 1990 to 2000 of 10 percent as reported by the census).

CONCLUSIONS

The development and use of the *luci2 Urban Simulation Model* provide a context for considering the role and importance of accessibility for urban development. In the development of the model, statistical models have been estimated that allow the assessment of the role played by accessibility in influencing various aspects of urban development. For the probability of residential development and population density models, accessibility to employment and (and changes in accessibility, for the former) were the most important predictors. For the probability of employment-related development and employment density models, accessibility to population was the most important predictor for the former and highly significant for the latter. Change in accessibility to population was the dominant predictor in the model of local-service employment change.

An important aspect of these results is that the models for probability of residential and employment-related development and the model for local service employment were models predicting change, estimated using historical data on such changes. This provides more direct evidence of the role played by accessibility for urban change than models that have only been estimated using cross-sectional data. Furthermore, the estimation results provide the additional important finding that for two of these models, measures of changes in accessibility—change in accessibility to employment for probability of residential development and change in accessibility to population for local-service employment change— are significant predictors. In thinking about the role of accessibility for urban development, it is thus important to consider accessibilities not only to levels of employment and population but also the changes in accessibilities or, equivalently, accessibilities to changes in the levels of employment and population.

With its ability to generate alternative scenarios, the *luci2* model also provides the opportunity to examine the effects of changes in the relative importance of accessibility to employment on patterns of residential development. Assuming that the model simulation is reasonable, increases in the importance of accessibility to employment results in new residential development being denser and clustered more closely to existing urban development, while decreases cause development to be less dense, more spread out, generating more urban sprawl. While this is certainly what one would expect, being able to observe the simulated patterns provides a more effective demonstration. Model output shows the scenario with greater importance of accessibility results in the conversion of 16 percent less land to residential uses than the scenario with reduced importance of accessibly. And while the results can only be considered speculative given the issues involved in the prediction of the journey to work, the predicted increase in the mean journey to work was far lower for the scenario assuming the increased importance of accessibility to employment.

Certainly the results have limitations and many improvements are possible. Within the context of simple urban simulation models such as the *luci2* model, better data could enhance the conclusions that could be drawn about the importance of accessibility. The land use data estimated from the satellite imagery have significant error. Better data would provide a better foundation for examining urban change. Finer spatial disaggregation may also enhance the estimation and simulation. The use of travel times rather than direct airline distances could significantly improve the measures of accessibility considered.

More sophisticated urban simulation models that model more of the details of the urban development process using an economic framework have the possibility of providing more detailed information about the mechanisms through which accessibility influences urban development. Of course, the models would have to be developed and estimated in a manner that would allow one to draw such inferences.

The results presented in this paper clearly demonstrate that accessibility to employment and population play a major role in shaping patterns of urban development. They also show that not only accessibility to levels of employment and population but also the changes in accessibility to those levels are important determinants of urban change.

NOTES

[1] The *LUCI* and *luci2* models can be downloaded from the *LUCI* website, which also includes additional information on the models, including the *luci2 Urban Simulation Model and the Central Indiana Implementation: Working Paper*, which provides more complete documentation on the development and structure of the model (http://luci.urbancenter.iupui.edu).

[2] A modification was made to the basic procedures described in the text to avoid the loss of data for grid cells in which the proportion of land converted was either 0 or 1 (for which the logit cannot be calculated). Following Wrigley (1985) citing Fox (1970), values of 0.5 (one half of a pixel) are added to both the numerator and denominator of the fraction used in computing the logit. A similar procedure must be followed to avoid weights of 0 for such cases as described by Wrigley (1985) using the method proposed by Berkson (1953, 1955).

[3] The possibility that widespread telecommuting could reduce the importance of accessibility as it has been defined raises the possibility that broader measures of accessibility that include accessibility to telecommunications might become more appropriate. Shen (2000) provides an example of the use of such broader measures of accessibility.

[4] This range was selected to provide simulations of development patterns that yielded maximum variations in those patterns while retaining reasonable expectations regarding the pattern of development. Significantly lower accessibility coefficients tended to disperse development across the entire area. Greater increases in the accessibility coefficients had little additional effect, continuing to cause new development to be clustered tightly around existing areas of development.

[5]This decline in the value of the accessibility coefficient over time, if it could be confirmed using better data, would itself be a quite interesting finding. A few studies have shown declines in accessibility coefficients over time. Mills (1972) showed the steady decline of population density gradients in urban areas in the United States, which would be directly related to the values of accessibility coefficients (to the CBD). Ottensmann (1975) provides estimates of the declines in accessibility coefficients for urban models over a half century for one metropolitan area.

[6]Using either a constant amount or a constant rate of decline in the journey-to-work accessibility coefficient resulted in predicted increases in the mean journey to work in the later simulation periods greater than were observed from 1990 to 2000. As an alternative, the model uses an inverse power function for predicting the decline of the accessibility coefficient with time, $\beta_t = \beta_0(1 + t)^{-p}$ with p calculated from the estimated accessibility coefficients for 1990 and 2000 and β_0 being the accessibility coefficient in 2000. This produces predictions of the mean journey to work comparable to the rate observed from 1990 to 2000 for the early simulation periods and somewhat smaller increases in later simulation periods. This was chosen as a reasonably conservative approach.

REFERENCES

Abraham, J. E. and J. D. Hunt (2002). Spatial market representations: concepts and application to integrated planning models. Prepared for presentation at the North American Meetings of the Regional Science Association International, San Juan, Puerto Rico. Downloaded from http://www.ucalgary.ca/~jabraham/Papers/ spatialmarkets/RSAISMD2.pdf on May 1, 2005.

Agarwal, C., *et al.* (2000). *A Review and Assessment of Land-Use Change Models*. Center for the Study of Institutions, Population, and Environmental Change, Indiana University, Bloomington and U.S.D.A. Forest Service, South Burlington, VT.

David Simmonds Consultancy (1999). *Review of Land-Use/Transport Interaction Models*. Department of the Environment, Transport and the Regions, London.

de la Barra, T. (1989). *Integrated Land Use and Transport Modelling*. Cambridge University Press, Cambridge.

de la Barra, T. (2001). Integrated land use and transport modeling: The Tranus experience. In *Planning Support Systems* (R. K. Brail and R. E. Klosterman, eds.), pp. 129-156. ESRI Press, Redlands, CA.

Berkson, J. (1953). A statistically precise and relatively simple method of estimating the bioassay with quantal response, based on the logistic function. *J. Amer. Stat. Assoc.*, **48**, 130-162.

Berkson, J. (1955). Maximum likelihood and minimum—2 estimates of the logistic function. *J. Amer. Stat. Assoc.*, **50**, 565-599.

Chapin, F. S., Jr. and S. F. Weiss (1962). *Urban Growth Dynamics in a Regional Cluster of Cities*. Wiley, New York.

Chen, J., *et al.* (2002). Assessment of the urban development plan of Beijing by using a CA-based urban growth model. *Photogram. Eng. Rem. Sensing*, **68**, 1063-1071.

Creighton, R. L. (1970). *Urban Transportation Planning*. University of Illinois Press, Urbana.

Fox, D. R. (1970). *The Analysis of Binary Data*. Methuen, London.

Fotheringham, A. S., and M. E. O'Kelly (1989). *Spatial Interaction Models*. Kluwer, Boston.

Goldner, W. (1965). The Lowry model heritage. *J. Amer. Planning Assoc.*, **37**, 100-110.

Guttenberg, A. Z. (1960). Urban structure and growth. *J. Amer. Inst. Planners*, **26**, 104-110.

Handy, S. L. and D. A. Niemeier (1997). Measuring accessibility: an exploration of issues and alternatives. *Environ. Planning A*, **29**, 1175-1194.

Hansen, W. G. (1959). How accessibility shapes land use. *J. Amer. Inst. Planners*, **25**, 73-76.

Hansen, W. G. and T. R. Lakshmanan. (1965). A retail market potential model. *J. Amer. Inst. Planners*, **31**, 131-143.

Huff, D. L. (1963). A probabilistic analysis of shopping trade areas. *Land Econ.*, **39**, 81-90.

Hunt, J. D. and D. C. Simmonds (1993). Theory and application of an integrated land-use and transport modelling framework. *Environ. Planning B*, **20**, 221-244.

Hunt, J. D. *et al.* (2001). Design of a statewide land use transport interaction model for Oregon. Oregon Department of Transportation, Eugene. Downloaded from http://egov.oregon.gov/ODOT/TD/TP/docs/TMR/GEN2/OG2D_WCTR.pdf on May 1, 2005.

Landis, J. D. (2001). CUF, CUF II, and CURBA: a family of spatially explicit urban growth and land-use policy simulation models. In: *Planning Support Systems* (R. K. Brail and R. E. Klosterman, eds.), pp. 157-200. ESRI Press, Redlands, CA.

Landis, J. D. and M. Reilly (2003). How we will grow: baseline projections of California's urban footprint through the year 2100. In: *Integrated Land Use and Environmental Models* (S. Guhathakurta, ed.), pp. 55-98. Springer, Berlin.

Landis, J. D. and M. Zhang. (1998a). The second generation of the California urban futures model. Part 1: model logic and theory. *Environ. Planning B*, **25**, 657-666.

Landis, J. D. and M. Zhang. (1998b). The second generation of the California urban futures model. Part 2: specification and calibration results of the land-use change submodel. *Environ. Planning B*, **25**, 795-824.

Lathrop, G. T., J. R. Hamburg, and G. F. Young (1965). Opportunity accessibility model for allocating regional growth. *Highway Research Record*, **102**, 54-66.

Li, X. and A. G. Yeh (2002). Urban simulation using principal components analysis and cellular automata for land-use planning. *Photogram. Eng. Rem. Sensing*, **68**, 341-351.

Lowry, I. S. (1964). *A Model of Metropolis*. Memorandum RM-4035-RC. Rand Corporation, Santa Monica, CA.

Mills, E. S. (1972). *Studies in the Structure of the Urban Economy*. Johns Hopkins University Press, Baltimore.

Muth, R. F. (1969). *Cities and Housing*. University of Chicago Press, Chicago.

Ottensmann, J. R. (1975). *The Changing Spatial Structure of American Cities*. Lexington Books, Lexington, MA.

Ottensmann, J. R. (2003). LUCI: Land Use in Central Indiana Model and the relationships of public infrastructure to urban development. *Pub. Works Mgmt. Policy*, **8**, 62-76.

Pijanowski, B. C. and D. G. Brown, B. A. Shellito, and G. A. Manik (2002). Using neural networks and GIS to forecast land use changes: a Land Transformation Model. *Computers, Environ. Urb. Systems*, **26**, 553-575.

Putman, S. H. (1983). *Integrated Urban Models*. Pion, London.

Putman, S. H. and S. Chan (2001). The METROPILUS planning support system: urban models and GIS. In: *Planning Support Systems* (R. K. Brail and R. E. Klosterman, eds.), pp. 99-128. ESRI Press, Redlands, CA.

Putman, S. H. and F. W. Ducca (1978). Calibrating urban residential models 2: empirical results. *Environ. Planning A*, **10**, 1001-1014.

Shen, Q. (2000). Transportation, telecommunications, and the changing geography of opportunity. *Urb. Geog.*, **20**, 334-355.

Silva, E. A. and K. C. Clarke (2002). Calibration of the SLEUTH urban growth model for Lisbon and Porto, Portugal. *Computers, Environ. Urb. Systems*, **26**, 525-552.

Southworth, F. (1995). *A Technical Review of Urban Land Use-Transportation Models as Tools for Evaluating Vehicle Travel Reduction Strategies*. U.S. Department of Energy, Washington, DC.

U.S. Environmental Protection Agency (2000). *Projecting Land-Use Change: A Summary of Models for Assessing the Effects of Community Growth and Change on Land-Use Patterns*. U.S. Environmental Protection Agency, Washington, DC.

Waddell, P. (2000). A behavioral simulation model for metropolitan policy analysis and planning: residential location and housing market components of UrbanSim. *Environ. Planning B*, **27**, 247-263.

Waddell, P. (2001). Between politics and planning: UrbanSim as a decision support system for metropolitan planning. In *Planning Support Systems* (R. K. Brail and R. E. Klosterman, eds.), pp. 201-228. ESRI Press, Redlands, CA.

Waddell, P. (2002). UrbanSim: modeling urban development for land use, transportation, and environmental planning. *J. Amer. Planning Assoc.*, **68**, 297-314.

Wegener, M. (1994). Operational urban models: state of the art. *J. Amer. Planning Assoc.*, **60**, 17-30.

Wrigley, N. (1985). Categorical Data Analysis for Geographers and Environmental Scientists. Longman, New York.

Access to Destinations
D.M. Levinson and K.J. Krizek (editors)
© 2005 Elsevier Ltd. All rights reserved.

CHAPTER 15

AN ACCESSIBILITY FRAMEWORK FOR EVALUATING TRANSPORT POLICIES

Frank Primerano and Michael A. P. Taylor, University of South Australia

INTRODUCTION

Most measures of accessibility lack the capability to evaluate the impacts of transport policies that can directly influence travel decisions. The major reason for this is that many measures of accessibility do not take into consideration enough of the factors that can be influenced by policies. Among such factors are those related to the attributes of individuals who undertake travel for the purpose of accessing activities. This paper argues that the focus of such measures should be on accessibility for individuals to activities rather than purely being a property of locations since the accessibility of a location does not necessarily reflect the true accessibility of an individual to that destination. Although accessibility to destinations is important it is only part of a person's accessibility. The focus should be on accessibility to activities, which can be satisfied by a number of destination alternatives.

The paper describes an accessibility framework that combines a number of classical measures of accessibility using discrete choice modelling. The framework considers travel patterns of individuals and households that are simulated through a system of behavioural models that resembles travel demand models commonly used in practice. In considering travel behaviour of people, the framework is able to assess the impact of transport and urban planning policies according to how such policies address the needs and preferences of individuals. The innovation of this framework is in allowing accessibility-based assessment and policy evaluation by considering together the urban transport system, the spatial arrangement and characteristics of land-uses, and importantly yet often neglected, the socio-economic characteristics of the population.

The accessibility framework is able to assess the effectiveness of both transport and land-use policies and show the distribution of policy benefits geographically and across socio-demographics. This paper will focus on demonstrating the transport policy evaluation capabilities of the accessibility framework through its application to evaluate the impact of the Adelaide-Crafers Highway on road users in metropolitan Adelaide. The results show how the benefits of the highway are geographically distributed throughout metropolitan Adelaide and that there are benefits to the residents in the suburbs of the Adelaide Hills which the highway services.

Based on the development and application of the accessibility framework, further research recommendations are discussed to improve the capability of accessibility for use as a tool by planners to identify levels of accessibility and to develop and evaluate policies targeted towards improving accessibility for people to their activities.

ACCESSIBILITY

Within transport planning, accessibility is generally defined as the ease with which desired destinations can be reached (Dalvi, 1978; Koenig, 1980; Niemeier, 1997). This section will attempt to attain a more specific definition of accessibility that will be more applicable to transport and land-use planning.

The connection and importance of accessibility within transport and urban planning is well acknowledged throughout the literature. Accessibility is considered as one of the most important factors within the location choice decision process shaping land-use patterns (Martinez, 2000; Wegener, 1996). For example, individuals will base their decisions of where to reside (or where to locate their businesses) on the ease of accessing the services (or clientele) they desire. Accessibility is a function of land-use patterns and of the transport system (Morris, Dumble and Wigan, 1979) making it useful in the planning and development of policies associated with transport networks and in determining land-use configurations within urban space. Therefore, accessibility provides a connection from the transport network to land-uses to determine the effect of the transport network on the arrangement of land-uses.

To facilitate decision-making in transport and land-use planning, this paper considers accessibility to be the *ease* for *people* to participate in *activities* from specific locations to a *destination* using a *mode* of transport at a specific *time*. The *ease* of participation in activities refers to any benefits or costs associated with travel. Such benefits and costs may encompass money, time, convenience and comfort to name a few.

Subject to the factors indicated in italics (i.e. people, activities, destinations, modes and time), 'ease' must be estimated in order to provide an indicator of accessibility. The above definition of accessibility acknowledges that:

- Individuals have varied socio-economic characteristics, behaviour and needs which influence their choices that in turn impact their accessibility;

- Accessibility is different for all activity types because of their location, importance, availability and their nature as acknowledged in research by Stopher, Hartgen and Li (1996), Bowman and Ben-Akiva (2001) and Hansen (1959);

- Location characteristics of destinations vary by their spatial separation from the location of individuals and by the characteristics of the destination itself;

- Each transport mode varies in relation to costs, benefits, characteristics and user perceptions.

- Time influences accessibility through the variation in the availability of activities, the attractiveness of areas and the state of the transport system throughout different times of the day and between different days of the week.

In summary, accessibility is more than just overcoming spatial separation between locations. It also acknowledges the differences between the people for whom the measure is calculated, the activities to which people need access, the properties of activity locations, the modes of travel that overcome the spatial separation between people and activities, and the influence of time on accessibility.

ACCESSIBILITY MEASURES

A significant amount of research focuses on advancing the methods used for calculating accessibility and how to identify and encourage its use in transport and urban planning. Some researchers have described these measures of accessibility as either *process* or *outcome* measures (Morris, Dumble and Wigan, 1978). Outcome measures are based on the actual use and level of satisfaction obtained from taking part in a given activity or service. This type of measure implies that accessibility should be gauged from the travel patterns of people, not just the presence of the facilities. In contrast, process accessibility measures are based upon the supply characteristics of an urban system and the characteristics of the population, implying that accessibility is the opportunity or potential to travel to selected activities and is independent of actual trip making.

A review of accessibility measures commonly in use is made in this section. The purpose of this section is to discuss these measures so as to identify their place within the accessibility framework discussed later in this paper.

Topological

Topological accessibility is defined as the *nearness* or *propinquity* between geographic locations (Jiang, Claramunt and Batty, 1999). Topological accessibility is traditionally the number of links connecting one vertex to another in a connected graph (Pirie, 1979). Based on deriving shortest path matrices (representing the connectivity between pairs of vertices within a network), a range of topological measures exist that either indicate accessibility between any two locations or provide an overall measure of accessibility for an entire network (Briggs, 1972).

Space-time framework

The space-time framework is a concept first developed by Hagerstrand (1970) that introduces the constraints of time with space to determine the behavioural possibilities of an individual (Jones, 1981; Miller, 1991). Space-time prisms are three-dimensional objects with an x-y plane representing space and a z-coordinate representing time. The space-time framework assumes that events undertaken by an individual have a spatial and temporal component and that an individual can only participate in activities at a single location and point in time (Miller, 1991).

Opportunity/impedance based

The *impedance-based* method is the most commonly used method for measuring accessibility and is one of the most researched and developed methods to date. In its most basic form (commonly called *relative accessibility*), it is calculated using the formula

$$A_{ij} = O_j f\left(C_{ij}\right) \tag{1}$$

where A_{ij} is the accessibility from zone i to zone j, O_j represents the opportunities present in j and $f(C_{ij})$ is the impedance function of generalised cost for travel from i to j.

The *integral* form of this measure is summed over all j destinations. Without the cost function, this measure is referred to as the *cumulative opportunities* measure and with the impedance function referred to as *cumulative opportunity weighted by impedance* or more commonly known as the *gravity-based model*. There are three branches of the gravity-based measure: *potential accessibility*, *behavioural utility*, and *consumer's surplus*.

Potential accessibility

The potential accessibility measure is derived from the singly-constrained gravity model used in travel demand models. The gravity model has analogy with Newton's law on gravity where in terms of transport, the number of trips made between two locations is proportional to their sizes and inversely proportional to their distance apart. This analogy was originally derived by Hansen (1959) where he discussed a simple land-use model based on accessibility to determine development and population growth in a region.

Weibull (1976) formulated a measure of accessibility by first specifying the desirable properties for the measure and then deriving the measure from these properties. Six axioms were defined, which were used to establish the mathematical form of the measure. The mathematical form derived and used to represent accessibility to jobs in Stockholm was

$$A_i = \sum_{j=1}^{n} f(C_{ij}) \frac{O_j}{\partial_j} \qquad (2)$$

where A_i is accessibility, $f(C_{ij})$ is the cost function, O_j is the opportunities in zone j and $\hat{\partial}_j$ is the demand potential formulated (in the travel to work case by two possible modes of transport) as

$$\partial_j = \sum_{k=1}^{n} \left[P^1(C_{kj}^1) h_k^1 + P^2(C_{kj}^2) h_k^2 \right] \qquad (1)$$

where $P^m(C_{kj}^m)$ is a non-increasing function in the range [0,1] for mode m where $P^m(0)$ equals one and approaches zero as the cost approaches infinity, and h_k^m is the number of individuals that live in zone k choosing mode m.

The capabilities of the gravity-based model have been extended by Shen (2000) to include non-spatial travel. Shen proportioned the total number of opportunities O into three groups, those accessed by: telecommunications ϕO; both telecommunications and transportation λO; and transportation only $(1-\phi-\lambda)O$. Two types of travellers were considered: those that have access to telecommunications δP; and those that only have access to transportation $(1-\delta)P$. For those that have access to both types of travel, accessibility is formulated as

$$A_i^{cv} = \sum_j \left(\frac{(1-\phi_j-\lambda_j)O_j f(t_{ij}^v)}{\sum_m \sum_k P_k^m f(t_{kj}^m)} \right)$$

$$+ \sum_j \left(\frac{\lambda_j O_j f(t_{ij}^v)}{\sum_m \sum_k [(1-\delta_k)P_k^m f(t_{kj}^m) + \delta_k P_k^m f(t_{kj}^{cm})]} \right) \qquad (4)$$

$$+ \sum_j \left(\frac{\phi_j O_j f(t_{ij}^{cv})}{\sum_m \sum_k \delta_k P_k^m f(t_{kj}^{cm})} \right)$$

where cv identifies those with the capability of using telecommunications (v was specified for those that can only use transportation), $f(t_{ij})$ is the impedance function of travel time from zone i to j, m is the mode of travel, and P_k^m is the number of people in zone k travelling by mode m. For those that do not have access to telecommunications, the last summation term in **Error! Reference source not found.** is omitted. Shen applied the accessibility model to employment accessibility in metropolitan Boston to investigate the spatial and social effects of the automobile and telecommunications by considering: the speed of access; the spatial distribution of jobs and services; the spatial distribution of people; and industrial restructuring (Shen, 2000).

Behavioural utility

Behavioural utility is based on the assumption that individuals are rational entities and will make choices to maximise their own satisfaction or in the case of choice modelling, maximise utility. The utility of an alternative is derived from the observable attributes (weighted by their contribution to influence a decision) and unobserved attributes (random variables estimated from a distribution representing the sampled population).

A measure of accessibility can be derived from the derivation of marginal choice probabilities in logit models of multidimensional choice (see Ben-Akiva and Lerman, 1985). This measure of accessibility, which is also called the *inclusive value* or *logsum*, has the form

$$V_n' = \ln \sum_{i \in C_n} e^{V_{in}} \tag{5}$$

where V_n' is the systematic component of the maximum utility for an individual n and V_{in} is the systematic component of each secondary choice i in the set of choices C_n. This measure represents in a single value the benefit an individual obtains from a set of alternatives. As an example, in a mode-destination choice this measure of accessibility will provide for each mode of travel a value of its worth by considering the utility derived by an individual for every destination that mode can or does visit.

It has been proven that the logsum measure satisfies the following two properties (Ben-Akiva and Lerman, 1985):

- *Monotonicity with respect to choice set size*, which means that an individual will be no worse off with any additional choices; and

- *Monotonicity with respect to the systematic utilities*, which means that if the utility of any of the alternatives increases then accessibility will not decrease.

An issue with the logsum technique is that there are no units associated with the value. This makes it difficult to interpret. Furthermore, the logsum value derived is only useful to that model and cannot be compared to values derived from any other model. Because of this, when using values derived from the logsum, one must take care to define the model specifications before interpreting the accessibility measures (Ben-Akiva and Lerman, 1985).

Economic based

Several measures, which include the Hicksian compensated variation, the Harburger excess burden and the Marshallian consumer's surplus have been derived from discrete choice models for application to economic welfare (see Small and Rosen, 1981). Consumer's surplus (on which this paper is focussed) is the benefit, in monetary terms that an individual receives from a consumption choice situation (Train, 2002). It can also be referred to as a measure of the willingness-to-pay for a commodity as it is the difference between what a person is willing to pay for a commodity and what they actually pay. Consumer's surplus is the extra value an individual receives above the purchase price. Consumer's surplus can be negative. When a change occurs (i.e. a price movement) the margin between what the person is willing to pay and what they actually pay changes. This difference represents the change in consumer's surplus.

When using the multinomial logit model, it is possible to use the inclusive value and a coefficient representing cost to estimate consumer's surplus as follows

$$E(CS) = \frac{1}{\alpha} \ln\left(\sum_{j=1}^{J} e^{V_j}\right) + \text{Constant} \tag{6}$$

where the logsum part is equivalent to **Error! Reference source not found.**, α represents the negative of the coefficient of time or cost from the deterministic component of the utility function and the unknown constant term represents the difference between the actual value of consumer's surplus and the estimated value (Train, 2002).

The estimated change in consumer's surplus is then formulated as

$$\Delta E(CS) = \frac{1}{\alpha}\left[\ln\left(\sum_{j=1}^{J^1} e^{V_j^1}\right) - \ln\left(\sum_{j=1}^{J^0} e^{V_j^0}\right)\right] \tag{7}$$

where the superscripts 0 and 1 represent before and after scenarios, the two logsums represent the inclusive values derived from the behavioural models under the two scenarios, and α represents the negative of the coefficient of time or cost within the behavioural model to give the estimated change in consumer's surplus a unit of measure (Train, 2002).

ACCESSIBILITY FRAMEWORK

From the review of accessibility measures, it is evident that accessibility is dependent on three components. These are:

- the Traveller (individual or group);
- the Transport system (mode, roads and traffic characteristics); and
- Land-use (characteristics of land-uses at origins and destinations).

Along with the three components of accessibility identified above, Guers and van Wee (2004) identified a temporal component reflecting the constraints of time on the availability of opportunities and the ability of individuals to access such opportunities. Time is not considered a component of accessibility explicitly here because the influence of time is determined by the three identified components of accessibility, i.e. the traveller, the transport system, and the activities available at locations.

The more advanced measures such as the gravity-based measure include characteristics of locations of activities that make them more attractive for visitors, however the properties of the activity itself and the importance of that activity to the individual also needs consideration. The behavioural-based measures consider the individual, however most measures only do so as a by-product of a travel demand model and so are very limited in what they can be used for in transport and urban planning. The behavioural-based measures provide a technique to ascertain the influence of factors on accessibility and dissect accessibility into its various components for analysis. Such a measure of accessibility provides the transport or urban planner with the ability to isolate components of accessibility that need to be influenced by policies to improve accessibility. The time-space prism concept is useful in accessibility as it considers both spatial and temporal separation. This is a step forward in terms of determining what is realistically accessible to an individual, however it still lacks behavioural foundations and does not fully consider characteristics of activities at locations.

Therefore there is not a single measure of accessibility that can cater for all the issues associated with transport and urban planning. This is where Geographical Information Systems (GIS) and Discrete Choice Modelling provide a framework to combine the strengths of these measures to address specific issues in planning.

Most data associated with accessibility have a spatial component providing information at various locations within an urban system. A GIS serves two main purposes within the framework:

1. Allows for the analysis and preparation of spatial data related to land-use and transport to develop the behavioural models;
2. Is used to analyse the urban system using the results derived from the measures of accessibility.

This section will discuss an accessibility framework by Primerano (2004) that is based on a system of behavioural models. The behavioural models within the accessibility framework were built from revealed travel patterns of individuals within households through the use of a travel diary survey. The following sections discuss the data requirements for this framework and provide a detailed discussion of the system of behavioural models developed.

Data requirements

Data related to the transport system provides information on the transport infrastructure, traffic characteristics and modal systems operating within the transport system. Transport-related information includes the road network, the public transport system (particularly the level of service), and provisions available for private motor vehicles (such as parking).

Information on land-use provides an indication of what and how much is offered for activities at various locations. Densities of opportunities offered in zones were calculated for land-use types that included population, employment, education enrolment places, retail, social and recreational facilities.

Revealed preference data collected from the 1999 Metropolitan Adelaide Household Travel Survey (MAHTS99) provides information on the socio-economic and travel patterns of the population in various areas of the urban space. MAHTS99 was conducted by Transport SA to gather information on the population's travel behaviour for the purpose of planning Adelaide's transport needs (Transport SA, 1999). The survey gathered information based around people's day-to-day activities over two consecutive days within the Adelaide Statistical Division. A sample of approximately 9,000 homes, representing two per cent of all private dwellings, was randomly selected, resulting in the collection of data reflecting the travel patterns of over 14,000 individuals. The location of the origin and destination of trips were identified by Transport Analysis Zones (TAZ) which were defined by state road transport authorities around Australia for use in the Australian Bureau of Statistics 1996 Journey To Work survey (Trewin, 2001).

Behavioural Models

The behavioural models incorporate into the accessibility framework the preferences and needs of individuals through their revealed travel patterns. Analysis of the MAHTS99 data revealed the travel behaviour characteristics of the population in metropolitan Adelaide and provided insight into:

- the relationships between decisions made by individuals;
- data preparation for development of the behavioural models; and
- the influence of variables on the decision-making process to aid devel   of the behavioural models.

The flow chart presented in Figure 1 shows the framework used to capture   ces individuals make that influence their accessibility to activities. The interconnectiv   en and the hierarchy of choice models within the framework is a reflection of the detai   s performed on revealed preference data collected from MAHTS99. The fran   structured in a similar manner to the traditional four-step travel demand model excep designed to determine accessibility rather than travel demand. This is the reason why frequency stage is replaced by an activity choice stage. Although not shown in Figur assignment was undertaken by calculating the shortest path between locations. More and spatially dissaggregate data than provided by MAHTS99 could allow further decision-making such as route choice to be modelled and incorporated within the framev

The choices are represented in the rectangular boxes, the properties of the travell represented by the oval shapes, the alternatives of a choice set are represented by the rounded edge rectangular boxes, and the procedures used to restrict the choice sets of individuals based on their characteristics or their situations anytime during the survey period are represented by the diamond-shaped boxes. Five types of travel choices were modelled, these were: activity choice; time period choice; trip-base choice; location choice and mode choice. All models are multinomial logit with exception of the mode choice models, which are nested logit.

The arrows in Figure 1 indicate the flow of information between modelled choices and attributes. From the lowest to the highest in the hierarchy of models, the upward flow of information (represented by dotted-lined arrows) is undertaken via the inclusive value (represented by **Error! Reference source not found.**) determined for the lower layer. The inclusive value represents in a single value the total user benefit to an individual given the alternatives available and the properties of factors that influence the choice of alternatives. Ultimately, this accessibility measure provides the benefit associated with participating in an activity. The more disaggregate models also provide the benefit an individual receives from participating in an activity, but at a finer detail. The downward arrows represent the trip choices made or attribute information, which transcend to the next model. For example, based on an individual's circumstances and any influential factors, the individual will select an activity and then will choose where, when and how to participate in that activity.

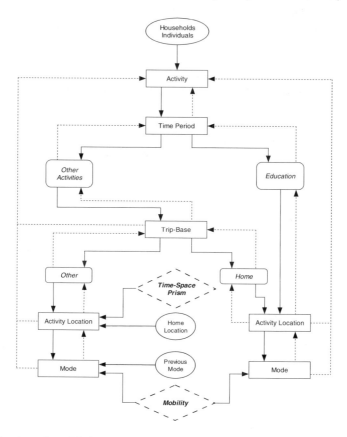

Figure 1 Behavioural model choice framework

The characteristics of the individual and their household are incorporated within the framework to consider: the differences in travel choices made by individuals; and the influence of other household members and the household's available resources on those travel choices. In addition, the choice of activity is also influenced by time of day, the possible trip-base from where travel to the activity can originate from, the possible location of such activities and the modal choice options available to the individual. To put it simply, the choice of activity is influenced by the benefits derived from all the options available to an individual to participate in an activity.

The time period choice model estimates the periods of departure time choice to an activity. This choice is influenced by the net benefit of the possible locations from which an individual can commence travel to the activity.

Trip chaining is considered by modelling the trip-base of trips. The choice is whether to participate in an activity directly from home or from another location. In determining accessibility, it is important to include trip chaining since depending on where one lives, the locations of an individual's frequent activities may impact accessibility to other activities. Modelling the trip-base considers the benefit of the location of the home relative to the location of activities and the benefit derived from linking trips to pursue activities from other locations. The choice of trip-base is dependent upon the opportunities surrounding the trip-base. Hence the inclusive value from the location choice model feeds back into the trip-base choice model.

The choice of location is highly influenced by the individual's ability to overcome the spatial separation between where they are currently and where they want to be. Hence, the decision of location choice for an activity is influenced by the mode choice alternatives available to the individual to overcome this separation.

The final decision to be made is to choose a mode of travel to get to the location of the activity. The mode choice is also influenced by the mobility options available. Essentially, this procedure determines what mode alternatives are actually available to the individual from the MAHTS99 data (Primerano, 2003).

The main flow of the decision process is described above, however there are variations to this depending on some of the decisions made along the way:

- The trip-base option for education-related travel was not modelled since the sample number of non-home-based education trips was small (six per cent of education trips and 0.4 per cent of total trips were non-home-based education trips). Hence, for education-related travel, the next decision is the choice of location for the activity with the time period choice model being influenced by the benefit of the possible educational opportunities available to an individual.

- Activity location choice models for trips originating away from the home include a variable that considers the location of the home, as it was assumed that individuals away from their home will choose activity locations that will get them closer to their home.

- The location choice sets for non-home-based travel were restricted using the space-time prism concept. It is considered that if an individual is away from their home, then their activity location options are limited in time and space by their current and next activity.

- Modal choice models for trips originating away from the home additionally consider the influence of the mode chosen for the previous trip since a person leaving their home using a particular mode of transport would most likely use that mode of transport for most of their other travels until they return home (Bowman and Ben-Akiva, 2001).

The accessibility measures within the framework

The accessibility measures reviewed were integrated within the accessibility framework in the following ways:

- Topological measures were included through the determination of travel distances, which were then used to calculate travel times and costs for the transport network. The shortest path along road links between zone centroids was calculated to determine these travel distances. This allows the framework to be applied to large real world transport networks rather than be confined to small hypothetical test networks.

- The space-time framework was used to determine the possible locations individuals could access given their current location, mobility resources, and time availability. This incorporates realism into accessibility calculations by only considering options that are truly available.

- Supply of opportunities and services provided by the transport system were considered to determine what land-uses are potentially accessible to individuals given their spatial arrangement.

- Behavioural choice models integrated the travel patterns and preferences of individuals within the framework to truly reflect the choices people make given their needs, characteristics and available resources.

- The inclusive values derived from the behavioural models were used to calculate the consumer's surplus to quantify accessibility for the evaluation of transport (demonstrated in the next section) and land-use policies.

Combining the measures within a framework reduces or even eliminates the weaknesses of each measure by using the strengths of other measures. By considering transport, land-use and socio-economic factors, the framework can be used to analyse the impact of each factor on accessibility, allowing for the development of policies that specifically target such factors.

EVALUATION OF THE ADELAIDE-CRAFERS HIGHWAY

The Adelaide-Crafers Highway was a South Australian road project funded by the Federal Government under the National Highways program. The new route shown in Figure 2 was opened on the 5 March 2000 (well after MAHTS99) providing road users with a direct route between Adelaide and the hills towns such as Stirling and Mt Barker in the Mt Lofty Ranges to the southeast of the city. The official name for the road is the Princes Highway and it is the major national road that connects Adelaide with Melbourne, Victoria. The new route, which also includes a tunnel through the hills provides a more gradual incline, is more direct (the old road was windy in parts) and is shorter in distance.

The benefits to road users include improved travel times, reduced fuel costs, and increased safety and reduced road crash costs (in particular through the bypass of Devil's Elbow, a former major crash site). The total length of the route improvement under the Adelaide-Crafers project was 8.3 km, a little over two kilometres shorter than the old route with estimated travel time reductions of five to ten minutes for the residents of the Adelaide Hills (approximately 10,000) and residents of more distant areas. Along with the improved road for motor vehicles, a bicycle path was also constructed for part of the new route. This bicycle path is mainly used by recreational riders but is much safer for commuting cyclists than the previous road. It is estimated that by 2006 the annual road user benefits will total $36 million with local businesses and residents accruing approximately $11 million of benefits each year. In addition, benefits were also expected for freight and commercial vehicle operators (Transport SA, 2003).

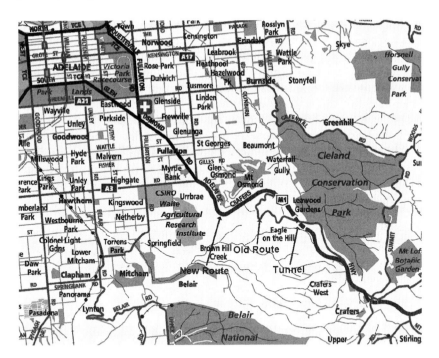

Figure 2. The Adelaide-Crafers highway including the old route

Implementation

Travel distances were updated for travel between Transport Area Zones (TAZ) in Stirling and all other TAZs in metropolitan Adelaide to reflect the distance reduction caused by the new highway. All travel distances to and from TAZs in Stirling were reduced by two kilometres except between TAZs within Stirling, which were left unchanged since although there would have been a reduction, it is uncertain how much that reduction would have been for travel within Stirling. Using the new travel distances, travel times and the fuel and taxi fare components of the travel costs were re-estimated. In cases where travel times were based on the start and end times of trips as given in MAHTS99, all such relevant travel times that were ten minutes or over were reduced by 7.5 minutes.

Results

The impact of the Adelaide-Crafers Highway on metropolitan Adelaide is demonstrated in Table 1. The figures in Table 1 include inclusive values (IV) (as formulated in **Error! Reference source not found.**) and estimates of the change in consumer's surplus (CS) (as formulated in **Error! Reference source not found.**). The new inclusive value, new rank and rank change represents the improvement in accessibility for that area with respect to metropolitan Adelaide due to the construction of the highway. The change in the inclusive value represents how much the inclusive value has changed for each area due to the highway. The change in consumer's surplus, represented for both time and money shows the time and money saved for each trip made by each resident with the introduction of the highway as compared to the scenario using the old route. The figures include an estimate of the change in consumer's surplus per year, which represents the distribution of cost savings from the highway over metropolitan Adelaide for all trips made by all residents per year. Consumer's surplus is used to evaluate the Adelaide-Crafers Highway development in this paper since infrastructure evaluation in Australia relies on the consideration of consumer surplus (Australian Transport Council, 2004).

Table 1 is sorted by the change in consumer's surplus per person per trip to show which residents benefit the most per capita from the highway. The reason for this is that it shows the true benefit derived from the policy on a per capita basis without the influence of other factors such as population size.

For metropolitan Adelaide as a whole, it is estimated that road users would have received cost savings of over $17 million dollars per year if the highway was opened in 1999. Compared to the estimate of $36 million per year by 2006 stated in Transport SA (2003), the estimate from the accessibility framework is reasonable considering (1) the assumptions made to produce the behavioural model estimates during implementation; (2) that trips that did not start or terminate within metropolitan Adelaide were not considered; and (3) inflation from 1999 to 2006.

Stirling benefits the most from the highway construction where each resident has on average a surplus benefit of over two minutes for every trip they undertake. This equates to an average surplus of just over 30 cents saved per trip per resident in Stirling. From the highway development it is estimated that residents of Stirling will accumulate a cost benefit of over $8 million per year. This estimate is comparable to the estimated $11 million worth of benefits per year by 2006 to residents and local businesses quoted in Transport SA (2003) given that:

- the accessibility framework excludes trucks as a mode of transport
- there is no assumed benefit for intrazonal travel; and
- no cost benefits were derived from the additional safety and comfort the highway provides road users (this issue is a further research topic discussed at the end of this paper).

Table 1 Benefits derived by road users from the development of the Adelaide-Crafers Highway if it were available in 1999

Local Government Area	New IV	New rank	Rank change	IV change	CS (min)	CS ($)	CS per year
Stirling	5.252	22	3↑	0.215	2.149	$0.315	$8,028,957
East Torrens	5.618	10	2↑	0.053	0.465	$0.056	$1,098,446
Adelaide	6.335	1	0	0.009	0.090	$0.012	$514,820
Unley	6.139	4	0	0.009	0.106	$0.016	$1,398,972
Burnside	5.765	7	0	0.007	0.073	$0.011	$1,151,929
St Peters	5.978	5	0	0.007	0.065	$0.010	$175,008
Mitcham	5.625	9	0	0.006	0.057	$0.009	$980,180
Walkerville	5.778	6	0	0.005	0.071	$0.010	$219,122
Glenelg	6.261	2	0	0.005	0.042	$0.005	$57,868
Payneham	5.554	13	0	0.003	0.039	$0.006	$104,735
Happy Valley	5.148	24	1↓	0.003	0.038	$0.005	$566,247
Enfield (Part A)	5.263	21	0	0.003	0.032	$0.005	$649,828
Munno Para	5.522	14	0	0.002	0.023	$0.004	$513,006
Enfield (Part B)	5.451	17	0	0.002	0.018	$0.002	$41,892
Kensington & Norwood	5.466	16	0	0.001	0.014	$0.002	$19,105
Noarlunga	4.821	28	0	0.001	0.010	$0.002	$240,436
Campbelltown	5.232	23	1↓	0.001	0.009	$0.002	$199,028
Marion	5.139	25	1↓	0.001	0.009	$0.002	$317,900
Tea Tree Gully	5.324	20	0	0.001	0.007	$0.001	$432,058
Willunga	4.572	29	0	0.001	0.006	$0.001	$36,610
Salisbury	4.944	27	0	0.000	0.005	$0.001	$307,995
Brighton	4.986	26	0	0.000	0.004	$0.001	$9,496
Hindmarsh & Woodville	5.469	15	0	0.000	0.004	$0.000	$49,209
West Torrens	5.635	8	0	0.000	0.003	$0.000	$31,551
Thebarton	5.600	11	1↓	0.000	0.003	$0.000	$4,156
Henley & Grange	5.355	19	0	0.000	0.003	$0.000	$11,849
Prospect	6.143	3	0	0.000	0.001	$0.000	$4,957
Port Adelaide	5.591	12	1↓	0.000	0.000	$0.000	$4,565
Elizabeth	5.411	18	0	0.000	0.000	$0.000	$893
Gawler	4.365	30	0	0.000	0.000	$0.000	$91
Metropolitan Adelaide	5.272			0.005	0.054	$0.008	$17,170,910

From the map in Figure 3, other areas that benefit greatly from the Adelaide-Crafers Highway are East Torrens, Burnside, Unley and Mitcham, which are close to where the highway commences out of metropolitan Adelaide. The majority of areas to benefit the least from the highway development are those areas from the southwest along the coast to the northwest of metropolitan Adelaide.

Figure 3. The distribution of benefit from the Adelaide-Crafers Highway to road users throughout metropolitan Adelaide

The accessibility web shown in Figure 4 shows the access of Stirling residents to activities as compared to that of the entire population of the metropolitan area. Each web spoke represents an activity, where the inclusive values for a specific area and for an entire metropolitan area (say) is located along this spoke. If the inclusive value for a specific area is located closer to the centre of the web than the inclusive value for the metropolitan area, then residents in that specific area have lower accessibility to that activity type than the average for the metropolitan area and vice versa. All inclusive values are normalised with the inclusive values of each activity derived for metropolitan Adelaide (with the normalised inclusive value for metropolitan Adelaide to every activity equalling one). Accessibility webs were derived from the use of similar graphs for fundamental analysis of public listed companies on financial stock markets, where the quality of a company as an investment is determined by comparing specific indicators (in place of activities in the case for accessibility webs) with those of the entire sector to which that company belongs (Primerano, 2004). Similarly, an accessibility web can be used to determine how accessible an area's inhabitants are to activities relative to an entire region. Furthermore, the accessibility web is activity specific, allowing policy-makers to assess the provision of opportunities for specific activities.

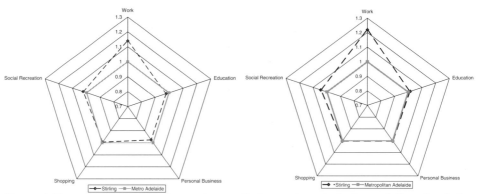

Figure 4. The accessibility webs for residents in Stirling without the highway (left) and with the highway (right) in 1999

The accessibility webs of Stirling in Figure 4 show that accessibility to every activity type modelled increased with education, personal business and shopping activities increasing just beyond the average for metropolitan Adelaide. In addition, Table 1 shows that the overall accessibility level of Stirling as compared to the other LGAs in metropolitan Adelaide increased three places to be just under the average accessibility level for metropolitan Adelaide.

Concluding comment

It should be noted that since the highway development, the Stirling Council amalgamated with three other local councils to form the Adelaide Hills Council. The Adelaide Hills Council along with neighbouring district council Mt Barker (which is the first district council outside of metropolitan Adelaide that the Adelaide-Crafers Highway passes through) have experienced new urban development. There is conjecture that the urban development growth experienced in Mt Barker is due to the Adelaide-Crafers Highway development.

The application of the accessibility framework to assess the impact of the Adelaide-Crafers Highway only considers the immediate benefits of the Highway and does not consider the long-term benefits through urban developments induced by the highway. Although not considered in this paper, it is possible to assess these long-term effects by incorporating the land-use changes within the framework. The accessibility framework considers the behavioural responses of individuals to changes in the transport system or land-uses, but does not model the impact of the transport system on the land-use configuration and vice versa. An estimate of such land-use changes could be determined using an interactive land-use/transport model with results incorporated within the accessibility framework through its activity location choice models. The capability of the accessibility framework to assess land-use policies has been demonstrated in Primerano and Taylor (2005).

CONCLUSIONS

Accessibility was defined as the ease for individuals to participate in activities. The ease is influenced by five factors, namely: the individual; activities; destinations; transport modes; and time. From the reviewed accessibility measures, it was found that no single measure can cater for all the issues associated with transport and urban planning. Discrete choice modelling and GIS were used to create a framework to bring together existing methods of calculating accessibility to address planning issues. The accessibility framework combines the strengths of some of the existing methods of calculating accessibility measures to develop a powerful and sophisticated accessibility framework for policy analysis and evaluation.

The accessibility framework determines the benefit or need for an individual or group of people to travel to an activity. The framework revolves around a hierarchical decision-making process that individuals undertake when deciding to participate in an activity. Binding the framework around behavioural models has the following benefits:

- allowing accessibility to reflect an individual's behavioural characteristics, which can be related to their socio-economic characteristics and decision-making when confronted with time, space and resource constraints;

- user benefit estimates obtained from available choice alternatives of individuals; and

- allowing the dissection of accessibility into its various disaggregate forms.

The framework was used to evaluate the impact of the Adelaide-Crafers Highway in terms of the distribution of benefit to road users in metropolitan Adelaide and to further investigate benefits to the residents of Stirling. This policy has shown that having a direct impact on the travel distance and time required to travel to activities has a strong impact on the accessibility of individuals to activities.

The accessibility framework is powerful in assessing the effectiveness of policies to change levels of accessibility in a manner intended by the policy-maker. The framework can be applied to policies based on modifying the characteristics of the transport system and those based on the adjustment of the characteristics and spatial arrangement of land-uses. This paper has demonstrated the ability for the accessibility framework to be applied to real world urban areas using detailed transport, land-use and socio-economic information.

This paper has shown the results from applying a framework to evaluate a transport policy taking into consideration people's sensitive to policy changes through the revealed travel preferences. In applying the framework to the Adelaide-Crafers Highway development, the results have shown how people will benefit through the increased opportunities the highway provides and in turn, how these opportunities influence their choices to maximise such benefits. For policy-makers with such objectives in mind, this accessibility framework can provide the required policy evaluation.

FURTHER RESEARCH

Through the development and application of the accessibility framework, a number of issues have surfaced. These are now discussed along with possible methods to further improve the capability of the accessibility framework for use as a tool used by planners to identify levels of accessibility and to better develop and evaluate policies.

The additional benefit derived from the increase in safety and comfort provided by the Adelaide-Crafers Highway was not included as part of the overall benefit gained by road users. Since MAHTS99 was a revealed preference survey, the data collected from MAHTS99 does not contain any information in regards to how individuals value safety and comfort of a journey. An additional stated preference survey is required where specific questions regarding issues of driver safety and comfort can be asked with results incorporated within the existing revealed preference survey.

During the development of the behavioural models, crude travel time estimates were used, These 'crude' estimates were due to the unreliable travel times often provided by household travel surveys like MAHTS99, the limited availability of intrazonal information and a lack of route choice information. Ideally, a sophisticated system is required to accurately calculate travel times for different times of the day, along various routes and among different mode types. More route choice information would also enable route choice behavioural models to be developed to add another decision choice dimension to the accessibility framework.

The travel distance and time adjustments made to simulate the highway were crude and without considering the benefits of intrazonal travel, time and cost savings were possibly underestimated. Evaluating a transport network upgrade such as the Adelaide-Crafers Highway using the accessibility framework would be ideally undertaken within a GIS where travel times and distances could be automatically updated along affected routes. In the case of the Adelaide-Crafers Highway, rather than the planner having to recalculate all of the travel times between all zones and incorporate the changes manually within the accessibility framework, these could occur automatically within a GIS that incorporates the highway. This will improve the usefulness of the accessibility-based planning tool for transport planners in their assessments of the effectiveness of transport network upgrades. Similarly so for urban planners where the benefits of different arrangements of land-uses can be assessed.

The policy evaluation undertaken in this paper only considered the short-term benefits offered by a transport infrastructure improvement. The paper identified that benefits of the long-term impacts of transport on land-use arrangements were not considered. Although it is possible to undertake such assessment, an appropriate integration of the accessibility framework with an interactive land-use/transport model is required to fully evaluate policies in the short and long terms.

To make the concept of accessibility more usable by transport and urban planners, an accessibility-based planning tool implemented within a GIS platform is required. Such a tool should allow planners to:

- Identify any transport and/or urban planning issues that negatively impact the social welfare of people and the viability of businesses;

- Create policies that target these issues and provide change according to the objectives of the policies; and

- Evaluate the effectiveness of policies through the spatial distribution of benefits throughout Adelaide and by showing that certain targets and objectives will be achieved.

ACKNOWLEDGEMENTS

The authors would like to thank the Department of Transport and Urban Planning for providing the travel diary survey data. We would also like to thank Planning SA and GISCA for providing spatial datasets and to the staff from the Transport Planning Agency of the Department of Transport and Urban Planning for their support. We would also like to acknowledge the funding support provided by the Australian Research Council.

REFERENCES

Australian Transport Council (2004). *National guidelines for transport system management in Austrlaia*, Vol. 3: Foundation Material. Department of Transport and Regional Services, Canberra.

Ben-Akiva, M. and S. R. Lerman (1978). Disaggregrate travel and mobility choice models and measures of accessibility. In: *Behavioural Travel Modelling* (D. A. Hensher and P. R. Stopher, ed.), pp. 654-679. Croom Helm, London.

Ben-Akiva, M. E. and S. R. Lerman (1985). *Discrete Choice Analysis: Theory and Application to Travel Demand*. MIT Press, Cambridge, Massachusetts.

Bowman, J. L. and M. E. Ben-Akiva (2001). Activity-based disaggregate travel demand model system with activity schedules. *Transportation Research Part A: Policy and Practice*, **35**, 1-28.

Briggs, K. (1972). *Introducing the New Geography - Introducing Transportation Networks*. University of London Press Ltd, London.

Dalvi, M. Q. (1978). Behavioural modelling, accessibility, mobility and need: concepts and measurement. In: *Behavioural Travel Modelling* (D. A. Hensher and P. R. Stopher, ed.), pp. 639-653. Croom Helm, London.

Guers, K. T. and B. van Wee (2004). Accessibility evaluation of land-use and transport strategies: review and research directions. *Journal of Transport Geography*, **12**, 127-140.

Hagerstrand, T. (1970). What about people in regional science? *Papers of the Regional Science Association*, **24**, 7-21.

Hansen, W. G. (1959). How accessibility shapes land use. *Journal of the American Institute of Planners*, **25**, 73-76.

Jiang, B., C. Claramunt and M. Batty (1999). Geometric accessibility and geographic information: Extending desktop GIS to space syntax. *Computers, Environments and Urban Systems,* **23,** 127-146.

Jones, S. R. (1981). Accessibility measures: a literature review. *Transport and Road Research Laboratory,* **Report 967.**

Koenig, J. G. (1980). Indicators of urban accessibility: Theory and application. *Transportation,* **9,** 145-172.

Martinez, F. J. (2000). Towards a land-use and transport interaction framework. *Handbook of Transport Modelling,* (D. A. Hensher and K. J. Button, ed.), Vol. 1, pp 393-407. Elsevier Science Ltd., Oxford.

Miller, H. J. (1991). Modelling accessibility using space-time prism concepts within geographical information systems. *International Journal of Geographical Information Systems,* **5,** 287-301.

Morris, J. M., P. L. Dumble and M. R. Wigan (1979). Accessibility indicators for transport planning. *Transportation Research,* **13A,** 91-109.

Niemeier, D. A. (1997). Accessibility: an evaluation using consumer welfare. *Transportation,* **24,** 377-396.

Ortuzar, J. D. D., F. J. Martinez and F. J. Varela (2000). Stated preferences in modelling accessibility. *International Planning Studies,* **5,** 65-85.

Pirie, G. H. (1979). Measuring accessibility: A review and proposal. *Environment and Planning A,* **11,** 299-312.

Primerano, F. (2003). Mobility considerations in restricting choice sets in modal choice models. *25th Conference of Australian Institutes of Transport Research,* Adelaide, Australia, 3-5 December.

Primerano, F. and Taylor M.A.P. (2005). Increasing accessibility to work opportunities in Metropolitan Adelaide. *Accepted on 14 April 2005* in the forthcoming *Journal of the Eastern Asia Society for Transportation Studies* **6.**

Small, K. A. and S. R. Harvey (1981). Applied welfare economics with discrete choice models. *Econometrica,* **49**(1), 105-130.

Stopher, P. R., D. T. Hartgen and Y. J. Li (1996). SMART: simulation model for activities, resources and travel. *Transportation,* **23,** 293-312.

Train, K. (2002). *Discrete Choice Methods with Simulation.* (Not yet published). Cambridge University Press.

Transport SA (1999). 1999 Metropolitan Adelaide Household Travel Survey - Interviewer's Manual. Government of South Australia, March.

Transport SA (2003). The Adelaide Crafers Highway Project http://www.transport.sa.gov.au/transport_network/projects/adel_crafers/index.asp. Transport SA. *Accessed* 9 March 2004.

Trewin, D. (2001). Statistical Geography Volume 2. *Census Geographic Areas Australia,* Australian Bureau of Statistics, 1 July.

Wegener, M. (1996). Reduction of CO_2 emissions of transport by reorganisation of urban activities. *Transport, Land-Use and the Environment* (Y. Hayashi and J. Roy, ed.) Kluwer Academic Publishers, Dordrecht, 103-124.

Weibull, J. W. (1976). An axiomatic approach to the measurement of accessibility. *Regional Science and Urban Economics,* **6,** 357-379.

Access to Destinations
D.M. Levinson and K.J. Krizek (editors)

CHAPTER 16

MODELING ACCESSIBILITY IN URBAN TRANSPORTATION NETWORKS: A GRAPH-BASED HIERARCHICAL APPROACH

Ahmed Abdel-Rahim, University of Idaho
Ayman M. Ismail, The Ohio State University

INTRODUCTION

Transportation systems are composed of a complex set of relationships between supply, (mainly the operational capacity of the network), demand, and the networks that support movements. The planning and design of urban transportation systems have been primarily based on engineering principles that address optimization of certain operational and safety indicators with little or no focus on the accessibility and mobility of system users at different parts of the network.

Accessibility is considered a combination of both impedance factors (time or cost of reaching a destination) and an attractiveness factor (the qualities of the potential destinations and the availability of desired services and activities). Hansen (1959) defined accessibility as the "potential for interaction". In the context of transportation planning, accessibility can be thought of as the ease with which desired destinations could be reached (Niemeier, 1997). Mobility, on the other hand, has been defined as the potential for movement, the ability to get from one place to another, and the ability to move around (Hansen 1959) and (Handy and Niemeier 1997). Mobility, with this definition, is related to the impendence components of accessibility. Decision makers need evaluation tools that allow them to determine how a particular decision or activity affects the network accessibility. Traditional level-of-service (LOS) measures in different transportation facilities have been used as measures of mobility. Accessibility, however, needs to be evaluated by examining factors that measure both the *availability* and *quality* of transportation services. Examples of such factors are presented in (BTS 1997), Litman (2003) and (Neimeier 1997).

Accessibility can be modelled and analyzed from different perspectives: a particular zone or area, a particular group of users, or a particular activity. Transportation projects that aim to improve travel time and increase speed for different travel modes can result in a proportionally larger increase in accessible areas, as shown in Figure 1. Policies to improve mobility will generally increase accessibility as well by making it easier to reach destinations. However, it is still possible to have cases where accessibility is not dependent on good mobility, for example, a community with severe congestion but where residents live within a short distance of all needed and desired destination. (Handy 2004).

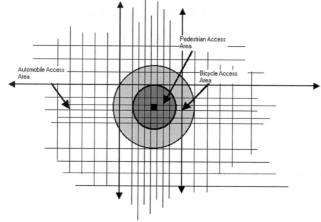

Figure 1, Accessible areas for different travel modes, (VPTI 2003)

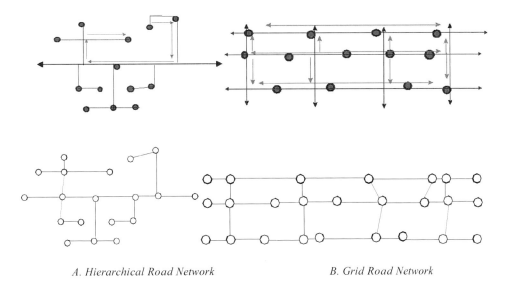

A. Hierarchical Road Network   B. Grid Road Network

Figure 2. Road networks and their Graph representation

The topological characteristics of the transportation network, the layout of the roadway network and different paths along it affects the overall network accessibility. Figure 2 illustrates how two types of road networks differ in their graphic representation. An Interconnectivity Index is commonly used to evaluate how well a roadway network connects destinations (Ewing, 1996). The Index is computed by dividing the number of roadway links by the number of roadway nodes in the network graph representation. A higher index indicates increased route choice, allowing for more direct connections and access between trip origins and destinations. When calculating the Interconnectivity Index, individual road segments and paths can be given level-of-service ratings for various types of travel. According to this method of calculation, the grid roadway network in Figure 2B has a higher Interconnectivity Index than the hierarchical road network (1.24 vs. 0.95).

Topological accessibility measures the nearness or propinquity between geographic locations (Jiang, Claramunt, and Batty, 1999). Pirie (1979) defined topological accessibility as the number of links connecting one vertex to another in a connected graph. A number of topological measures exist based on deriving the shortest path matrices that indicate the least number of links required to be passed to reach another vertex (Briggs, 1972).

Researchers have used several forms of accessibility measures. In most cases, measures of accessibility include both an *impedance factor*, reflecting the cost of travel, and an *attractiveness factor*, reflecting the qualities of the potential destination (Handy 2004). One of the widely used accessibility models used is the opportunity/impedance based model introduced by Hansen (1959). The model was the base for several cumulative opportunities and gravity-based accessibility models. Hansen's model, in its most basic form, is calculated using the formula:

$$A_i = \sum O_j f(C_{ij})$$

Where:

A_i is the Accessibility of point i
O_j is the Opportunities at point j, and
$f(C_{ij})$ is a function of the generalized travel cost from i to j

A graph-based approach to model accessibility in urban transportation networks is presented in this paper. Three different graph connectivity-based measures are used to analyze the network structure efficiency and connectivity providing topological accessibility measures. A network-wide functionality-based measure was used to assess the network accessibility based on the demand exchanged between different Origin-Destination (O-D) pairs (opportunities) and the quality of service on routes connecting the O-D pairs using different modes of travel (impedance). The network functionality measures provide opportunity/impedance accessibility measures for the network or for subnetworks within the network. The model provides decision makers with a tool that allows them to model and evaluate the accessibility of transportation networks under different scenarios.

MODEL OVERVIEW

A graph is a symbolic representation of a network. It implies an abstraction of the reality representing the structure of a network. A transportation network can be represented as a graph G with a set of verticies (nodes) v connected by edges (links) e.

$$G = (v, e) \tag{1}$$

A node v is a terminal point or an intersection point of a graph. It is the abstraction of a location such as a city, a division, or a road intersection. An edge e is a link between two nodes representing an abstraction of a physical roadway infrastructure or a transit line supporting movements between nodes. Travel routes or paths consist of a sequence of links that are travelled in the same direction. Finding all the possible paths in a graph is a fundamental attribute in measuring accessibility and traffic flows in a network.

In typical transportation networks, the edges are not merely binary entities, either present or not, but have associated weights that represent their strengths relative to one another (i.e. their capacity or quality of service). A weighted network can be represented mathematically by an adjacency matrix with entries that are not simply zero or one, but are equal, instead, to the weights on the edges:

A_{ij} = weight of edge connecting i to j

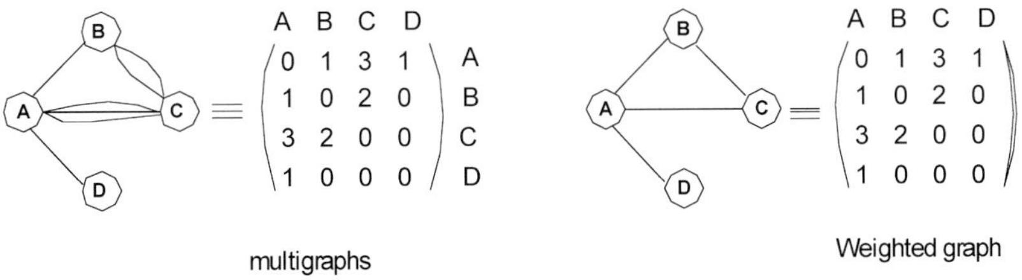

Figure 3. Adjacency matrices for a weighted graph and multigraphs

Figure 3 shows the adjacency matrices for a weighted graph and a multigraph (a graph representing a network with multiedges). As can be seen in the figure, the two networks have the same adjacency matrix. Newman (2002) suggests that weighted graphs can be analyzed by mapping them onto unweighted multigraphs. That is, every edge of weight n is replaced with n parallel edges of weight 1 each, connecting the same vertices. The adjacency matrix of the graph remains unchanged and any techniques that can normally be applied to unweighted graphs can now be applied to the multigraph as well.

GRAPH INDICES: CONNECTIVITY-BASED MEASURES

Several measure and indices can be used to analyze the network efficiency and connectivity (Kansky 1963). These measures can be used to express the relationship between values and the network structures they represent and to compare different transportation networks under different structure scenarios. For the purpose of this paper, three indices that measure the network connectivity were used, (Rodrigue 2004):

Iota Index: measures the ratio between the network and its weighed vertices. It divides the length of a graph ($L(G)$) by its weight ($W(G)$). The lower its value, the more efficient the network is. The weight of all nodes in the graph ($W(G)$) is the summation of each node's order (o) multiplied by 2 for all orders above 1. The order of the node (o) is the number of links attached to the node.

$$\varsigma = \frac{L(G)}{W(G)} \qquad W(G) = 1, \forall o = 1 \qquad W(G) = \sum_e 2*o, \forall o > 1 \qquad (2)$$

Beta Index (Interconnectivity Index): measures the level of connectivity. It is expressed by the relationship between the number of links (e) over the number of nodes (v). Higher values of interconnectivity index indicate higher number of paths possible in the network.

$$\beta = \frac{e}{v} \qquad (3)$$

Gamma Index: measures of connectivity that considers the relationship between the number of observed links and the number of possible links. The value of gamma is between 0 and 1.

$$\gamma = \frac{e}{3(v-2)} \qquad (4)$$

FUNCTIONALITY BASED MODELING

In order to reduce the complexity of the transportation system graph model $G = (v, e)$, it is viewed as a collection of essential mobility-based functionalities on paths connecting different O-D pairs using different modes. It is assumed that the network is composed of, or capable of engaging, "n" essential functionalities. Each functionality (F_k^{rs}) represents demand between O-D pair r and s using travel mode k. It is also assumed that the network consists of N components defined by the component set C, composed of all nodes and edges in the network. $C = \{e, v\}$. Each functionality (F_k^{rs}) is executed by a component set C_{rsk}^p representing the components of different paths that connect nodes r and s using mode k.

It will be useful to express the system essential functionalities in terms of the set of components it needs. Let C_{rsk}^p be the components sets associated with functionality (F_k^{rs}),

It will be useful to express the system essential functionalities in terms of the set of components it needs. Let C_{rsk}^p be the components sets associated with functionality (F_k^{rs}), where p represents the number of alternative paths in which (F_k^{rs}) can be executed. If $p = 1$ then the functionality can be executed through a unique path and O-D pair r and s are connected by only one route using mode k. This is typical for public transit mode. If $p > 1$, then the component set C_{rsk}^p has p subcomponent sets and the functionality can be executed through "p" alternatives paths. Thus, p in the component set C_{rsk}^p denotes the number of alternative paths that connect O-D pair r and s using mode k. Component sets can be used to define functional primitives as will be described below.

Given a component set C_{rsk}^p, let $V(C_{rsk}^p)$ be a function that defines a quantitative measure for the quality of transportation service on the components set C_{rsk}^p (nodes and edges). Then, $V(C_{rsk}^p)$ is the product of the quantitative measures of its components (edges along the path), in this case:

$$V(C_{rsk}^p) = \prod_{C_i \in C_{rsk}^p} V(C_i) \qquad (5)$$

The definition is analogous to the definition of the reliability of a series, which is defined as the product of the reliabilities of the components of the series.

Having several alternatives, yet functionally equivalent, functionality F_k^{rs} allows for a choice based on certain selection criteria. Recall that C_{rsk}^p is the component set of functionality F_k^{rs} with p alternative paths. We can now define a numerical value (v) that represents the quality of the transportation services that serve functionality F_k^{rs}

$$v = V(C_{rsk}^p) \tag{6}$$

and,

$$v_k^{rs} = \max(V(c_{rsk}^1), V(c_{rsk}^2), \ldots, V(c_{rsk}^p)) \tag{7}$$

let $S(v)$ be a selection function that maps v to a specific functionality F_k^{rs} then:

$$S(\max(V(c_{rsk}^1), V(c_{rsk}^2), \ldots, V(c_{rsk}^p))) \tag{8}$$

The functionality of the network (or part of the network with) can be assessed using the equation:

$$F_{network} = \sum_{i=1}^{n} ((F_k^{rs})(v_k^{rs}))_i \tag{9}$$

MODEL IMPLEMENTATION

In order to demonstrate the model, we will consider the roadway network presented in Figure 4 for the Sixth of October City, a new city with a population of around 350,000 people located 40 km southwest of Cairo, Egypt. A graph representation of a subset of its road network is presented in Figure 4 and Figure 5. The existing network consists of 22 nodes and 29 links (v=22 and e=29). Due to the unavailability of detailed operational characteristics data, a hypothetical LOS data for both automobile and public transit were generated according to the Highway Capacity Manual (HCM) guidelines for urban arterials and transit facilities. The LOS data were converted to numerical weights (A=5, B=4, C=3, D, 2, and E and F=1). The adjacencies matrices for both automobile and transit are presented in Figure 6. A hypothetical O-D matrix was used to determine the functionality set for both automobile and transit F_1^{rs} and F_2^{rs}, respectively.

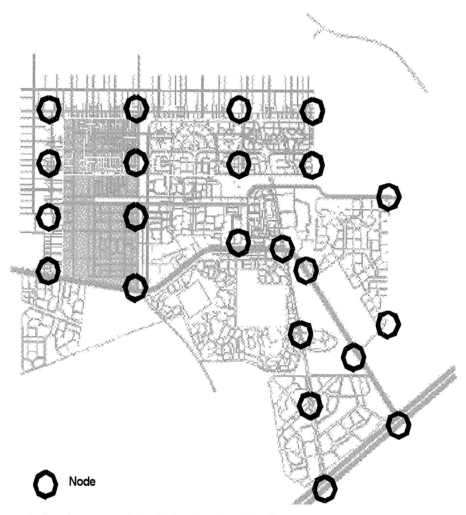

Figure 4 - Roadway network for Sixth of October City, Egypt

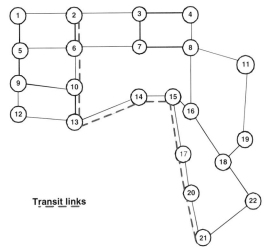

Figure 5. Graph network representation

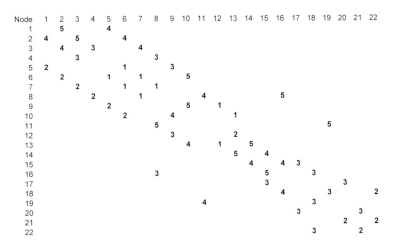

Figure 6-a. Adjacency matrix for automobile use

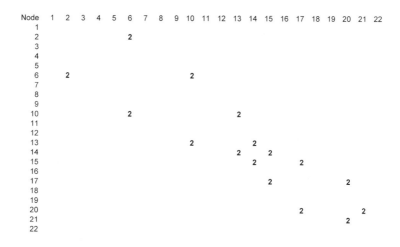

Figure 6-b. Adjacency matrix for transit use

ANALYSIS AND RESULTS

The graph representation of the network was analyzed under four different scenarios: 1) existing conditions, 2) the addition of a new transit line connecting node 22 to node 2 through node 4 and improving the LOS of the transit service to C, 3) adding a new highway between nodes 7 and 14, and 4) improving the LOS on all links that has LOS of 2 or less to 4. A Visual Basic script was used to determine the network wide functionality under different scenarios. First, the O-D data were used to generate a set of functionalities for both automobile and transit users. For each functionality, the components set for each connection alternative was identified. To simplify the analysis, a maximum of three connection alternatives were considered for each functionality ($p \leq 3$). LOS data were used to determine the quality of service measure (v_k^{rs}) associated with each functionality. These data were then used to determine the network wide functionality index using equation (9).

Results from the analysis are presented in Table 1 and Table 2 for network wide and node specific measures, respectively. The network-wide indices and measures yielded consistent results, with the new transit line and LOS improvements, increasing network connectivity and functionality by an average of 22.5 %. The graph-based network connectivity indices (Beta, Gamma, and Iota) provide a topological accessibility assessment of the network hierarchy, structure efficiency, and connectivity for the network under different alternatives. The node specific measures revealed a more detailed picture of benefits per zones/areas. Nodes' order data, (Table 2), provide more detailed information on the effect of proposed alternatives on a zone-by-zone basis. The data shows that the new highway alternative provided marginal topological accessibility improvement compared with the other two alternatives. Network-wide functionality index revealed similar conclusion. The new highway construction provided 6% improvement in the network wide functionality index, compared to 20% and 22% for LOS improvements and the new transit line, respectively. The new transit line provided connectivity and LOS improvements ranging from 13% to 109% for 15 of the 22 zones, with an average improvement 31%. Improving the LOS provided connectivity and LOS improvements ranging from 7% to 140% for 15 of the 22 zones, with an average improvement 44%. The service improvement data, together with cost data, can provide decision makers with good assessment of the network-wide or zone-by-zone impact of different alternative on network accessibility.

The graph-based connectivity indices are simple measures and do not require extensive computation. However, they primarily provide measures for network topological accessibility. The inclusion of the LOS data in the weighted network graph provides an assessment of the impedance side of accessibility measures. However, these connectivity-based indices do not include any measures for the opportunity side of the accessibly measures. The network wide functionality index, while requiring extensive computational procedure, provides opportunity/impedance accessibility measures for the entire network or for part of the network by taking into account both O-D and LOS data.

Table 1. Network-wide measures under different scenarios

MOE	Existing	Transit	% Improvement	new highway	% Improvement	LOS improvements	% Improvement
v	22	22		22		22	
e	214	262		224		263	
Beta Index	9.727	11.909	22.430	10.182	4.673	11.955	22.897
Gama Index	0.282	0.345	22.430	0.295	4.673	0.346	22.897
Iota Index	5.841	4.771	-18.321	5.580	-4.464	4.753	-18.631
Network-wide functionality (10^6)	6.120	7.470	22.059	6.490	6.046	7.380	20.588

CONCLUSION

This paper presented a graph-based approach to model accessibility in urban transportation networks. The example presented in this paper demonstrated the effectiveness of the proposed model in analyzing transportation networks from both the *connectivity*, as analogous to topological accessibility, and *functionality* as analogous to impedance/opportunity accessibility perspectives. The model provides decision makers with a tool that allows them to model and evaluate the accessibility of transportation networks under different scenarios. Three different graph connectivity-based measures were used to analyze the network hierarchy, structure efficiency, and connectivity. A network-wide measure was used to assess the network functionality based on the demand between different O-D pairs and the quality of service on routes connecting the O-D pairs using different travel modes.

Table 2. Nodes' order under different scenarios

Node	Existing	Transit	% Improvement	new highway	% Improvement	LOS improvements	% Improvement
1	15	15	0.0	15	0	17	13.3
2	28	33	17.9	28	0	30	7.1
3	21	30	42.9	21	0	23	9.5
4	11	23	109.1	11	0	13	18.2
5	13	13	0.0	13	0	23	76.9
6	25	29	16.0	25	0	41	64.0
7	10	10	0.0	20	100	24	140.0
8	24	33	37.5	24	0	32	33.3
9	18	21	16.7	18	0	23	27.8
10	29	33	13.8	29	0	34	17.2
11	18	18	0.0	18	0	18	0.0
12	7	7	0.0	7	0	15	114.3
13	26	30	15.4	26	0	34	30.8
14	26	30	15.4	26	0	26	0.0
15	31	35	12.9	31	0	31	0.0
16	24	36	50.0	24	0	24	0.0
17	20	24	20.0	20	0	20	0.0
18	22	33	50.0	20	-9	22	0.0
19	15	15	0.0	25	67	15	0.0
20	19	23	21.1	19	0	21	10.5
21	15	18	20.0	15	0	21	40.0
22	11	15	36.4	12	9	18	63.6

The proposed measures were used to analyze a subset of the roadway network ($v=22$ and $e=29$) for a city of population of around 350,000 people located near Cairo, Egypt. The graph representation of the network was analyzed under existing conditions and three different transportation improvement scenarios. The network-wide indices and measures introduced in this paper yielded consistent results. The node specific measures revealed a more detailed picture of benefits per zones/areas. These network-wide and zone specific measures, together with cost data, can provide decision makers with good assessment of the network-wide or zone-by-zone impact of different alternative on accessibility.

REFERENCES

Briggs, K. (1972). *Introducing the New Geography - Introducing Transportation Networks*, University of London Press Ltd, London.

BTS (1997). Mobility and Access; Transportation Statistics Annual Report 1997, Bureau of Transportation Statistics, pp. 173-192.

Ewing, R. (1996). *Best Development Practices; Doing the Right Thing and Making Money at the Same Time*, Planners Press, 1996.

Handy, S. and D. Niemeier (1997). Measuring Accessibility: An Exploration of Issues and Alternatives. Environment and Planning A, Vol. 29, pp. 1175-1194.

Handy, S. (2004). *Planning for Accessibility: Definitions, Measures and Benefits, Access to Destinations: Rethinking The Transportation Future of Our Region*, University of Minnesota President's 21st Century Interdisciplinary Conference Series, November 8-9, 2004

Hansen, Walter G. (1959). How Accessibility Shapes Land Use. Journal of the American Institute of Planners, **25**, 73-76.

Jiang, B., C. Claramunt and M. Batty (1999). Geometric Accessibility and Geographic Information: Extending Desktop GIS to Space Syntax. *Computers Environment and Urban Systems*, **23**, 127-146.

Kansky, K. (1963). *Structure of Transportation Networks: Relationships Between Network Geography and Regional Characteristics*, University of Chicago, Department of Geography, Research Papers 84.

Litman, Todd (2003). Measuring Transportation: Traffic, Mobility and Accessibility, *ITE Journal*, **73(10)**, 28-32.

Neimeier, D. (1997). Accessibility: An Evaluation Using Consumer Welfare. *Transportation*, **24(4)**, 377-396

Newman, M. E. J. (2002). The Structure and Function of Networks, *Computer Physics Communications,* **147**, 40-45.

Pirie (1979). Measuring accessibility: a Review and Proposal. *Environment and Planning A*, **11**, 299-312.

Rodrigue, J.-P. *et al.* (2004). Transport Geography on the Web, Hofstra University, Department of Economics & Geography. http://people.hofstra.edu/geotrans, Accessed September 2004.

Victoria Transport Policy Institute VTPI (2003). Accessibility: Defining, Evaluating and Improving Accessibility. http://www.vtpi.org/tdm/tdm84.htm, accessed October 2004.

Access to Destinations
D.M. Levinson and K.J. Krizek (editors)
© 2005 Elsevier Ltd. All rights reserved.

361

CHAPTER 17

ACCESSIBILITY AND SPATIAL DEVELOPMENT IN SWITZERLAND DURING THE LAST 50 YEARS

M. Tschopp, Ph. Fröhlich, and K. W. Axhausen, Institute for Transport Planning and Systems, ETH Zurich

INTRODUCTION

Transport systems have been built primarily to expand the reach of both people and industry. One measure of the resulting spatial impacts is the change in *accessibility*, which measures what can be reached with what effort. Accessibility is both the primary service provided by transport infrastructure and the link between transport infrastructure and land use. It can measure the spatial impact of newly built transport infrastructure and show the attractiveness of a region's location.

The link between accessibility improvement and economic and population growth, at least change, is a key tenet of regional science, transport and planning, while the literature acknowledges that accessibility is only a sufficient and not necessary condition for growth (Vickerman, 1992 or Banister and Berechman, 2000). The previous empirical work trying to document this link (see for example Lutter, 1980; Kesselring, Halbherr and Maggi, 1982; Seimetz, 1987; Aschauer, 1989; Fernald, 1998; Holtz-Eakin, 1994; Munnell, 1990; Nadiri, 1998; Boarnet and Haughwant, 2000; Shirley and Winston, 2004) suffers a number of short comings: the use of large spatial units, such as US states or UK counties; the reliance on short study periods, typically ten to twenty years; omission of railway services and finally in most cases the approximation of the services delivered by the transport system by the value of the public capital stock (See Axhausen, 2004 for a detailed critique).

The first part of this paper provides a general survey of population growth and its spatial distribution in Switzerland. The second part analyzes the development of spatial accessibility and models its impacts on demographic and economic change using a multi-level regression approach.

This research is based on a data set with unique spatial detail and historical depth: extensive structural data on each of the about 2896 Swiss municipalities and transport networks for the last fifty years (1950-2000). The socio-economic data were collected from Swiss censuses and restructured to refer to the year 2000 geographies throughout (Tschopp, Frey, Reubi, Keller and Axhausen, 2003). In 2000, Switzerland's 2896 municipalities cover the entire country. In the last 5 decades more than 300 mergers have changed these boundaries. The matching network models were built by Fröhlich, Frey, Reubi and Schiedt, 2003.

SETTLEMENT PATTERNS IN SWITZERLAND

During recent decades Switzerland had one of the highest growth rates in Western Europe (see Haug, 2002). The development was characterised by a continuous population growth of different intensities. Periods of strong growth can be observed around 1900, the time after the World War II until 1970, and the last two decades until 2000. From 1850 to 2000 population almost tripled to 7,200,000 inhabitants. The patterns of this growth are discussed below.

A useful indicator of the distribution of the population between municipalities is its Gini coefficient, measuring the concentration displayed in a Lorenz curve (see Figure 1). The Lorenz curve displays the relative concentration of inhabitants, implying the dispersion of population between municipalities. The municipalities are ranked by size, with the smallest assigned rank 1. To make the curves comparable, the x-axis is scaled as percentage of the varying number of municipalities. The x-axis shows the cumulative share of the population. The closer the curve lies to the diagonal the more equal is the population spread over the municipalities. If the curve equals the diagonal, then all municipalities have the same size. The Gini Index is a measure of concentration (Bökemann and Kramar, 1980). It measures the ratio of the area between the diagonal and the curve to the total area underneath the diagonal: It ranges from zero for equality to one for an extremely concentrated pattern.

The Gini Index is increasing over the decades (Figure 1) and reaches 0.7, its highest value, in 1970. More than 66% of the population lived in the 10% largest towns. In the last three decades a continuous decline of this value to 0.67 indicates a reversal of this trend. Winners were medium sized municipalities, while the largest municipalities lost inhabitants. Figure 2 compares Gini Indices over time for different Cantons (Cantons are the basic elements, roughly comparable in their function and law setting powers to federal states in the United States). Zurich is an urban, highly industrialised Canton, Solothurn a peripheral area between large agglomerations and Graubünden a rural and alpine Canton. All three cantons show a three-part pattern: From 1850 to 1910 a strong increase, then a less steep continuation and finally a trend reversal form 1960 for urban areas onwards. The more rural and mountainous the areas areas are, the later Gini Indices are falling.

It is possible to express this change as changes in the coefficients of Zipf's law (1949). Eeckhout's recent paper (2004) casts doubt on the usual way of estimating the rank-size rule. Future research will test his conclusion about the log-normal distribution of settlement size with our data going back to 1850.

Figure 1. Lorenz curves based on municipal population size in Switzerland

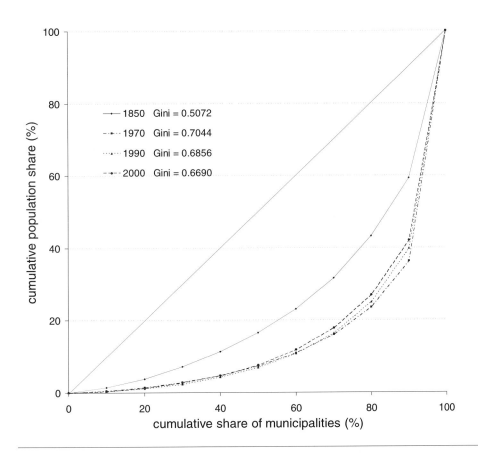

Figure 2. Development of the Gini index of selected Swiss Cantons

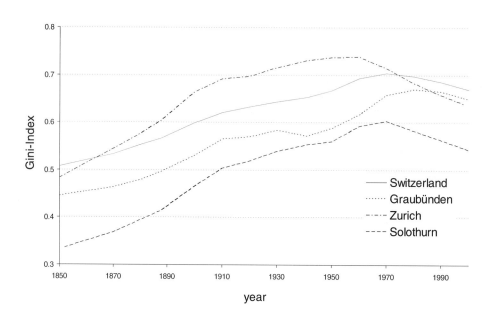

If we focus on the municipalities and therefore take a more local point of view we can see obvious differences in the spatial patterns over the decades (see Figure 3). Indicated with a dot are all municipalities growing for the first time more than one standard deviation above the average growth rate in Switzerland in their decade. If municipalities grew above the average in more than one decade, the first decade is marked. Squares indicate the core municipality of the five largest agglomerations: Zurich, Geneva, Basle, Bern and Lausanne.

In the years after 1950 we see a first wave of suburbanisation around Zurich, Basle and Geneva. In this time period an inner circle around the city centers is growing rapidly. During the next decades a further growth in the urban areas is obvious, now population at the outskirts of the agglomerations is increasing (simultaneous with the decline of the Gini index in Figure 2). Finally, between 1990 and the year 2000 the growth zones of the agglomerations (especially between Zurich and Basle and Lausanne and Geneva respectively) are growing together.

If a municipality grows this fast in more then one decade, then it is noted with the first decade of such growth.

Figure 3. Fast growing municipalities (+1 std. dev above the national mean) in different decades

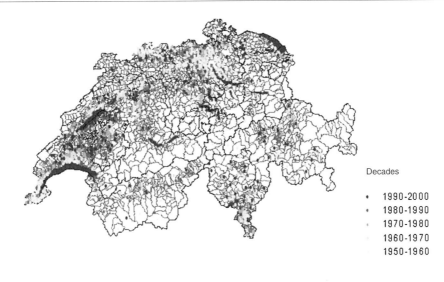

Decades

• 1990-2000
• 1980-1990
• 1970-1980
  1960-1970
  1950-1960

ACCESSIBILITY DEVELOPMENT

In this section we shift the focus from settlement patterns to question how accessibility has changed over the same period. Accessibility is defined as in Geurs and Ritsema van Eck, 2001:

> "...the extent to which the land-use transport system enables [groups of] individuals or goods to reach activities or destinations by means of a [combination of] transport mode[s]."

which is operationalized using the log sum term of a simple destination choice model:

$$A_i = \ln \sum_j O_j f(C_{ij})$$

where

A_j	is the Accessibility at point i
O_j	is the number of Opportunities at point j
$f(C_{ij}) = \exp(-\beta * c_{ij})$, where β is the distance weight, c_{ij} travel time between municipalitiy i and municipality j

Figure 4 shows road and railroad accessibility values for every municipality in Switzerland for 1950 and 2000. To show the differences between the localities clearly, no logs were taken. The parameter β of the generalised costs in accessibility formula is assumed to be 0.2 and kept constant over time. This value was estimated for 1960 and 1970 for Switzerland (Schilling, 1973) and recent literature also suggests a value in that range. For a discussion of other possible weighting functions see e.g. Kwan, 1998. It is realistic to assume that the parameter has changed over time reflecting the change in the residential locations and in the cost of travel relative to other household budget categories, such as housing and food, but the relevant estimates are not yet available.

Ideally one would calculate the travel times between individual street addresses, as there is a large amount of spatial variability in travel times and costs. This is currently not feasible due to the very large computing times required. Locations are therefore aggregated into zones with an associated center of gravity, which represents all activity opportunities and also is the source of all travel leaving the zones. In the road-based model the centroids of the Swiss zones (municipalities) have connectors to the next road network node with a speed of 15 km/h. In the railroad network model the connector from the centroid of a zone to the next station is a proxy for the way in which passengers reach the station. In the last 50 years several lines and therefore their stations have been closed. Hence, this development had to be considered for the connectors in the different decades. The connectors are classified by type of zones, distance and if the municipality has its own station or not. It is assumed that half of the passengers are using the urban public transport or walk and the other half come by bicycle, car or are taken by car there. The assumed speeds for the different connector types are the following: for Swiss municipalities with their own station 6 km/h airline (as-the-crow-flies) distance, for municipalities without their own station and under 10 km road distance to the station 12 km/h, for municipalities without their own station and over 10 km road distance 25 km/h, for major cities and airline distance under 1 km 6 km/h, for Swiss major cities and airline distance over 1 km 10 km/h. These speeds were checked against known bus timetables for a number of areas and decades (Bodenmann, 2003). The shortest time path travel times on the network between the municipalities were calculated using mean speeds that differ by link type and decade. For the year 2000 this compared well with the results of the assignment of the new national demand matrix to the same network (See Fröhlich and Axhausen, 2004).

Figure 4. Comparison of the accessibilities between road and railroad, between 1950 and 2000 respectively (variables: population; travel time) (no logs taken)

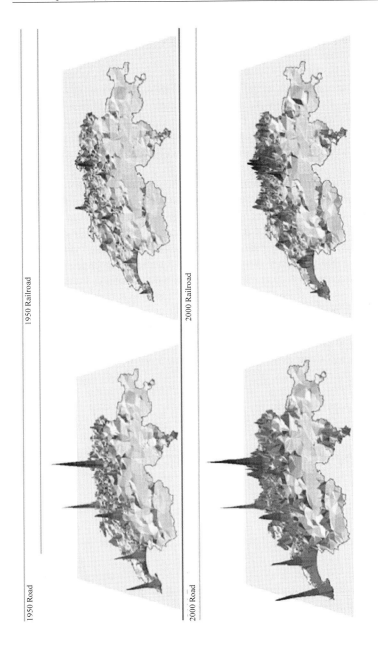

If we focus on road network accessibility in the year 1950 the major urban areas of Zurich, Bern, Basle and Geneva had a clear absolute accessibility advantage over the other parts of Switzerland. The only band of high accessibility is the Mittelland, the Swiss lowlands, spreading between Bern and Zurich. With the exception of the Rhône valley and southern Ticino, large parts of the mountain and alpine regions have very low accessibility values.

In 2000 the locations within the highest quintile of road network accessibility values are concentrated in a circular area around Zurich including the whole agglomeration and the northern part of Lake Lucerne. Around Bern, a cross-shaped area with municipalities of high accessibility can be seen, leading from Biel to Thun and from Solothurn to Fribourg. Around Lake Geneva, the distribution has changed from two main peaks, Geneva and Lausanne, to a more homogeneous appearance, with two additional peaks at Vevey and Nyon.

Focusing on the alpine range no similar development can be found. Those areas could not develop in the same way as the former rural areas the low lands in between the agglomerations. The map for the railroad accessibility shows in 1950 clear advantages of urban areas as well as generally low levels in rural and alpine areas. Comparing to road network accessibility the advantages are more concentrated on the major railroad axes (see for example the major railroad cross in Olten). In the year 2000 the pattern remained but on a higher level. Especially in the urban Zurich area, a more homogeneous appearance between town and countryside can be observed.

The investment into the motorway system since the 1960s has allowed the municipalities in the mayor agglomeration to improve their situation substantially, while the major cities were able to maintain their absolute, but not relative positions. The true peripheries have not benefited in relative terms.

ACCESSIBILITY AND ITS SPATIAL IMPACT

So far the expansion of population and the development of accessibility during the last decades were analysed separately. In this chapter those two aspects shall be linked. The question we raise now has two parts:

1. To what extent can we explain population change by changed accessibilities?

2. Can differences be seen between the different regions and if yes, where are they?

Multilevel modelling tries to combine an individual level representing disaggregate behaviour with a macro-level model representing contextual (in our case: spatial) variations in behaviour. The point of multilevel modelling is that a statistical model should explicitly recognise a hierarchical structure where one is present (Fotheringham, 2000). By focusing attention on the levels of hierarchy in a dataset, multilevel modelling enables the researcher to understand where and how effects are occurring. This approach has obvious appeal in our case, as the municipalities are grouped in cantons or can be classified by the location relative to the major centers. The formulation of the multilevel regression model is:

$$y_{ij} = \beta_{0ij} x_0 + \beta_{1j} x_{1ij}$$

where

$$\beta_{0ij} = \beta_0 + u_{0j} + e_{0ij}$$

and

$$\beta_{1j} = \beta_1 + u_{1j}$$

with:

y relative population growth
$\beta_{0,1}$ parameter
x_0 constant
x_1 absolute change of accessibility
u residual (departure of the j-th Canton's intercept (slope respectively) from the overall value)
e residual (departure of the i-th municipality's actual score from the predicted score)
i level 1 (municipality)
j level 2 (Canton)

In effect, instead of calculating one regression line, 23 regression lines are calculated, one for each Canton. Population change is explained by the change in accessibility (see Figure 5). The models were estimated using ML Win (Rasbash *et al.*, 2000).

We notice a strong link between population growth and accessibility. Nevertheless there are big differences between the Cantons.

If we focus on the regression results (Figure 6) an obvious pattern of intercepts and slopes can be seen. The intercepts of four Cantons (dark) are significantly above the average of all 23 Cantons, while five Cantons have a significantly steeper slope (light). Interestingly the Cantons with steep slopes have small intercepts and vice versa. See Figure 7 for the locations of these two groups: one urban (Basle, Geneva and Zürich plus suburban Aargau), the other covering the peripheral areas in the Alps and the Jura mountains. For the urban areas there is not much evidence that accessibility change is associated with strong population growth. In rural and alpine areas the situation is completely different: At a lower level absolute level, further accessibility growth is strongly associated with healthy population growth.
Comparing the response among the Cantons, there are no big differences between the impacts of railroad network and road network accessibility change. As seen before several rural Can-

tons show a significantly higher value for the slope and for the intercept (urban Cantons) respectively (Figure 8).

Figure 5. Road network accessibility and population growth 1950–2000 (all Cantons)

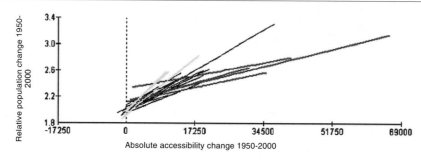

Figure 6. Intercepts and slopes of the regression lines (all Cantons)

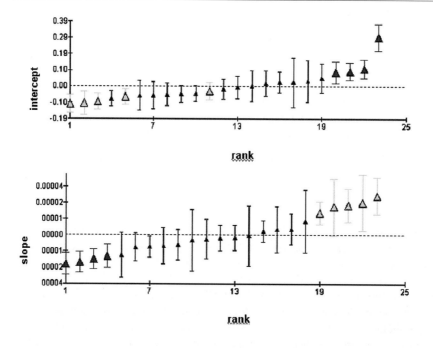

Figure 7. Cantons by values of the regression parameters

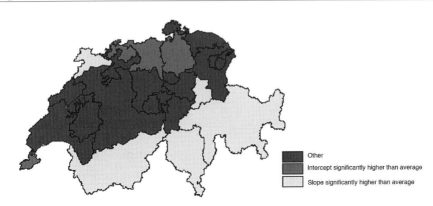

Other

Intercept significantly higher than average

Slope significantly higher than average

Figure 8. Railroad network accessibility and population growth 1950–2000 (all Cantons)

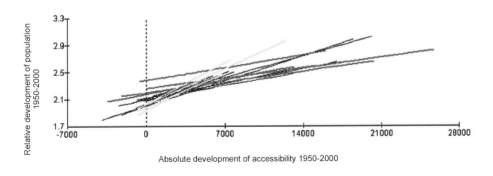

So far we focused only on accessibility and its impact on population development. For a wider set of models see Table 1 linking accessibility and spatial development. Three aspects are of special interest:

First, as shown before, a connection between accessibility development for both road and railroad and population development is obvious. Second the effect of population size in 1950 on the further population development of municipalities is negative (what we have already seen with the falling Gini Index from 1960 onwards). Smaller municipalities were in the last 50 years more successful in attracting population. Beneficiaries were the small, well accessible areas in the outskirts of the agglomerations. Third the number of employees reacts much quicker with aspect to accessibility than population does. Working places are much more sensitive to an accessibility change. During the last 50 years, areas with large accessibility advantages were able to gain a substantial amount of new workplaces.

Table 1. Switzerland: Accessibility and growth

dP =	0	S.E	1	S.E	2	S.E.	-2*loglikeihood
f(dA$_{road}$)	101.55	4.78	0.01	0.00			34395.17
f(dA$_{road}$, dA$_{railroad}$)	98.49	3.32	0.10	0.01	0.09	0.02	34300.79
f(P$_0$, dA$_{road}$)	103.61	4.96	-0.01	0.00	0.01	0.00	34359.36
f(P$_0$, dA$_{railroad}$)	133.95	4.97	0.00	0.00	0.02	0.00	34680.25
dE =							
f(dA$_{road}$)	204.20	42.16	0.04	0.01			50638.05

Where:

dP   Population change between 1950 and 2000

P$_0$ Population 1950

dE   Employment change between 1950 and 2000

dA   Accessibility change between 1950 and 2000

DISCUSSION AND OUTLOOK

The population in Switzerland has almost tripled since 1850. In the beginning the growth was limited to the large towns in the Mittelland. Industrialisation, reduction of agricultural employment, the transition to industrial mass production had concentrated the work places and thus the population. This process of concentration ended during the boom phase between 1960 and 1970. The Gini indices (all municipalities) show a striking trend reversal from 1970 onwards.

The massive increase of road accessibility can be explained with the construction of the Swiss motorway network during the last 50 years. Beneficiaries of the motorway network are not primarily the urban centers but the areas in between, where accessibility improved tremendously. Peripheral, alpine regions are less or not affected at all from this trend. In the same time period only a few new railroad lines were constructed. But with the implementation of synchronised timetables and with a higher utilisation of the railroad system, waiting times declined while connections improved which led to shorter travel times.

The income growth after World War II led to an individualisation and mass mobilisation of the society. Suddenly the car was affordable to a wide range of population groups. The building boom of single family houses, which grew continuously inside the agglomerations, was a consequence. That social change was followed by a massive extension of traffic infrastructure and road based accessibility increased enormously in the Swiss Mittelland. Figures 1 to 3 demonstrate this change in society over the decades: A migration into cities and a tremendous augmentation of population in the town centers is followed by suburbanisation. The results in Table 1 show that population does not react very quickly in an already dense transport system (which exists in these agglomerations (see Figure 7)) to further accessibility growth, work places are still responding to accessibility even in urban areas. In those areas work- and living places can be further apart than previously. Meanwhile workplaces were concentrating during the last 50 years on the inner circles of the five major agglomerations, while population migrated to the outskirts of those areas. The consequence is a general dispersion of the agglomerations, which seem to grow together (e. g. the agglomerations Basle and Zurich or the metropolitan areas Lausanne and Geneva) and hence longer travel times for commuters.

The development of the population in space and time is characterised by a continuous dispersion during the last five decades. Strongly connected to this development is increased land consumption, distances covered, spatial dispersion of traffic, and thus motorised individual traffic.

The mobility of people, the reduced impedance of space in the Swiss Mittelland led to more interactions. The short distances in Switzerland, the federal structures expedite those trends and led to an abolishment of the former division of town and countryside.

The paper reports for Switzerland the development of transport infrastructure and therefore accessibility and socio-economic variables consistent with a high spatial resolution and over a long span of time. Nevertheless the present work is only the beginning, as the causal relations between accessibility and spatial organisation are not trivial[1]. Useful models should be able to address the following:

- The calculation of accessibility requires an estimate of the generalised cost of travel. Is the development of travel time or the development of travel costs (e. g. gas prices, car prices) more suitable to explain spatial development?

- Does accessibility follow spatial and economical development or is the causal relation reversed? Can both effects occur? Are they synchronous or time lagged?

- What influence have other variables (e. g., starting conditions of a region, structural changes of the economy, competition situation on the world market, amount of taxes, subsidies etc.)?

- What impacts has the vicinity in time and space on the developments of neighbouring settlements? (In modelling terms: what impacts have spatial and temporal (error) correlations?)

ACKNOWLEDGEMENTS

The work is part of the project "Development of the Transit Transport System and its Impact on Spatial Development in Switzerland," conducted within the framework of Action 340 of the European Co-operation in Scientific and Technical Research (COST) consortium "Towards a European Intermodal Transport Network: Lessons from history."

REFERENCES

Aschauer, D. (1989). Is public expenditure productive?, *Journal of Monetary Economics*, **23(2)**, 177- 200.

Axhausen, K. W. (2004). Does road investment produce accessibility gains at a constant rate? A comment on Shirley and Winston (2004), *Arbeitsberichte Verkehrs- und Raumplanung*, **228**, Institut für Verkehrsplanung und Transportsysteme (IVT), ETH Zürich, Zürich.

Banister, D. and J. Berechman (2000). *Transport Investment and Economic Development*, UCL Press, London.

Boarnet, M. and A. Haughwout (2000). *Do highways matter? Evidence and policy implications of highways' influence on metropolitan development*, Departments of Urban and Regional Planning and Economics, University of California, Irvine.

Bodenmann, B. (2003). Zusammenhänge zwischen Raumnutzung und Erreichbarkeit: Das Beispiel der Region St. Gallen zwischen 1950 und 2000, MSC thesis, Institut für Verkehrsplanung und Transportsysteme, ETH Zürich, Zürich

Bökemann, D. and H. Kramar (2000). Auswirkungen von Verkehrsinfrastrukturmassnahmen auf die regionale Standortqualität, *Schriftenreihe*, **109**, Bundesministerium für Verkehr, Innovation und Technologie, Wien.

Eeckhout, J. (2004). Gibrat's Law for (All) Cities, *The American Economic Review*, **94(5)**, 1429-1451.

Fotheringham, A. S., C. Brunsdon and M. Charlton (2000). *Quantitative Geography*, The Cromwell Press Ltd, Trowbridge.

Fernald, J. G. (1998). Roads to prosperity? Assessing the link between public capital and productivity, *American Economic Review*, **89(3)** 619-638.

Fröhlich, Ph. and K.W. Axhausen (2004). Sensitivity of accessibility measurements to the underlying transport network model, *Arbeitsberichte Verkehrs- und Raumplanung*, **245**, IVT, ETH Zürich, Zürich.

Fröhlich, Ph., T. Frey, S. Reubi and HU. Schiedt (2003). Entwicklung des Transitverkehrs-Systems und deren Auswirkung auf die Raumnutzung in der Schweiz (COST 340):

Verkehrsnetz-Datenbank, *Arbeitsbericht Verkehrs- und Raumplanung*, **208**, IVT, ETH Zürich, Zürich.

Geurs, K.T. and J.R. Ritsema van Eck (2001). Accessibility measures: review and applications, *RIVM report*, **408505006**, National Institute of Public Health and the Environment, Bilthoven.

Haug, W. (2002). Räumliche und strukturelle Bevölkerungsdynamik der Schweiz 1990-2000 Bundesamt für Statistik, Neuchâtel.

Holtz-Eakin, D. (1994). Public sector capital and the productivity puzzle, *Review of Economics and Statistics*, **76(1)** 12-21.

Kesselring, H., P. Halbherr and R. Maggi (1982). *Strassennetzausbau und raumwirtschaftliche Entwicklung*, Paul Haupt, Bern.

Kwan, M. (1998). Space-time and integral measures of individual accessibility: A comparative analysis using a point-based framework, *Geographical Analysis*, **30(3)** 1991-216.

Lutter, H. (1980). Raumwirksamkeit von Fernstrassen, *Forschungen zur Raumentwicklung*, **8**, Bundesforschungsanstalt für Landeskunde und Raumordnung, Bonn.

Munnell, A. H. (1990). How does public infrastructure affect regional economic performance? *New England Economic Review*, **5**, 11-33.

Nadiri, M. I. (1998). *Contributions of highway capital to output and productivity growth in the U.S. economy and industries*, report prepared for the Federal Highway Administration Office of Policy Development, Washington, D.C.

ORL-Institut (1971). Landesplanerische Leitbilder der Schweiz, Schlussbericht Bd I-IV, Schriftenreihe zur Orts-, Regional- und Landesplanung, Nr. OA-D, Institut für Orts-, Regional- und Landesplanung, ETH Zürich, Zürich.

Rasbash, J., W. Browne, H. Goldstein, M. Yang, I. Plewis, M. Healy and G. Woodhouse (2000). *A user's guide to MLwiN*, Institute of Education, London.

Schilling, H. R. (1973). Kalibrierung von Widerstandsfunktionen, *Studienunterlagen*, Lehrstuhl für Verkehrsingenieurwesen, ETH Zürich, Zürich.

Shirley, C. and C. Winston (2004). Firm inventory behavior and the returns from highway infrastructure investments, *Journal of Urban Economics*, **55(2)**, 398-415.

Seimetz, H.-J. (1987) *Raumstrukturelle Aspekte des Fernstrassenbaus*, Geographisches Institut der Johannes Gutenberg-Universität, Mainz.

Tschopp, M., T. Frey, S. Reubi, P. Keller und K. W. Axhausen (2003). Raumnutzung in der Schweiz: Eine historische Raumstruktur-Datenbank, *Arbeitsberichte Verkehrs- und Raumplanung*, **165**, IVT, ETH Zürich, Zürich.

Vickerman, R. W. (1991). *Infrastructure and Regional Development*, Pion Limited, London.

Zipf, G. (1949). Human Behaviour and the Principle of Least Effort, Addison Wesley, New York.

NOTES

[1] See for example Aschauer, 1989; Banister and Berechman, 2000; Boarnet and Haughwout, 2000; Bökemann and Kramar, 2000; Kesselring, Halbherr and Maggi, 1982; Lutter, 1980; Seimetz, 1987; Vickerman, 1991).

Access to Destinations
D.M. Levinson and K.J. Krizek (editors)

CHAPTER 18

PARKING AND ACCESSIBILITY

Erik Ferguson

INTRODUCTION

Parking and accessibility each has been the subject of considerable prior research, but only rarely in conjunction with one another. The parking problem was first identified as such by McClintock (1930).[1] Walking distance (to and from available parking) was identified as a critical aspect of the parking problem by Eno (1942),[3] Mogren and Smith (1952), Burrage and Mogren (1957), Gittens (1965), and others, but has been downplayed or ignored in more recent studies.[4]

Accessibility, measured in terms of travel distance or time, has long played a critical role in models of location, land use, urban development, and travel demand. Linking origins to destinations (trip distribution) is fundamentally a function of accessibility, whether measured in terms of gravity, intervening opportunities, or any other evaluation criterion. Regional accessibility measures rarely take parking cost or availability explicitly into account, however, apparently assuming that parking is negligible as a factor limiting highway accessibility.

How important is the largely unobserved and unexplored relationship between parking and accessibility? Parking is critical to accessibility whenever a vehicle is used to make a trip, and the driver does not act as a chauffeur. Accessibility is critical to parking whenever drivers have more than one choice regarding parking options at their final destination. In modern suburban settings, an ample supply of free parking is critical to automobile accessibility. In traditional downtown areas, parking supply generally is more limited and more expensive, making accessibility critical in the location of parking facilities.

Key parking and accessibility concepts are defined in the first section of this paper. This is followed by a brief exposition on the policy environment in which parking and accessibility operate today, namely government regulation through local zoning ordinances. Travel demand

and parking are compared and contrasted. Parking characteristics are discussed in more detail. A simple parking choice model is proposed, which explicitly addresses tradeoffs between walking distance and parking price.

Parking policy is discussed in relation to activity scheduling, accessibility modelling, location and land use. A second model is introduced, which explicitly addresses tradeoffs between maximum walking distance and minimum parking requirements measured in terms of maximum building size, serving as a measure of centrality, density of development, and revealed accessibility. Policy and research recommendations follow the exposition of these complex interrelationships between parking, travel, accessibility and the location of activities across time and space.

Definitions

Parking refers to facilities used to store vehicles. There are two types of parking: on and off-street. Parking requirements vary depending on the number, type and size of vehicles to be stored. More vehicles require more space, as do larger vehicles.[5] Vehicles with much different operating or user characteristics sometimes require separate facilities. Individual parking spaces generally conform to a standard size, although smaller spaces for compact cars and larger spaces for oversized vehicles are not uncommon.

Parking requirements vary over time as a function of vehicle trip purpose. There are three major divisions of parking based on functional characteristics (McClintock, 1925; Mueller and Reber, 1930): stopping, standing and parking. Stopping refers to the loading and unloading of passengers or freight. Standing refers to vehicle stacking in queues, including drive-in[6] and drive-thru[7] applications. Parking refers to the short or long term storage of unattended vehicles. This paper focuses on parking requirements, as opposed to stopping and standing requirements. Stopping and standing require less total space, but need much better access than parking.

Accessibility may refer to persons, vehicles, or both. Personal accessibility is a function of personal abilities and disabilities, and is closely tied to the concept of mobility. An average person, drawn at random from the general population, experiences a typical level of accessibility, given a specific transportation system configuration.[8] The very young, the very old, and those who suffer from mental, physical or emotional handicaps may be less mobile, and therefore would experience lower levels of accessibility. Active, healthy individuals in the prime of life are often more mobile, and would experience correspondingly higher levels of accessibility. Income and education may operate in a similar fashion to influence personal accessibility.

Vehicular accessibility refers to mode-specific differences in accessibility. Vehicular accessibility varies as a function of both transportation system and vehicle characteristics. Highway accessibility usually is best on a regional basis, but may fall as a result of peak period congestion on heavily travelled roads. Transit accessibility usually varies more from one part of a region to another, depending on route coverage and system design. Bicycle and pedestrian accessibility typically operate over shorter distances, but may exceed transit or highway accessibility in selected high density, multiuse environments.

Zoning for Parking

Zoning for parking[9] has been identified as a major advantage for automobile traffic, and a corresponding constraint on transit use (Buchanan, 1963; Shoup and Pickrell, 1978; Shoup, 1995 and 1999; Willson, 1995 and 2000). Parking pricing has been suggested as a better way to allocate curb space (Vickrey, 1954; Shoup 1999), reduce traffic congestion (Roth, 1965; Kulash, 1974), and encourage commuters to consider alternatives to driving alone (Shoup, 1982; Shoup and Willson, 1992). Nonetheless, zoning for parking remains firmly entrenched in planning practice today (Ferguson, 2004).

Modern zoning ordinances may include some or all of the following provisions: number, size, and location of parking spaces, and exceptions to the rule. Minimum automobile parking requirements form the core of most local zoning ordinances. Maximum automobile parking requirements and minimum bicycle parking requirements are recent additions to a small but growing number of ordinances. Both of these policy innovations are still far from being ubiquitous in planning practice today (Davidson and Dolnick, 2002).

A majority of American cities with populations exceeding 10,000 zoned for parking by 1960 (Ferguson, 2003). It was necessary, even at that time, to specify both the number and size of required parking spaces, in order to ensure full regulatory compliance. Schrader (1964) compared the number and size of required off-street parking spaces in 20 exemplary zoning ordinances adopted between 1960 and 1963. Of 30 separate land use categories included in his study, only one (hospitals) had minimum parking requirements identified in all 20 of these ordinances. All 20 ordinances specified the minimum area required to adequately provide one parking space. The average parking dimensions required by zoning in 1964 were 2.6 meters (width) by 6.0 meters (length) by 2.1 meters (height). The average area required to provide one parking space in 1964 was 16.5 square meters, exclusive of the area required for aisles (internal circulation) and driveways (ingress and egress).

Today, typical required parking dimensions are 2.4 x 6.7 meters for parallel parking and 2.7 x 5.5 meters for angle and perpendicular parking, with minimum aisle widths of 3.7 to 7.6 meters for one-traffic traffic, and 6.7 to 7.6 meters for two-way traffic.[10] The average parking space today consumes a minimum of 30 to 32 square meters of space, which includes the area required for the parking space itself, and adequate aisles for internal traffic circulation, but nothing else.[11]

The location of parking is obviously important in terms of its effects on accessibility. Most zoning ordinances require that all parking must be provided on the same site as the building to be served. Many further specify a maximum allowable distance between required parking and the buildings or activities to be served. Zoned parking is rarely allowed off-site, and even then, never more than 100-200 meters from the building served. Exceptions may be allowed under extenuating circumstances, especially in traditional downtown areas. The fact remains that ample parking, located on site, within a short walking distance from the building served, virtually guarantees that automobile access will equal or more likely exceed comparable transit and pedestrian access for most new developments built in the U.S.

Exceptions to the number, size, and location of parking spaces required under zoning may be allowed under certain circumstances. Joint use and shared parking are related but distinct concepts. Joint use parking refers to parking areas shared by multiple buildings located on separate sites, typically in a downtown setting. Shared parking refers to parking areas shared by multiple activities located on a single site, typically in a suburban setting. Full or partial waivers of parking requirements may be offered as an incentive for downtown redevelopment, historic preservation, affordable housing, mixed-use development, transit-oriented development, or transportation demand management (TDM) program development. In lieu fees are sometimes required in exchange for reduced parking requirements.

PARKING AND TRAVEL DEMAND

The relationship between parking and accessibility is complex. It may be useful in explaining this relationship to pass it through the interpretative lens of travel demand modelling, to which both are related in a much more obvious and direct manner. The traditional four-step travel demand model includes four sequential steps: trip generation, trip distribution, mode choice, and route selection.

Measures of accessibility are commonly used in modelling trip distribution, while parking cost occasionally enters mode choice models as an explanatory variable. Trip generation may be modelled as a function of socio-economic factors (family income, vehicle ownership, etc.) and/or land use (residential, commercial, etc.). Trip generation is a function, at least indirectly, of the location of activities. The location of activities (and/or the distribution of land uses) generally is modelled using some type of gravity formulation, which is fundamentally predicated on measures of accessibility. Parking often is cited as a critical element in real estate development decisions, or more generally the location of activities (e.g., limited or high cost parking as a constraint, and ample or free parking as a facilitating factor).

Multinomial logit regression may be used to model mode choice. Market shares for various modes of transportation are then estimated as a function of the generalized cost of travel, including some or all of the following factors: modal attributes (time, cost, etc.), user attributes (income, auto ownership, etc.), and system characteristics (capacity, etc.).

Travel cost generally is limited to out-of-pocket expenses such as bus fares, road tolls and parking fees. Where roads and parking are provided free of charge, only public transportation realizes a penalty for out-of-pocket travel expenses.[12] Travel time may be separated into in-vehicle travel time (IVTT) and out-of-vehicle travel time (OVTT), with IVTT generally imposing a lesser penalty than OVTT on modal preferences.

Consider the following personal transportation modes: private automobile, public transit, and non-motorized. In standard travel demand models, IVTT is estimated for the first two modes, but not the last one, which is exposed to the elements throughout the trip.[13] OVTT is measured for the last two modes, but not the first one. Motor vehicle users do experience some OVTT, however, depending on the actual distance between their chosen parking spaces and their final origins and destinations. The implicit assumption in most travel demand models is that motor vehicle OVTT is negligible, at least in relation to public transit and/or non-motorized OVTT. This assumption may be correct in many cases (e.g., virtually all U.S. suburban developments built after 1950), but it cannot be correct in all cases, and is therefore an egregious example of systematic model misspecification on a global (or regional) scale.

Travel time and cost (and their multiple variations and extensions) are the principal components of most measures of accessibility and most multinomial mode choice models as well. Thus, mode choice is in fact primarily a function of modal accessibility, with or without parking availability and cost included as an additional operational constraint on motor vehicle travel.

If looking for a parking space requires any time at all, route selection may be affected by the availability and cost of parking, consumer knowledge of such parking availabilities and costs, and the use of such information in choosing exactly when, where, and how to make a final approach through a specific road network to any particular activity-based destination.

These are just a few examples of the extent to which both parking and accessibility are embedded in virtually all aspects of travel demand, and should be incorporated into models intended to capture critical elements of travel behaviour and the travel choice mechanisms associated with such behaviour.

Parking Characteristics

Parking may not appear in travel demand models as often as it should, but parking has itself been modelled in many different ways. Models of parking demand are more common than models of parking supply, perhaps because parking supply is more subject to government oversight and regulatory control. On-street parking is only sufficient to meet a small part of the overall demand for parking, and is heavily regulated as a result. Most private parking is mandated by local zoning ordinances, and is thus insulated, wholly or partially, from the direct influence of market forces. Public off-street parking frequently is provided at cost in response to localized parking shortages, but makes up only a small part of total parking supply.

The demand for parking is a *doubly indirect* demand, in that it arises solely from the following dually contingent circumstances: the need to link spatially separated activities (travel demand derives indirectly from activity and/or land use demand); the need to store vehicles used in travel on a temporary basis (parking demand derives indirectly from travel demand)[14]

Parking spaces and parking facilities (lots, garages) may be treated as unitary objects (non-differentiable bulk commodities) for the purposes of analysis, or grouped into categories based on user, vehicular, modal, or temporal restrictions: personal, vehicular, modal, and temporal.

Personal restrictions include spaces reserved for specific individuals or classes of individuals (e.g., handicapped parking). Vehicular restrictions include spaces reserved for larger, smaller, or specialized vehicles (e.g., emergency vehicles). Modal restrictions include spaces reserved for carpools and vanpools (a hybrid category, combining personal and vehicular attributes). Temporal restrictions include hourly, daily, weekly, and seasonal variations in parking rules and regulations, particularly with respect to curb parking.

The desirability of any particular parking space or facility, holding user, vehicular, modal, and temporal restrictions constant, is a function of the following factors: access distance and/or time, out-of-pocket parking fees (if any), and amenity values.

The single most important of all of these factors arguably is the first.[15] Holding all other factors constant, the nearest parking space to one's final destination is the most desirable one. Other factors may enter the equation, depending on tradeoffs between parking accessibility, cost, safety and security, and cover. Avoidance of other parked cars and proximity to the exit are positive features of increasing walking time or distance that are attractive to a small minority of parking users, including those with particularly nice vehicles, and those who are in a bigger hurry to get away at the end of the day.

On-street parking makes up only a small part of total parking supply, but is heavily regulated, largely because on-street parking frequently is closer to building entrances than parking spaces located in large parking lots or structures. Temporal restrictions of one form or another are used almost universally in the regulation of on-street parking. Parking meters frequently are used to control on-street parking where demand is particularly high.

Public parking also makes up only a small part of total parking supply, and is highly concentrated in a few specific areas, namely traditional downtown areas of older and larger urban communities. Public parking is more likely than private parking to take the form of multilevel garages and to impose parking charges because of the high density locations where it is most likely to be found. Public parking may be provided by municipalities, parking authorities, commercial operators or some combination of all three.

Private parking is the most common type of parking found in most communities, and constitutes the vast majority of all parking facilities in existence today. Private parking may be provided in direct response to market demands, but more commonly is provided in response to local regulatory requirements in zoning for new development.

Parking space requirements in local zoning ordinances are often considered to be the single greatest contributor to low density development and urban sprawl, but this may not always be true. Of equal importance from the perspective of accessibility is the common requirement that all required parking be provided on the same parcel of land on which the building to be served is constructed.[16] This restriction virtually guarantees that all required parking will be a short walking distance from the building. In actual practice today, most new buildings are set back from the street, with most of the required parking located between the building and the street. This further guarantees that all required parking will be a shorter walking distance to the building than any comparable transit or pedestrian access from the street might be.[17]

Parking demand may be further decomposed into the following major elements: location, duration, accumulation, and turnover. Parking location and duration are related, in that the shorter the duration of parking and associated activities is, the more proximate the location of parking must be to attract vehicles, drivers and/or passengers. Parking accumulation and turnover likewise are related, in that the more rapidly parking spaces turn over, the lower the resulting parking accumulation will be.

Parking generation refers to maximum parking accumulation, generally without reference to alternative modes of transportation, transit or pedestrian accessibility, parking charges or access restrictions, etc. Parking generation thus provides an upper limit on the amount of parking necessary to serve the needs of particular activities and land use types.

In general, the following rules govern observed variations in parking demand:

1. The more time spent in any particular activity (or, more generally, the greater the economic or social benefit associated with that activity), the longer the associated parking duration must be, the greater the associated parking accumulation may be, and the lower the associated parking turnover will be. Parkers generally are willing to walk longer distances and/or pay higher total parking fees in order to access parking facilities serving such activities.

2. The less time spent in any particular activity (or, more generally, the lower the economic or social benefit associated with that activity), the shorter the associated parking duration must be, the lesser the associated parking accumulation may be, and the higher the associated parking turnover will be. Parkers generally are unwilling to walk longer distances, but may be willing to pay higher marginal parking fees in order to access parking facilities serving such activities.

If a larger share of the total parking supply were publicly provided at competitive market rates, parking could be treated as an activity in and of itself, with its own accessibility needs and requirements. In certain situations, such as older, more traditional, high density commercial centres, this is at least partially the case.

Shoup (1999) illustrates the efficacy of parking pricing through development of a model of parking choice that includes total walking time and marginal parking price as its principal explanatory variables. According to Shoup's model, the optimal walking distance to parking under market pricing might be as high as 400 to 800 meters.

Most parking today is provided free of charge, negating the potential of Shoup's market-based approach to parking provision. HRB (1971) showed that walking distance to parking increased with parking duration, itself a function of activity duration. WSA (1965) showed that walking distance increased with city size, itself a reasonable proxy for parking cost. The most illuminating results come from BPR (1956), however, which showed that prior to the implementation of zoning for parking on a widespread, long term basis, walking distances varied by parking location as shown in Table 1. Among curb parkers, those who parked illegally walked the shortest distances, followed by those who paid to park and those who parked for free legally. This phenomenon clearly adheres to the Shoup model, which predicts a trade-off between walking distance and parking payment.[18]

Table 1. Mean walking distance by city size and parking type, 1956

City Population	N Cities	Mean Walking Distance (Meters)						Percentage of Parkers Who	
		Curb			Off-Street				
		Illegal	Pay	Free	Pay	Free	All		
<25k	4	42				48	61	11.2%	
25-50k	8	56	93	128	150	81	87	8.0%	3.9%
50-100k	3	62	94	138	179	78	105	6.2%	5.2%
100-250k	6	59	118	131	201	86	130	7.0%	8.9%
250-500k	4	82	161	147	260	111	177	9.3%	29.9%
500k+	7	62	157	187	244	107	172	13.6%	24.8%
Mean		61	125	146	207	85	122	9.2%	14.5%

Source: Gittens 1965
Data: BPR 1956

Note: these results are based on a compilation of downtown parking studies drawn from across the U.S.

Table 2. "Acceptable" walking distance (meters) by parking access and level of service

Parking Access[3]	[Pedestrian] Level of Service (LOS)[1,2]			
	A	B	C	D
Enclosed pedestrian sidewalk[4]	300	600	900	1,200
Covered pedestrian sidewalk	150	300	450	600
Uncovered pedestrian sidewalk	120	240	360	480
Through surface parking lot	105	210	315	420
Through parking garage[5]	90	180	270	360

Source: Modified from Smith and Butcher (1994)

[1] The LOS concept employed here is borrowed from civil engineering principles applied to streets, highways and traffic intersections.
[2] LOS in traffic engineering is based on traffic delay, with A = no delay (free flow traffic) and F = (unacceptable) traffic gridlock.
[3] Smith and Butcher refer to these as "parking conditions", but these are in fact pedestrian, not vehicular, access amenity attributes.
[4] Sidewalks provide an exclusive pedestrian right-of-way, whereas generic parking facilities operate under mixed flow traffic conditions.
[5] Parking garages generally have narrower traffic lanes and shorter lines-of-sight, providing fewer pedestrian amenities as a result.

Those who paid for parking off-street walked longer distances than any curb parkers, suggesting that commercial off-street parking facilities were farther away, but cheaper, than comparable curb spaces in 1956. Of greatest interest today is the fact that those who parked off-street for free walked shorter distances than those who paid for off-street parking, an apparent contradiction of the Shoup model. The only way this could happen is if "free" off-street parking was somehow different than the paid kind, especially in terms of its proximity to the buildings, activities, and land uses served. One plausible explanation is that zoned parking was preferentially provided at no cost to the user even as long ago as 1956. These data provide some idea of how ample parking under zoning transformed American cities, by eliminating parking pricing as a factor in travel decision making, and shortening automobile access times through shorter walking distances as well.

Smith and Butcher (1994) propose level-of-service (LOS) standards for parking facilities based on parking conditions (pedestrian amenities associated with parking access) and walking distance to the activity served (an obvious measure of parking accessibility), with higher LOS standards associated with shorter walking distances and improved pedestrian amenities (Table 2). Their proposed standards are conceptually intriguing, but are backed up by very little hard data on actual parking access patterns, and do not reference parking pricing at all.

A PARKING MODEL

For any vehicle trip in which the vehicle must be stored, a travel party may choose to pay for parking closer to the destination, or park free further away (Shoup, 1999; Arnott and Inci,

2005).[19] The maximum distance a travel party is willing to walk in order to avoid paying for parking is:

$$D_w \quad = \quad C_p * (T_a + T_w) / (C_w / S_w) * 1000$$

Where

D_w	=	maximum walking distance (meters)
C_p	=	marginal parking cost (\$/hour)
T_a	=	time spent engaged in activity a (hours)
T_w	=	time spent walking to and from activity a (hours)
C_w	=	marginal value of walking time (\$/hour)
S_w	=	average walking speed (kilometers per hour)

And

$$T_w \quad = \quad 2 * N * D_w / S_w$$

Where

N =   number of persons in the travel party[20]

Substituting for T_w

$$D_w \quad = \quad C_p * (T_a + 2 * N * D_w / S_w) / (C_w / S_w) * 1000$$

Rearranging terms

$$D_w \quad = \quad (C_p / C_w * T_a * S_w) / (1 - 2 * N * C_p / C_w) * 1000$$

Willingness to walk to avoid paying for parking increases with marginal parking rates, time spent engaged in activities (at a single location), and average walking speed. Willingness to walk to avoid paying for parking decreases with the average value of walking time and the number of persons travelling together in one vehicle.

The model is illustrated for one person travelling alone in Figure 1, assuming an average walking speed of 5 kilometers per hour, and an average value of walking time of \$15 per hour. In this example, a traveller may be willing to pay up to \$5 per hour for parking, but only for activities of very short duration. Figure 2 illustrates model results for a party of two with the same average walking speed and value of walking time. In this example, the travel party will not pay more than \$3.75 per hour for parking, even for activities of very short duration.

Figure 1. Marginal Parking Cost Vs. Walking Distance by Activity Duration for One Person Travelling Alone

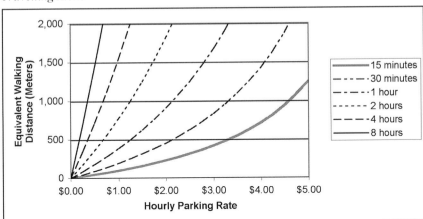

Figure 2. Marginal Parking Cost Vs. Walking Distance by Activity Duration for Two Persons Travelling Together

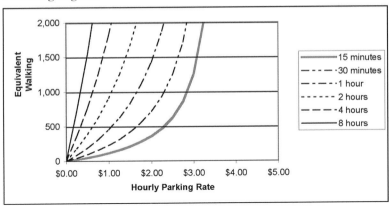

Many parking facilities charge a fixed daily rather than a variable hourly rate for parking. In such cases, parking duration does not affect parking cost directly, and

$$D_w = (C_p / C_w * S_w) / (1 - 2 * N * C_p / C_w) * 1000$$

Where

C_p = total parking cost ($/day)

Figure 3 illustrates model results for parties of different size, irrespective of activity duration, with the same average walking speed and value of walking time as before. Since parking fees in excess of $3.00 per day are not uncommon in large activity centres, especially in older, higher density urban areas, the value of walking time must be higher than $15.00 per hour, or free parking is not a readily available option in such areas.

Marginal pricing may discourage long term parking, especially where free parking is within reasonable walking distance. Fixed pricing reduces this threat, but may discourage short term parking.

Parking and Activity Scheduling

Consider the value of choosing to engage in an activity at a particular location in time and space as a function of rationally expected benefits and costs:

Figure 3. Fixed Parking Cost Vs. Walking Distance by Party Size

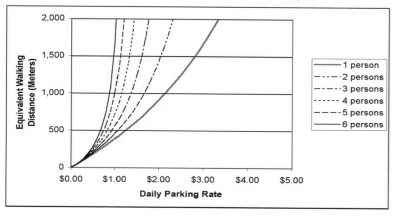

$$V_a \quad = \quad f(B_a - C_a)$$

Where

V_a = Net benefit or value of activity a
B_a = Benefit of activity a
C_a = Cost of activity a
And

B_a = $f(U_a, T_a)$
C_a = $f(C_t, C_x)$

And

U_a = marginal utility of engaging in activity a ($/hour)

T_a = time spent engaged in activity a (hours)
C_t = generalized travel cost
C_x = all other activity-related costs (entry fees, supplies and materials, etc.)

The marginal utility of engaging in any activity a may be either fixed (working for wages) or variable (most other activities), but is always positive for $0 < t < T_a$. B_a increases monotonically with T_a. T_t and C_t may vary slightly as a function of time (time of day, day of week, seasonally, etc.) due to variable traffic and parking conditions, but are generally fixed in relation to any specific activity length, T_a. Other activity-related costs (not discussed here) may include admission fees (for some recreational and entertainment activities), tools of the trade (for some work activities), and housing costs (for some home-based activities), among many others (Kurani and Lee-Gosselin, 1996).

Figure 4. Traditional Time-Space Prisms

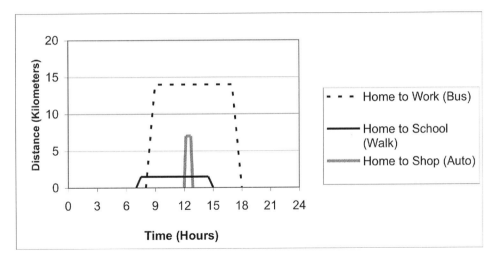

Figure 4 shows traditional time-space prisms for various types of activities. Horizontal lines represent activities over time. Slanted lines represent travel. The slope of each line represents the speed of travel. The height of each line represents distance travelled. Shopping trips typically involve more travel per time spent engaged in the activity than most other types of trips.

Figure 5 zooms in on the parking portion of a trip. Searching for a parking space may require anywhere from several seconds to several minutes, and typically requires slowing the vehicle down to avoid missing a good spot. Walking to the final destination from parking also requires a certain amount of time, and takes place at an even slower pace.

The decision to engage in any activity *a* may be viewed as a probabilistic function of its expected value in relation to all other possible activities. Activity choice and duration may be influenced by access cost (measured in time and money), which includes parking cost for auto travel choices (Jara-Diaz and Guerra 2003). Fixed parking fees may influence the choice and/or location of activities more than marginal parking fees do, especially with respect to shorter activities. Marginal parking fees may influence the duration of activities and parking location decisions more than fixed parking fees do.

The effect of various parking pricing strategies on total parking cost is shown in Table 3. Fixed parking prices are most attractive to long term parkers. Variable parking prices are most attractive to short term parkers. Hybrid parking prices are most attractive to medium term parkers. Daily parking price maximums ensure that people whose activities take longer than expected (including those who leave their cars unexpectedly overnight) are not unduly penalized by marginal parking rates. Early bird specials are designed to attract long term parkers on a preferential basis, signalling a competitive market. Significant price discounts on monthly or annual parking permits versus average daily charges serve the same basic purpose.

Figure 5. Parking in Time and Space

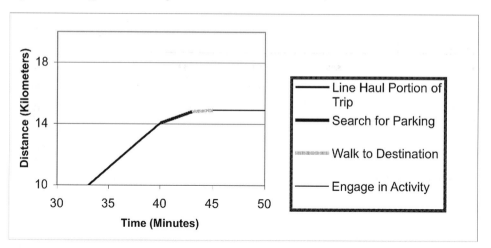

Activities vary in terms of scheduling and schedule flexibility across time (start time, end time, and duration) and space (location and travel). Schedule flexibility varies with activity type. Some activities, such as work, childcare and education are associated with rigid scheduling requirements. Other activities, including shopping, recreation and entertainment, are associated with more schedule flexibility. In fact, however, most activities exhibit a wide range of scheduling requirements, from very rigid to very flexible (Table 4).

The more rigid the schedule, the less tolerant a traveller will be to any and all sources of uncertainty, including the need to search for a parking space, and the time required to walk from any particular parking space to their final destination. More rigid schedules should be associated with a greater willingness to pay for more proximate parking. More flexible schedules should be associated with a greater interest in variable versus fixed parking prices. More uncertain schedules (e.g., frequent overtime) should be associated with a greater desire for capped parking rates as a form of risk management.

Some additional perspective on how maximum acceptable walking distance to parking varies by activity type is shown in Table 5. Various activity-user interfaces are compared in terms of maximum acceptable walking distance, activity duration and activity frequency, all rated on a relative scale from high to low. In general, it seems that maximum acceptable walking distance increases with activity duration, and decreases with activity frequency. Maximum acceptable walking distance also decreases with user income or social status, both of which are proxies for the marginal value assigned to time by specific individuals or groups.

Table 3. Parking pricing strategies

	Parking Rate Structure				
	Fixed Price	Variable Price		Hybrid	
		No Cap	With Cap	No Cap	With Cap
Parking Duration (Hours)	X Dollars per Day	Y Dollars per Hour		X Dollars for the First Hour, Y Dollars Thereafter	
1	$10	**$2**	**$2**	$4	$4
2	$10	**$4**	**$4**	$5	$5
3	$10	**$6**	**$6**	**$6**	**$6**
4	$10	$8	$8	**$7**	**$7**
5	$10	$10	$10	**$8**	**$8**
6	$10	$12	$10	**$9**	**$9**
7	**$10**	$14	**$10**	**$10**	**$10**
8	**$10**	$16	**$10**	$11	**$10**
9	**$10**	$18	**$10**	$12	**$10**
10	**$10**	$20	**$10**	$13	**$10**

Note: the lowest parking prices for any given parking duration are shown in **boldface** type.

Table 4. Schedule flexibility by activity type

Activity		Schedule			
		Very Rigid	Rigid	Flexible	Very Flexible
Employment	Temporal	Wages	Salaries	Paid overtime	Unpaid overtime
	Spatial	National headquarters	Regional office	Satellite office	Work at home
	Functional	Production	Customer service	Marketing and sales	Research and development
Education	Temporal	Lecture	Lab	Study group	Independent study
	Spatial	Graduate school	College	Secondary school	Elementary school
	Childcare	Preschool	Daycare	Babysitter	Home alone
Shopping	Temporal	Formal auction	Special sales event	With friend	By self
	Spatial	Real estate	Durable goods	Nondurable goods	Convenience goods
Recreation	Temporal	League, tournament play	Informal match play	Casual pickup game	Solo workout
	Spatial	Field	Court	Street	Backyard
Entertainment	Temporal	Live performance	Television broadcast	Theatrical film release	Video rental
	Spatial	Live performance	Theatrical film release	Video rental	Television broadcast
Personal business	Temporal	Site visit	Preset appointment	Same day appointment	Walk-in appointment
	Spatial	Specialized facility	Generic store, office	Telephone call	Internet access

Source: loosely based on Olaru and Smith (2003)

Potential behavioural responses to parking policy changes are shown in Table 6. Parking price increases or supply reductions may induce changes in travel behaviour, activity scheduling or both. Travel behaviour responses are limited to changing where one parks or the mode of travel one uses to access a particular destination. Activity scheduling responses include changing the start, end (duration), and/or location of an activity, or eliminating the activity altogether. The most likely responses to small shifts in parking policy are no change at all, small changes in parking location, and small changes in activity schedules. Major shifts in parking policy may produce any or all of the behavioural responses shown in Table 6.

Table 5. Maximum walking distance by activity-user interface

Activity-User Interface	(Maximum) Walking Distance	(Typical) Activity Duration	(Typical) Activity Frequency
Airports - annual vacation	very high	very high	very low
Airports - occasional business travel	high	high	low
Airports - frequent flyers	medium	medium	medium
College - administration	low	varies	high
College - faculty and staff	medium	varies	high
College - students	high	varies	high
Employees - full-time	medium	high	high
Employees - part-time	low	medium	medium
Entertainment - major cultural events	very high	varies	low
Entertainment - movies	medium	medium	medium
Recreational - daily workout routines	low	low	high
Recreational - major sporting events	very high	medium	low
Religious - regular services	medium	medium	medium
Religious - holiday services	high	high	low
Residential - apartments, condos	medium	high	high
Residential - single family, attached	low	high	high
Residential - single family, detached	very low	high	high
Restaurants - fast food	low	very low	high
Restaurants - general	medium	low	medium
Restaurants - quality	low	medium	low
Retail - convenience store	very low	very low	high
Retail - grocery store	low	low	medium
Retail - shopping mall	medium	medium	low
Services - medical/dental	low	medium	low
Services - professional	low	medium	medium
Special - emergency services	very low	varies	low
Special - loading and unloading	very low	varies	high
Special - people with disabilities	very low	varies	high

Table 6. Behavioral response to parking policy changes

Increase Price

	Response Category	Alternatives	Explanation
1	Pay increased price	No behavioral change	The original parking space or facility is worth the increased price
2	Change parking location	Legal or illegal	Legal choices are socially desirable, illegal choices represent spillover parking
3	Change mode of travel	Alternative modes of travel	The marginal value of auto (or bicycle) travel must be low for this to happen
4	Change activity start time	Earlier or later	Either to take advantage of parking pricing that varies by time of day, if applicable
5	Change activity duration	Shorten or lengthen	Shorter to reduce total cost under marginal pricing, longer to reduce average cost under fixed pricing
6	Change activity location	Closer or farther	Closer to reduce time cost of activity-travel combinations, farther to reduce monetary cost
7	Eliminate activity altogether	Choose another activity or do nothing	The marginal value of the activity must be low for this to happen

Reduce Supply

	Response Category	Alternatives	Explanation
1	Increase parking search time	Circulate or camp out	Circulate to find any spot, or wait for a spot to open up at a particular location
2	Change parking location	Legal or illegal	Same as above
3	Change mode of travel	Alternative modes of travel	If there is no place to park, a mode change may be required
4	Change activity start time	Earlier or later	Either to avoid peak parking demand as capacity is reached
5	Change activity duration	Shorten or lengthen	Only indirectly as a result of start time changes combined with rigid scheduling requirements
6	Change activity location	Closer or farther	Either to reduce time cost of searching for a parking space
7	Eliminate activity altogether	Choose another activity or	Same as above

Reduce Price

	Response Category	Caveats	Explanation
1	Opposite increase price	As long as supply is sufficient	Overpricing can be as harmful as underpricing
2	Same as reduce supply	If demand approaches capacity	Underpricing leads to traffic and/or parking congestion

Increase supply

	Response Category	Caveats	Explanation
1	Opposite reduce supply	If demand approaches capacity	Undersupply can be as harmful as oversupply
2	No effect, or harmful effect	If capacity is already sufficient	Excess supply may result in reduced pedestrian access for all modes, including auto

PARKING AND ACCESSIBILITY

A general model of accessibility is:

$$A_i = \sum O_j f(C_{ij})$$

where

A_i	=	a measure of accessibility for object i (a person, place or thing)
O_j	=	a measure of opportunities available to object i at object j
C_{ij}	=	a measure of separation (generalized cost) between object i and object j

Accessibility is a relative measure, and cannot be estimated directly. It can be estimated indirectly as a function of trip attractions between activities across a transportation network (Harris 2001). Accessibility increases with the number of opportunities available at particular locations, and decreases with the generalized cost of linking separate locations through the medium of transportation technologies.

Opportunities are reciprocal in nature. Workers need access to jobs. Employers need access to the labour market. Shoppers need access to stores. Retailers need access to customers. Pupils need access to schools. Colleges need access to students. And so on. Opportunities may be measured in terms of travel (person trips, vehicle trips, mode choices) or activities (purpose of trip, activities engaged in, activity outcomes). For example, variations in retail opportunities might include the number of shoppers attracted (person or vehicle trips, with or without purchases), the number of dollars spent by shoppers (total, per shopper, or as a percentage of shopper income), and retail income generated (gross and/or net mall or store revenue; store profitability, total or by retail category).[21]

Generalized cost functions in accessibility models may take many different forms, the two most common being the gravity form (x^{-b}) and the exponential decay form (e^{-x}). The exponential decay form is preferred today, given its association with the multinomial logit model used in discrete choice modelling (Krizek, 2003).

The basic form of a generalized cost function is:

$$C_{ij} \quad = \quad f(T_{ij}, M_{ij}, X_{ij})$$

Where

T_{ij}	=	Time costs
M_{ij}	=	Monetary costs
X_{ij}	=	All other costs

Table 7. Generalized travel costs by mode of travel

Mode	Time		Information (Knowledge)	
	In-Vehicle	Out-of-Vehicle	Alternative Routes	Variable Conditions
Auto	Line haul	To and from parking	Streets, highways	Traffic, weather
Transit	Line haul	To and from stop, station	Routes, schedules	Traffic, weather
Bicycle	Line haul	To and from parking	Streets, bike paths	Traffic, weather
Walk	n/a	To and from destination	Streets, sidewalks	Traffic, weather
Parking	Search time	To and from destination	Facilities, spaces	Occupancy, rates

Mode	Money		Transaction	
	Variable	Fixed	Intramodal Transfers	Intermodal Transfers
Auto	Fuel, tolls	Vehicles, facilities	Wait time	Time, money
Transit	Fares, passes	Vehicles, facilities	Wait time	Time, money
Bicycle	n/a	Vehicles, facilities	n/a	Time, money
Walk	n/a	Facilities	n/a	Time, money
Parking	Fees, permits	Facilities	n/a	Shuttle service

Mode	Amenity (Actual, Perceived)		Uncertainty (Risk)	
	Safety, Security	Comfort, Convenience	Weather	Traffic
Auto	Accidents, crimes	Privacy, luxury	Minor	Major
Transit	Accidents, crimes	Hands free	Medium	Medium
Bicycle	Accidents, crimes	Health	Major	Minor
Walk	Accidents, crimes	Health	Major	Minor
Parking	Accidents, crimes	Cover	Minor	Major

Generalized travel costs include time, money and any and all other factors that may influence individual travel decisions. Other costs may include amenity values, information and transaction costs, and risk and uncertainty. Representative examples of generalized cost components are shown for auto, transit, bicycle and walk modes of travel in Table 7. There is some duplication across modes and cost categories, but each is important enough to warrant separate consideration. Time and money costs are self-explanatory. Amenity values are positive or negative aspects of travel modes, which do not translate directly into time or money costs.

Information costs assume limited consumer knowledge of available transportation options. Transaction costs include time, money and other costs associated with intramodal and intermodal transfers. Intramodal transfers include carpool and vanpool assembly, bus transfers, train changes, etc. Intermodal transfers include trips that involve the use of more than one transportation technology for the line haul portion of the trip, such as rail to bus, auto to rail, bike to bus, etc. Uncertainty includes the risk associated with predicting travel times and costs under variable operating conditions.

Generalized parking costs are compared with generalized travel costs in Table 7. Time spent searching for a parking space after a destination (or its near vicinity) has been reached is in-vehicle time. Time spent walking between parking and activities is out-of-vehicle time. Parking fees are parking costs. Amenity values associated with parking may include covered vs. uncovered parking, lighting, drainage, visibility, surveillance, etc. Information costs can be significant in choosing where to park, especially in competitive markets with high demand, limited supply and variable parking rates (Sun *et al.*, 2005). Remote parking facilities with shuttle service, carpool park-and-ride lots and transit parking facilities are examples of parking transaction costs.

Generalized parking costs may be combined with generalized auto or bicycle costs to provide a more balanced estimate relative to generalized transit costs. Generalized parking costs may also be considered in isolation. In this case, local access to parking is the thing measured, rather than regional access to the highway system.

Parking opportunities may be combined with generalized parking costs in measuring the parking accessibility of different land uses or locations. Parking opportunities may be measured at the aggregate level as the number of parking spaces in each parking facility located within a particular study area. Parking accessibility increases with the number of parking spaces available, and decreases with generalized costs associated with such parking opportunities.

The desirability of particular parking spaces or facilities may be measured as a function of access to activities that require (or at least may use) such parking. In this case, traditional measures of land use opportunities (residential, commercial, industrial, etc.) are combined with generalized parking costs to determine the land use accessibility of parking facilities (a proxy for parking needs, parking demand, and/or commercial parking market potential) at a given location.

Parking influences accessibility directly and indirectly. Parking affects accessibility directly in terms of the relative ease with which activities can be linked through the intermediate office of a place to store vehicles used to transport people or freight across a transportation network. Parking affects accessibility indirectly in terms of consumer attitudes and opinions regarding the quality, quantity and cost of parking associated with activities at particular locations across time and space.

Generalized parking costs may be treated as a subcomponent of generalized travel costs associated with those transportation technologies that require parking facilities in order to function properly. In an automobile-oriented society, where a large percentage of the population is automobile-dependent (or transit-deprived), there is little loss of generality in developing a model of parking accessibility for captive automobile users. In this case, the accessibility of a specific type of activity (shopping, for example) may be viewed as a function of parking availability and/or cost alone. Opportunities in such a model include all available parking spaces within a reasonable walking distance. Maximum walking distance increases with out-of-pocket costs in a closed parking system, but decreases in an open system.[22]

Location and Land Use

Local zoning ordinances typically specify the number, size and location of parking spaces required for all new developments within the community. These rules virtually guarantee that person trips attracted to new developments will be more accessible via automobile than by any other means. When parking is provided free of charge, and located a shorter walking distance from the street than any other mode, a clear advantage is obtained.

The value of this accessibility advantage will vary depending on several factors: walking distance, walking speed, and value of time. In general, the value of walking time to parking is smaller than the opportunity cost (potential market value) of parking space consumed (Shoup 1999). Average walking distances to parking in downtown areas vary from as low as 60 meters in the smallest communities to more than 275 meters in the largest ones (Weant and Levinson 1990, Box 1993).

The average walking distance to parking in the U.S. is about 120 meters. Assuming an average walking speed of 5 kilometers per hour and an average value of walking time of $15 per hour, the average value of one round trip to and from parking is $0.36. Because most parking is provided free of charge, walking distance is the sole determinant of parking value in most communities.

The average cost of providing one commercial parking space in a surface lot is on the order of $500-$2,000 per year, which includes the opportunity cost of land, plus amortized capital, operating, and maintenance expenses. Assuming the parking space is occupied continuously throughout the year, the average value of such a parking space is $0.06-$0.23 per hour. Most parking spaces are used considerably less than all of the time. Parking spaces in high demand areas generally are occupied only 25-50% of the time, and in areas with low demand or excessive supply, even lower occupancy rates may be expected. The market value (opportunity cost) of high demand suburban parking may be as high as $1.00 per hour or more.

If parking prices were higher, and the supply of parking was lower, walking distances to parking would increase as consumers sought the most efficient combination of parking attributes to meet their own particular transportation needs. Under present circumstances, the value of walking time to parking exceeds the market price of parking, because most parking is provided free of charge. Parking walk time accessibility thus determines parking utilization rates, even though the true market value of parking is considerably higher than the value of walking time in most instances.

Parking as a Constraint on Development

Parking can only influence the location of activities negatively in situations where parking pricing and/or supply act as a direct deterrent to travel. The easiest solution to this type of problem would seem to be an ample supply of free parking, but this solution is not without its own set of costs (Shoup, 2005). How much parking is too much? When does parking act as a deterrent to development?

Ferguson (2005) proposes the following "law of centrality" in a sea of parking:

$$A_f = \frac{\pi W^2}{\left(\sqrt{1+P_r} - 1\right)^2}$$

where

A_f	=	maximum floor area (a measure of centrality, or maximum revealed accessibility)
W	=	maximum walking distance (a valid measure of parking cost in the absence of parking pricing)
P_r	=	minimum parking ratio (parking area ÷ floor area, a measure of the minimum parking opportunities required under most current zoning, lending and development industry practices)

Centrality (maximum building size) decreases with the amount of parking required by zoning and/or industry practice, and increases with maximum walking distance to parking in response to zoning requirements and/or customer preferences (Table 8). Average parking requirements for shopping centres have steadily declined over the last fifty years, as average shopping centre size increased (Figure 6).

Automobile ownership rates have skyrocketed over the last fifty years. Parking requirements in zoning ordinances also increased over the last fifty years, but much less rapidly (Ferguson, 2004). The unprecedented decrease in parking requirements observed for major regional shopping centres over the last fifty years can only be explained as the result of pedestrian access limitations imposed by the vast amount of space required to provide an ample, free supply of parking (Ferguson, 2005).

Table 8. Maximum floor area (square meters) by pedestrian access and parking requirements

Maximum Walking Distance (Meters)	Minimum Parking Ratio (Parking Area / Floor Area)					
	1	2	3	4	5	6
30	17,011	5,446	2,919	1,910	1,389	1,078
60	68,044	21,785	11,675	7,641	5,557	4,310
90	153,099	49,016	26,268	17,192	12,502	9,698
120	272,177	87,140	46,698	30,564	22,226	17,241
150	425,276	136,156	72,966	47,757	34,729	26,940
180	612,397	196,065	105,071	68,770	50,009	38,793

Source: Ferguson (2005)

Parking is not merely an issue in disaggregate discrete choice travel behaviour and activity analysis, but has clearly discernible effects on macroeconomic issues associated with the location of large-scale economic activities such as regional shopping malls. In a completely automobile-dependent society, parking supply requirements limit the allowable density of development as a fundamental constraint. Even where alternative modes are in use, parking may serve as a major deterrent to both allowable development densities and the use of alternative modes. This is a direct function of the gross physical requirements frequent automobile use imposes on spatial resource allocation at the macroscopic level.

Figure 6. U.S. Shopping Centre Size and Parking Supplied, 1949-1999

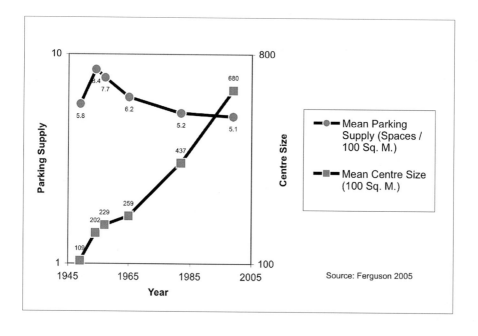

Source: Ferguson 2005

CONCLUSIONS

Parking and accessibility are inexorably linked in an automobile-oriented society, particularly when parking oversupply conditions force the marginal price of parking to zero in most instances. Minimum parking requirements in zoning contribute to this phenomenon, as do parking size and location requirements. Parking is often neglected in travel demand modelling, but this can be justified (at least technically) as a function of parking access (which is generally better than comparable transit or pedestrian access) and cost (which is generally nothing).

In the absence of parking pricing, accessibility to adjacent land uses is paramount in determining the desirability and use of parking. Curb parking is heavily regulated, due to its relative scarcity, close proximity to adjacent land uses, and resulting desirability to a wide range of motor vehicle drivers and their passengers. Although parking is rarely considered in modelling accessibility as a determinant of location patterns and land use intensity, this omission distorts the large-scale effects excessive parking requirements may have on macroscopic urban development patterns and the location of activities across time and space.

Research recommendations include the need for more data and a better understanding of the incidence and dynamics of:

1. Walking distance to parking (by activity type, duration and location)
2. Parking price schedules (fixed, variable, hybrid and other)
3. Parking policy regimes (on-street, off-street, public, private, commercial, etc.)

Policy recommendations include the following:

1. Zoning and regulation (review zoning requirements for parking, reduce or eliminate as necessary to encourage more efficient travel markets)

2. Markets and pricing (encourage formation of parking markets in more urban areas, and larger sections of existing urban areas, through on-street parking pricing and off-street parking regulatory reform)

3. Education and awareness (inform public concerning the true cost of ample, free parking, and the availability of other transportation and parking alternatives)

REFERENCES

Arnott, R. and E. Inci (2005). *An integrated model of downtown parking and traffic congestion*. Working Paper 57. Boston, MA: Boston College, Department of Economics.

Box, P. C. (1993). Parking systems and loading facilities. In: *Transportation Planning Handbook* (J.D. Edwards, Jr., ed.), pp. 179-200. Prentice Hall, Englewood Cliffs, NJ.

Buchanan, C. (1963). *Traffic in Towns*. Ministry of Transport, London

Bureau of Public Roads (1956). *Parking Guide for Cities*. U.S. Department of Commerce, Washington, DC.

Burrage, R. H. and E. G. Mogren (1957). *Parking*. Eno Foundation, Saugatuck, CT.

Davidson, M. and F. Dolnick (2002). *Parking Standards*. PAS Report 510/511. Chicago, IL: APA.

Eno Foundation (1942). *The Parking Problem: A Library Research*. Eno Foundation, Saugatuck, CT.

Ferguson, E. (2003). Zoning for parking as policy innovation. *Transportation Quarterly* **57(2)**, 47-55.

Ferguson, E. (2004). Zoning for parking as policy process: a historical review. *Transport Reviews,* **24(2)**, 177-194.

Ferguson, E. (2005). The law of centrality in a sea of parking. *ITE Journal on the Web* (forthcoming)

Gillen, D. W. (1977). Effects of parking costs on urban transport mode choice. *Transportation Research Record,* **637**, 46-51.

Gittens, M. J. (1965). Parking. In: *Traffic Engineering Handbook* (J. E. Baerwald, ed.), pp. 459-500. Institute of Traffic Engineers, Washington, DC.

Harris, B. (2001). Accessibility: concepts and applications. *Journal of Transportation and Statistics* **4(2/3)**, 15-30.

Highway Research Board (1971). *Parking Principles*. Special Report 125. HRB, Washington, DC.

Jara-Diaz, S. R. and R. Guerra (2003). *Modeling activity duration and travel choice from a common microeconomic framework*. Presented at the 10th International Conference on Travel Behavior Research, Lucerne, Switzerland.

Krizek, K. J. (2003). Operationalizing neighborhood accessibility for land use—travel behavior research and regional modeling. *Journal of Planning Education and Research,* **22** 270-287.

Kulash, D. (1974). *Parking Taxes for Congestion Relief: A Survey of Related Experience*. Urban Institute, Washington, DC.

Kurani, K. S. and M. E. H. Lee-Gosselin (1996). Synthesis of past activity analysis applications. *Activity-Based Travel Forecasting Conference Proceedings*, June 2-5.

Lambe, T. A. (1969). The choice of parking location by workers in the Central Business District. *Traffic Quarterly* **23(3)**, 397-411.

LeCraw, C. S. (1946). *An Economic Study of Interior Block Parking Facilities*. Eno Foundation, Saugatuck, CT.

McClintock, M. (1925). *Street Traffic Control*. McGraw-Hill, New York.

McClintock, M. (1930). *A Report on the Parking and Garage Problem of the Central Business District of Washington, DC*. National Capital Park and Planning Commission, Washington, DC.

Mogren, E. G. and W. S. Smith (1952). *Zoning and Traffic*. Eno Foundation, Saugatuck, CT.

Mueller, W. and A. Reber (1930). Parking and garaging of road vehicles. Report 6-S, Part 2 (Switzerland). *Proceedings of the Sixth International Road Congress*. Paris, France: Permanent International Association of Road Congresses, pp. 13-14.

Olaru, D. and B. Smith (2003). *Modeling daily activity schedules with fuzzy logic*. Presented at the 10th International Conference on Travel Behavior Research, Lucerne, Switzerland.

Roth, G. J. (1965). *Paying for Parking*. Hobart Paper 33. Institute of Economic Affairs, London.

Schrader, J.G. (1964). *Off-Street Parking Requirements*. PAS Report 182. ASPO, Chicago, IL.

Shoup, D. C. (1982). Cashing out free parking. *Transportation Quarterly,* **36(3)**, 351-364.

Shoup, D. C. (1995). An opportunity to reduce minimum parking requirements. *Journal of the American Planning Association,* **61(1)**, 14-28.

Shoup, D. C. (1999). The trouble with minimum parking requirements. *Transportation Research A,* **33(7-8)**, 575-599.

Shoup, D. C. (2005). *The High Cost of Free Parking*. Planners Press, Chicago, IL.

Shoup, D. C. and D. H. Pickrell (1978). Problems with parking requirements in zoning ordinances. *Traffic Quarterly,* **32(4)**, 545-563.

Shoup, D. C. and R. W. Willson (1992). Employer-paid parking: the problem and proposed solutions. *Transportation Quarterly,* **46(2)**, 169-192.

Smith, Mary S. and T. A.Butcher (1994). How far should parkers have to walk? *Parking,* **33(8)**, 29-32.

Sun, Z., T. Arentze, and H. Timmermans (2005). *Modeling the impact of travel information on activity-travel rescheduling decisions under conditions of travel time uncertainty*. Presented at the annual meeting of the Transportation Research Board, Washington, DC, January.

Vickrey, W. (1954). The economizing of curb parking space. *Traffic Engineering,* **25(2)**, 62-67.

Weant, R. A. and H. S. Levinson (1990). *Parking*. Eno Foundation, Westport, CT.

Wilbur Smith and Associates (1965). *Parking in the City Center*. Automobile Manufacturers Association, New Haven, CT:.

Willson, R. (1995). Suburban parking requirements: a tacit policy for automobile use and sprawl. *Journal of the American Planning Association,* **61(1)**, 29-42.

Willson, R. (2000). Reading between the regulations: parking requirements, local perspectives, and public transit. *Journal of Public Transportation,* **3**, 111-128.

NOTES

[1] The parking problem, as seen from the perspective of McClintock and his colleagues in 1930, was one of insufficient parking supply in a society experiencing rapid growth in automobile ownership and use. The parking problem, as seen today is quite the reverse, with parking oversupplied and underpriced in a largely saturated market (per capita) for auto ownership and travel.

[3] The seminal report published and nominally authored by the Eno Foundation in 1942 in point of fact seems to have been initiated and principally authored by Miller McClintock, whose name was removed from the final report due to a sudden falling out with Eno Foundation chairman and founder William Phelps Eno just prior to publication.

[4] Box (1993) identifies BPR (1957), WSA (1965), and HRB (1971) as the primary national studies of average walking distances to parking available. More detailed case studies of individual cities also exist (Lambe 1969, Gillen 1977), but even these are neither common nor particularly up to date.

[5] The key variable in determining the physical space required to store a vehicle is its footprint: width times length. Vehicle height is of negligible importance in surface lot design, but critical in the design of multistory parking structures. Vehicle weight is important only in the case of very large, very heavy, oversized vehicles.

[6] Drive-in theaters and restaurants were once quite popular in the U.S., but are increasingly difficult to find today.

[7] The drive-thru concept is increasingly popular today, especially for fast food restaurants, drycleaners, and banks. Gasoline stations, car washes, quick lube, and emissions testing facilities operate on similar principles. Private residential communities, military and industrial installations, recreational and parking facilities, toll roads and bridges, and any other activity or land use with gate-controlled access also fall into the drive-thru category, at least in terms of operating characteristics.

[8] Accessibility, by definition, is a relative measure. There are no absolutes in either the measurement of accessibility or its use in modeling.

[9] Zoning for parking is the practice of requiring all new developments in a community to meet minimum parking requirements as a condition of development approval. Such requirements are included in zoning ordinances as a matter of public policy, promoted by the federal government since at least 1940, adopted at the local level by city and county government almost universally since 1970, and implemented into action by the development community at large (Ferguson 2003).

[10] Recommended (and required) parking space dimensions vary from as low as 7'6'' by 16' for small foreign cars, to 10' by 20' for short-term retail parking.

[11] If the size and shape of the area available for parking is sub-optimal in terms of the most efficient parking plan for each type of parking layout, more space is required. Driveways connecting to streets and highways, pedestrian walkways connecting to streets and buildings, landscaping, lighting, and drainage will increase space requirements above and beyond those listed here as well.

[12] Monthly bus passes, like annual motor vehicle registration fees, stretch the definition of marginal cost to the limit.

[13] One might argue that bicycle travel is less onerous than walking, jogging or running. Then again, there is a much wider continuum of travel choices, if one considers motorized bicycles, mopeds and motorcycles each as a separate form of transportation technology and mode choice. Shades of blue and red buses, which when blended together produce a lighter shade of purple. Which is to say that, elimination of irrelevant alternatives aside, who is to say which

modes are distinguishable, one from another, and which are not? The choice is not always as obvious or discrete as one might like.

[14] Travel demands served by non-vehicular transportation modes such as walking, jogging or running do not impose any parking demands. Similarly, travel demands in which the vehicle operator is not engaged in the activity otherwise being served by the vehicle trip in question also impose no parking demands, other than the temporary need for a place to load or unload passengers (drop-off locations, bus stops, etc.).

[15] Most parking is provided free of charge, making out-of-pocket costs irrelevant in practice for most consumers.

[16] Many local zoning ordinances go even further to specify exactly how close required parking must be to the activity, land use or building served (e.g., within 300 feet).

[17] The larger the building, the larger the surrounding parking lot, and the further away pedestrian sidewalks and transit bus stops must be. LeCraw (1946) proposed that required parking should be placed on the interior of blocks, to minimize disruption of street traffic, including pedestrian and transit access.

[18] The fact that illegal curb parkers walked the shortest distances does not contradict th model, but rather confirms it, once the effects of enforcement (the probability of being and fined for violating local parking regulations) is considered.

[19] Loading and stacking facility requirements are often included with parking requirement zoning ordinances. Such facilities most often are located directly in or adjacent to parking The demand for loading and stacking space is much different than the demand for long or sh term parking, however, and would require a different model based on different assumptions with different explanatory variables. Parking models might further identify parking demand i terms of the specific spatial dimensions of individual parking spaces required to store vehicles of different lengths, widths, heights and axle weights. Parking prices rarely discriminate on the basis of vehicle size or storage space requirements, and most parking facilities divide parking areas into individual spaces of roughly equal size. Compact parking spaces rarely entitle compact cars to a lower parking rate. Buses do, however, generally pay a higher premium to enter parking areas at major recreational facilities.

[20] For convenience, it is assumed here that all persons in the travel party have the same value of travel time and average walking speed. This is always true for one person, but not necessarily true for larger parties. A decision rule would be needed to evaluate the heterodox case, more specifically, whether to use the shortest, longest, or average expected walking distance to choose among various parking alternatives.

[21] These are all direct measures of operational accessibility. When individuals choose to engage in particular activities at particular locations starting and ending at particular times, they have chosen to access those activity-time-space combinations over all others. Shoppers are related to shopping trips. Dollars spent are related to shoppers. Income generated is related to dollars spent. Profit is related to income generated. From a business point of view, retail profit is the key variable of interest. From a retail customer point of view, the utility of shopping (including the marginal value of purchased goods, if any) is paramount. From a land use planning perspective, the location of activities is of greatest concern. From a traffic engineering

perspective, vehicle trip generation is the key. All of these are but partial reflections of accessibility in action.

[22] In a closed parking system, auto users attracted to a particular location must travel to that location regardless of parking price changes, and will walk further to avoid parking fees as necessary to minimize their generalized parking costs. In an open system, auto users attracted to a particular location may choose a different location with lower generalized parking costs instead, if the monetary price of parking at their original preferred location rises.

INDEX